PATENT IT YOURSELF
Second Edition

By Patent Attorney David Pressman

Edited by Attorney Stephen Elias

Illustrations by Linda Allison

Please read this: We have done our best to give you useful and accurate information in this book. But please be aware that laws and procedures change constantly and are subject to differing interpretations. If you are confused by anything you read here, or if you need more information, check with an expert. Of necessity, neither the author nor the publisher of this book makes any guarantees regarding the outcome of the uses to which this material is put. The ultimate responsibility for making good decisions is yours.

NOLO PRESS • 950 Parker Street, Berkeley, CA 94710

PRINTING HISTORY

Nolo Press is committed to keeping its books up-to-date. Each new printing, whether or not it is called a new edition, has been completely revised to reflect the latest law changes. This book was printed and updated on the last date indicated below. You might wish to call Nolo Press, (415) 549-1976 to check whether there has been a more recent printing or edition.

New *printing* means there have been some minor changes, but usually not enough so that people will need to trade in or discard an earlier printing of the same edition. Obviously, this is a judgment call and any change, no matter how minor, might affect you. New *edition* means one or more major—or a number of minor—law changes since the previous edition.

FIRST EDITION	October 1985
Second Printing	March 1986
Third Printing	October 1986
Fourth Printing	April 1987
Fifth Printing	October 1987
SECOND EDITION	November 1988
Second Printing	June 1989
Third Printing	July 1990
Editor	Stephen Elias
Illustrations	Linda Allison
Production	Stephanie Harolde
	Michael Sigal
Book Design	Toni Ihara
Printing	Delta Lithograph

ISBN 0-87337-075-9
Library of Congress Card Catalog No.: 85-63111
The title "Patent It Yourself!" is a registered trademark, California Reg. Nr. 74002.

recycled paper

ACKNOWLEDGEMENTS

First Edition: My deep thanks go to my clients and other inventors whose creativity and genius I so greatly admire and envy. Your valuable and helpful suggestions for expanding and improving my earlier book have been incorporated in this expanded work. I have been deeply gratified to see the results of the do-it-yourselfers who have followed my instructions and hope that this opus will spur you to even greater heights. Also thanks to librarians par excellence, Mrs. Margreta J. Nisbett and Dr. Mary-Jo Di Muccio of the Patent Information Clearinghouse in Sunnyvale, California, for their generous help. And lastly, thanks to the staff at Nolo Press, including Steve Elias, Stephanie Harolde, Keija Kimura, and Toni Ihara, for their helpful suggestions, insightful queries, and extensive efforts which accomplished more than I could do alone.

David Pressman, 1985 October

Second Edition: I iterate my thanks to all of those listed above; the unprecedented success of the first edition has confirmed the value of your help. For this second edition, I am again deeply grateful to all of my clients, readers, other inventors, Roberta, Steve Elias, Jake Warner, and others at Nolo who have given me many helpful suggestions and ideas for improving the first edition. They have been incorporated in this, the second edition. I continue to welcome your invaluable feedback and sugestions. Also this edition is significantly expanded to include additional forms, insights I have learned in the past three years during media interviews, lecturing, and patent prosecution, fee and timing tables, charts, legal examples, and most importantly, several checklists which should prove invaluable to the do-it-yourselfer. Finally, I have been able to solve the sexist personal pronoun problem thanks to a helpful article in Ms. *Magazine* by Marie Shear. I hope that the result meets your needs even more than before and engenders success in your inventive activities. I continue to welcome your invaluable feedback and suggestions.

David Pressman, 1988 October

RECYCLE YOUR OUT-OF-DATE BOOKS & GET 25% OFF YOUR NEXT PURCHASE!

Using an old edition can be dangerous if information in it is wrong. Unfortunately, laws and legal procedures change often. To help you keep up to date we extend this offer. If you cut out and deliver to us the title portion of the cover of any old Nolo book we'll give you a 25% discount off the retail price of any new Nolo book. For example, if you have a copy of TENANT'S RIGHTS, 4th edition and want to trade it for the latest CALIFORNIA MARRIAGE AND DIVORCE LAW, send us the TENANT'S RIGHTS cover and a check for the current price of MARRIAGE & DIVORCE, less a 25% discount. Information on current prices and editions is listed in the NOLO NEWS. Generally speaking, any book more than two years old is of questionable value. Books more than four or five years old are a menace. This offer is to individuals only.

OUT OF DATE = DANGEROUS

patent 6/89

14—YOUR APPLICATION CAN HAVE CHILDREN

15—AFTER YOUR PATENT ISSUES: USE, MAINTENANCE AND INFRINGEMENT

16—OWNERSHIP, ASSIGNMENT, AND LICENSING OF INVENTIONS

BOOKS OF USE AND INTEREST

GLOSSARY

APPENDIX

ABOUT THE AUTHOR

David Pressman is a member of the Pennsylvania (inactive), California, and Patent And Trademark Office bars. He's had over 27 years' experience in the patent profession, as a patent examiner for the U.S. Patent Office, a patent attorney for Philco-Ford Corp., Elco Corp., and Varian Associates, as a columnist for EDN Magazine, and as an instructor at San Francisco State University. He's also an inventor, with two patents issued. When not writing, dabbling in electronics, programming, or inventing he practices as a patent lawyer in San Francisco. Originally from Philadelphia, he has a BS in EE from Penn State University and a JD from George Washington University. A member of Mensa, he is also active in the general semantics and vegetarian movements.

INTRODUCTION

Here's a book that allows you, the inventor, to patent and commercially exploit your invention by yourself. It provides:

- Step-by-step guidance for obtaining a U.S. patent, together with tear-out forms that are necessary for each step of the process;

- An overview of the procedures and requirements for getting patent protection abroad and concrete suggestions for finding the necessary resources to help you do this;

- An overview of the alternative and supplementary forms of protection available for inventions, such as trade secreting, copyrighting, trademarks, and using unfair competition law; and

- Detailed information and advice on how to commercially evaluate and market your invention.

By following the instructions set out here, you'll not only save healthy attorney fees, but you'll be personally involved with every step of the patenting process. This is a wise way to proceed since you know your invention better than anyone else, and assuming you're willing to wade through a fair number of patent technicalities, you're the best person to patent it.

I think of the book as a great equalizer since it provides the know-how to enable the garage-shop or basement do-it-yourselfer to get as good a patent as a large corporation. It provides the legal tools necessary for inventors (whether large or small) to provide first class legal protection for their work. And it especially gives the small inventor the tools to competently and efficiently protect an invention, whether or not he or she can afford a patent attorney.

In this view, it's almost a universal misconception that one must use a patent attorney to get a valid patent. This isn't true. First, the laws contain absolutely no requirement that one must have a patent attorney to file a patent application,

deal with the Patent Trademark Office (PTO) concerning the application, or to obtain the patent. In fact, PTO regulations [MPEP, Section 707.07(j)] specifically require patent examiners to help inventors in *pro se* (no lawyer) cases. Second, and perhaps more persuasive, it's a simple fact that many hundreds of patent applications are filed and successfully prosecuted each year by *pro se* (no attorney) inventors.

The quality of a patent is mainly dependent upon two basic factors:

(1) whether the patent application contains a full, clear, and accurate description of the invention, and

(2) whether the reach of the patent is as broad as possible, given the state of prior developments in the field.

Fortunately, it takes no special legal expertise to do an excellent job for either of these factors.

Even if you do choose to work with an attorney, or have one available to you through the process, you'll find that this book allows you to take an active role in the process, do a better job of monitoring your attorney (no trivial consideration), and greatly add to your understanding of the ways in which the law is willing to protect your invention. Indeed, many corporate legal departments have used the first edition of this book to educate their inventors and support personnel to deal with patent attorneys and to protect their inventions more effectively.

The big question is, of course, even though it's possible to prosecute your own patent application, should you do so on your own or hire an expert? After all you probably hire people to do all sorts of things for you, from fixing your car to cleaning the chimney, that you could in theory do yourself. Of course a powerful incentive for patenting it yourself is the amount of money expert help costs. Or put another way, even though most car mechanics make a pretty good living, most of them can't afford to belong to the same country club as patent attorneys.

The best answer to this question may be to do some of both. Using this approach the diligent inventor will do much of the patent work herself, only consulting with an attorney at his hourly rate if snags develop, or to check the patent application before submission.

The book is organized primarily for chronological use, starting with an overview of the patent process and then sequentially covering the steps most inventors will take to protect and profit from their inventions. I strongly recommend that you first read the book all the way through, skimming lightly over the many chapters that actually tell you how to do things.

In this way you'll first get an overview of the patent forest before you return and deal with the individual steps (trees) necessary to fully protect your invention.

Throughout the book I refer to a number of forms and in many instances reproduce them in the text. A tear-out version of each is also located in the Appendix for your use. I recommend that you make photocopies so you'll have ample spares for drafts and extra copies for your records.

Also throughout the book I refer to various statutes and governmental administrative rules, mostly in the patent area. I use standard forms of legal citation; these are interpreted as follows:

35 USC 102 = Title 35 of the U.S. Code, Section 102

37 CFR 1.111 = Title 37 of the [U.S.] Code of Federal Regulations, Section 1.111

Title 35 of the U.S. Code (USC) comprises all of the federal statutes patents and Title 37 of the U.S. Code of Federal Regulations (CFR) comprises all of the federal administrative rules issued by the Patent and Trademark Office and Copyright Office which deal with patents, trademarks, and copyright matters. Part 1 of 37 CFR is concerned with patents. Thus

Patent Rule 111 = 37 CFR 1.111.

Both the U.S. Code and the CFR are available in any law library and as indicated in the Bibliography.

Good luck and successful inventing!

Chapter 1

Introduction to Patents and other Intellectual Property

IN THIS CHAPTER I'll first introduce you to the world of patent law. Each of the patent-related items discussed here will be amplified in subsequent chapters, as they relate to the actual process of obtaining and profiting from a patent. I also present an overview of the other forms of "intellectual property" (trademarks, copyright, trade secrets, etc.) potentially available to you. Although you may think that a patent is the only form of protection for your creation, you should at least consider the alternatives, some of which can be used in addition to or in lieu of a patent to gain the maximum or most cost-effective offensive rights ("protection") for your creation.

A. What Is a Patent and Who Can Apply for It

WHAT IS A PATENT? It's a right granted by the government to a person or to a legal entity (partnership or corporation).

What is the nature of the patent right? A patent gives its holder the right to exclude others from making, using, or selling the invention "claimed" in the patent deed for seventeen years, provided certain fees are paid. (See Chapter 9 for more on claims.) This right to exclude others is exercised by filing a patent infringement lawsuit in federal court.

Who can apply for a patent? Anyone, regardless of age, nationality, mental competency, or any other characteristic, so long as he or she is a true inventor of the invention. Even dead or insane persons may apply through their personal representative (see Chapter 16 for more on patent ownership).

A patent is like property and can be sold outright for a lump sum, or it can be licensed in return for royalty payments. More on this in Ch. 16.

HELPING HANDS HAT
No. 3,169,192.

B. Types of Patents

THERE ARE THREE types of patents—utility patents, design patents, and plant patents. Let's briefly look at each.

- **Utility Patents:** As we'll see in Chapters 7 to 9, a utility patent, the main type of patent, covers inventions which function in a unique manner to produce a utilitarian result. Examples of utility inventions are Velcro hook and loop fasteners, new drugs, electronic circuits, and automatic transmissions. To get a utility patent, one must file a patent application which consists of a detailed description telling how to make and use the invention, together with claims (formally-written sentence fragments) which define the invention, drawings of the invention, formal paper work, and a filing fee. Again, only the actual inventor can apply for a utility (or any other) patent.

- **Design Patents:** As discussed in more detail in Chapter 10, a design patent (as opposed to a utility patent) covers the unique shape or design of an invention. Thus if a lamp, a building, a computer case, or a desk has a truly unique shape, its design can be patented. However the uniqueness of the shape must be purely ornamental or aesthetic; if it is functional, then only a utility patent is proper, even if it is also aesthetic. A good example is a jet plane with a constricted waist for reducing turbulence at supersonic speeds: although the shape is attractive, its functionality makes it suitable only for a utility patent.

A useful way to distinguish between a design and a utility invention is to ask, "Will removing or smoothing out the novel features substantially impair the function of the device?" If so—as in the jet plane with the narrowed waist—this proves that the novel features have a significant functional purpose, so a utility patent is indicated. If not—as in Dizzy Gillespie's trumpet with its upwardly-bent bell, where bending back the bell to a horizontal position won't affect the trumpet's function—a design patent is indicated. Another useful question to ask is, "Is the novel feature(s) there for structural or functional reasons, or only for the purpose of ornamentation?"

The design patent application must consist primarily of drawings, along with formal paper work, and a filing fee.

- **Plant Patents:** A plant patent covers asexually or sexually reproducible plants, such as flowers.

C. The Novelty and Unobviousness Requirement

WITH ALL THREE types of patents, the patent examiner in the Patent and Trademark Office (PTO) must be convinced that your invention satisfies the "novelty" requirement of the patent laws. That is, your invention must be different from what is already known to the public. Any difference, however slight, will suffice.

Novelty, however, is only one small hurdle to overcome. In addition to being novel, the examiner must also be convinced that your invention is "unobvious." This means that at the time you came up with your invention, it would have been considered unobvious to a person skilled in the technology (called "art") involved in your creation. As we'll see in Chapter 5, unobviousness is best shown by new and unexpected, surprising, or far superior results, when compared to the state of the art in the particular area you're concerned with.

D. How Long Do Patents Last?

HOW LONG CAN YOU, the patent owner, exclude others from infringing the exclusive rights granted by your patent? In the U.S., utility and plant patents are granted for a period of 17 years (assuming maintenance fees are paid), while design patents are granted for 14 years.[1] The 14- or 17-year monopoly starts when the patent issues, usually several years after it is filed. From the date of filing to issuance (termed "pendency period") the inventor has no rights. However, when and if the patent later issues, the inventor will obtain the right to prevent the continuation of any infringing activity that started during the pendency period. Patents aren't renewable, and once patented, an invention may not be repatented.

I provide a time chart in the Appendix to indicate these and other pertinent times.

---

[1]The terms of patents for certain product whose commercial marketing has been delayed due to regulatory review (e.g., drugs, food additives) can be extended beyond 17 years under new statutes (35 USC 155-156).

E. Patent Filing Deadlines

AS WE'LL SEE in more detail in Chapter 5, in the United States you must file your patent application within one year after you first commercialize or publish details of the invention. However most foreign countries don't have this one-year grace period, so there's some disadvantage if you sell or publish before filing. For this reason, your safest route is to file your U.S. patent application before you publish or commercialize your invention.

F. Patent Fees

HOW MUCH WILL IT COST to protect your invention? Assuming you use this book and don't use any patent attorneys or agents, and not including costs of drawing, typing, photocopying, and postage, the only fees you'll have to pay are government fees.

The amounts of these fees are listed on the PTO Fee Schedule in the Appendix. As indicated in the Schedule, most PTO fees are two-part: large entity and small entity. The large-entity fees must be paid when the inventor has agreed to assign (legally transfer) the invention to a large business (over 500 employees), or any individual(s) or organization which has (or is under an obligation to) license or assigned the invention to a large business. The small-entity fees, which are half the large-entity fees, can be paid by all other inventors, namely private inventors, or inventor-employees of non-profit organizations or small businesses (less than 500 employees). The names of these fees and the circumstances when they're due are as follows:

- **Utility Patents:** To file a utility patent application, you must pay a *Utility Patent Application Filing Fee* and to have the PTO issue your utility patent, you must pay a *Utility Patent Application Issue Fee.* To keep the patent in force for its full 17-year term, you must pay the PTO three maintenance fees, as follows:

  Maintenance Fee I, payable 3.0 to 3.5 years after issuance.

  Maintenance Fee II, payable 7.0 to 7.5 years after issuance.

  Maintenance Fee III, payable 11.0 to 11.5 years after issuance.

- **Design Patents:** To file a design patent application, you must pay a *Design Patent Application Filing Fee* and to have the PTO issue your design patent, you must pay a *Design Patent Application Issue Fee.* The law doesn't require maintenance fees for design patents.

- **Plant Patents:** To file a plant patent application, you must pay a *Plant Patent Application Filing Fee* and to have the PTO issue your plant patent, you must pay a *Plant Patent Application Issue Fee.* Again, the law doesn't require maintenance fees for plant patents.

G. The Scope of the Patent

THE PATENT RIGHT extends throughout the entire U.S., its territories, and possessions. It's transferrable by sale or gift, including by will or by descent [under the state's intestate succession (no-will) laws]. The patent rights can also be licensed, i.e., you can own the patent and grant anyone else, including a company, the right to make, use, or sell your invention for fees, called "royalties" (more on licensing in Chapter 16). As mentioned, the patent right is granted by the Federal Government, acting through the Patent and Trademark Office (a division of the Department of Commerce), in Arlington, Virginia.

H. How a Patent Can Be Lost

THE PATENT RIGHT isn't an absolute monopoly for a seventeen-year period. It can be lost if:

- Maintenance fees aren't paid;

- Some prior art references (earlier patents or other publications) is uncovered which shows that the invention of the patent wasn't new or wasn't different enough when the invention was made;

- The patent owner engages in certain defined types of illegal conduct, i.e., commits anti-trust or other violations connected with the patent; and

- The patent asset is seized or lost in a general legal proceeding such as in a bankruptcy or by creditors who levy on your assets. In this case the patent right will continue to exist, but you will lose your ownership of it.

In short, the patent monopoly, while powerful, is defeatable and is limited in scope and time.

I.   How Patent Rights Are Asserted

THE PATENT GRANT gives its owner the right to file and maintain a lawsuit against any person or legal entity (infringer) who makes, uses, or sells the claimed invention, or an essential part of it. If the patent owner wins the lawsuit, the judge will issue an injunction (a signed order) against the infringer, ordering the infringer not to make, use, or sell the invention any more. Also, the judge will award the patent owner damages (money to compensate the patent owner for loss due to the infringement). The amount of the damages is often the equivalent to a reasonable royalty (say 5%), based on the infringer's sales. However, if the patent owner can convince the judge that the infringer acted in bad faith (e.g., infringed intentionally with no reasonable excuse), the judge can triple the damages and make the infringer pay the patent owner's attorney fees.

Since the patent defines the invention monopoly very precisely, the patent owner can use the patent only against supposed infringers who make, use, or sell things or processes which fall within the defined monopoly. Thus not everyone

who makes something similar to your invention will be an infringer; you can validly sue only those whose products or processes fall within the claims in your patent. The claims are legal definitions of your invention—see Chapters 9, 13, and 15 for more on claims.

J.   What Can't Be Patented

YOU CAN'T PATENT any process that can be performed mentally, or which can be performed with very simple implements, such as a paper and pencil. The reason is that the law doesn't wish to limit what people can do essentially with their bodies and brains. The same rule applies to abstract ideas which aren't reduced to hardware form, naturally-occurring articles, computer programs which aren't related to computer hardware, business forms and other printed matter, methods of doing business, scientific principles in the abstract (without hardware), and atomic energy inventions. See Chapter 5.

Note: Computer programs per se can be protected by copyright (and sometimes by trade secret) law. See Salone, *How to Copyright Software* (Nolo Press).

The Life of an Invention

Although most inventors will be concerned with the rights a patent grants during its monopoly or in-force period (from the date the patent issues until it expires 17 years later), the law actually recognizes five "rights" periods in the life of an invention. These five periods are as follows:

1. **Invention Conceived But Not Yet Documented:** When an inventor conceives of an invention, but hasn't yet made any written, signed, dated, and witnessed record of it, the inventor has **no rights whatsoever**.

2. **Invention Documented But Patent Application Not Yet Filed:** After making a proper, signed, dated, and witnessed documentation of an invention, the inventor has valuable rights against any inventor who later conceives of the same invention and applies for a patent. The invention may also be treated as a "trade secret" (i.e., kept confidential), which gives the inventor the legal right to sue and recover damages against anyone who "steals" the invention.

3. **Patent Pending—Patent Application Filed But Not Yet Issued:** During this patent pending period, the inventor's rights are the same as in Period 2 above. A patent application gives an inventor no rights whatsoever—only the hope of a future monopoly, which commences only when a patent issues. However most companies which manufacture a product which is the subject of a pending patent application will mark the product "patent pending" in order to warn potential copiers that if they copy the product, they may have to stop later (and thus scrap all their molds and tooling) if and when a patent issues. The Patent and Trademark Office (PTO) by law must keep all patent applications preserved in secrecy. The patent pending period usually lasts from one to three years.

4. **In-Force Patent—Patent Issued But Hasn't Yet Expired:** After the patent issues, the inventor (or patent owner if the inventor has assigned [legally transferred] ownership to the patent) can bring and maintain a lawsuit for patent infringement against anyone who makes, uses, or sells the invention without permission. The patent's in-force period lasts for 17 years.

5. **Patent Expired:** After the patent expires (17 years after issuance, or sooner if a maintenance fee isn't paid), the patent owner has no further rights, although infringement suits can be brought for any infringement that occurred during the patent's in-force period.

With respect to designs, as explained, the PTO won't grant design patents on:

- Any design which has significant functional utility (use a utility patent); or

- Ornamentation that is on the surface only rather than forming an integral part of the invention (use copyright).

K. Some Common Patent Misconceptions

OVER THE YEARS that I've practiced patent law, I've come across a number of misconceptions that laypersons have about patents. As part of my effort to impart what a patent is, I want to clear up a few of the most common here at the outset.

▲

Common Misconception: A patent gives one the right to practice an invention.

Fact: If you come up with an invention, you may practice (make, use, and sell) it freely, with or without a patent, provided that it's not covered by another's "in force" patent, i.e., a patent which is within its 17-year term (a non-expired patent).

▲

Common Misconception: Once you get a patent, you'll be rich and famous.

Fact: A patent is like a hunting license: it's useful just to go after infringers. If the invention isn't commercialized, the patent is usually worthless. You won't get rich or famous from your patent unless you also get the invention into commercial use.

▲

Common Misconception: *If a product has been patented, it's bound to be superior.*

Fact: *Although Madison Ave. would like you to believe this, in reality a patent merely means the invention is significantly different, not necessarily superior.*

L. How Intellectual Property Law Works to 'Protect' Inventors

WHILE MANY PEOPLE speak of a patent as a form of "protection," the fact is a patent is an offensive weapon, rather than "protection," which is a defensive shield. To properly benefit from a patent, as we'll see in Chapter 14, the patent owner must sue or threaten to sue any trespasser on the right. The patent doesn't provide any "protection" in its own right. Although the word "protection" is in common usage for all types of intellectual property, it's more accurate to say that a patent (as well as a copyright, trade secret, and trademark) gives its owner "offensive rights" against infringers. In other words the patent, copyright, trade secret or trademark provides a tool with which you can enforce a monopoly on your creation. The distinction between protection (a defense) and offensive rights is as important in intellectual property law as it is in football or basketball: while a good defense may be valuable, you'll need a powerful offense to win the game or stop the infringer.

To help you keep this distinction in mind, I try consistently to use the term "offensive rights" instead of "protection" However, if I slip up from time to time, please remember that by protection I only mean that inventors have the right to affirmatively come forward and invoke the court's help in preventing infringement by others.

M. Alternative and Supplementary Offensive Rights

AS YOU PROBABLY REALIZE, there are several alternative and often overlapping ways to acquire offensive rights on intellectual property. Let's think of these as different roads to the same destination. While the immediate filing of your patent application is one of these roads, it is only one. The purpose of this chapter is to provide you with a map to the other roads and to help you decide which is the best way to travel, given your circumstances.

The value of your invention can sometimes be better protected by using one of the other intellectual property protection schemes and can almost always be enhanced by simultaneously using one or more of these other forms of protection, such as unique trade dress, a good trademark, copyright-protected labels and instructions, and by maintaining later improvements as a trade secret.

N. Intellectual Property— The Big Picture

"INTELLECTUAL PROPERTY" (sometimes called intangible property) refers to any product of the human intellect, such as an idea, invention, expression, unique name, business method, industrial process, or chemical formula, which has some value in the marketplace. Intellectual property law, accordingly, covers the various legal principles which determine:

- Who owns any given intellectual property;
- When such owners can exclude others from commercially exploiting the property; and
- The degree of recognition that the courts are willing to afford such property (i.e., whether they will enforce the owner's offensive rights).

In short, intellectual property law determines when and how a person can capitalize on a creation.

Over the years, intellectual property law has fallen into several distinct subcategories, according to the type of "property" involved:

- **Patent Law** deals with the protection of the mental concepts or creations known as inventions—an example is the flip-top can opener. As indicated earlier, we have three types of patents: utility, design, and plant.

- **Trademark Law** deals with the the degree to which the owner of a name used in marketing goods or services will be afforded a monopoly over the use of the name (i.e., offensive rights against others who try to use it). Examples of trademarks are *Ivory, Coke, Patent It Yourself,* and *Nolo;*

- **Copyright Law** grants authors, composers, programmers, artists and the like the right to prevent others from copying or using their original creations without permission. Copyright law gives me offensive rights against anyone who copies this book.

- **Trade Secret Law** deals with the acquisition of offensive rights on private knowledge which gives the owner a competitive business advantage— e.g., manufacturing processes, magic techniques, and formulae. The method of producing the laser light shows and fireworks are trade secrets;

- **Unfair Competition Law** affords offensive rights to owners of mental creations that don't fall within the rights offered by the four types of law just discussed, but which have nevertheless been unfairly copied by competitors. Thus, "trade dress" (e.g., *Kodak's* yellow film package), a business name (e.g., *Proctor and Gamble Co.*), a unique advertising slogan (e.g., "Roaches check in but they don't check out"), or a distinctive packaging label (e.g., *Duracell's* copper top energy cells) may all enjoy offensive rights under unfair competition principles.

Having covered patent law earlier in this chapter, let's now wade a little deeper into the other forms of intellectual property law.

O. Trademarks

THIS IS THE MOST FAMILIAR branch of intangible property law. Everyone sees, uses, and makes many decisions on the basis of trademarks daily. For instance you probably decided to purchase your car, your appliances, much of the packaged food in your residence, your magazines, your computer, your watch, etc., on the basis of their trademarks, at least to some extent.

1. Trademarks Defined

In its most literal meaning, a trademark is any word or other symbol which is consistently attached to goods to identify and distinguish them from others in the marketplace. In other words, a trademark is a brand name.

Many patented goods or processes are also protected by trademarks, e.g., Xerox photocopiers have many patents on their internal parts, and also are sold under the well known *Xerox* trademark. Without the patents, people could copy the internal parts, but *Xerox* would still have a monopoly on its valuable and widely-recognized trademark.

The term "trademark" is also commonly used to mean "service marks." These are marks (words or other symbols) that are associated with services offered in the marketplace. The letters *NBC* in connection with the broadcast network is one example of a service mark. Another is the emblem used by *Blue Cross-Blue Shield* for its medical/insurance services. Other forms of marks commonly included within the term "trademark" are "certification marks" (the identifying symbol or name of an independent group, board, or commission which judges the quality of goods or services—e.g., the Good Housekeeping seal of approval), and "collective marks" (an identifying symbol or name showing membership in a organization—e.g., the

FDIC's symbol to show that deposits in a bank are insured).

An important third category of business identifier which is often confused with trademarks is called a "trade name." In the law, trade name is the word or words under which a company does business, while a trademark is the word or other symbol under which a company sells its products or services. To understand this better, let's use *Procter & Gamble* as an example. The words *Procter & Gamble* are a trade name, while *IVORY* is a trademark, i.e., as a brand name for *Procter & Gamble's* white soap. However, the media often refer to trademarks as trade names.

The trademark aspect of the word *Ivory* enjoys offensive rights under both federal and state trademark laws. The *Procter & Gamble* trade name, however, enjoys offensive rights primarily under state law (corporation registrations, fictitious name registrations, and unfair competition law). However, a federal law (17 USC § 1125) can also be used to slap down a trade name infringement as a "false designation of origin."

2. The Characteristics of a Trademark (with Offensive Rights)

Briefly, a trademark may or may not qualify for legal offensive rights depending on how distinctive (or strong) the law considers the trademark. Trademarks which are arbitrary (e.g., *Elephant* floppy disks), fanciful (e.g., *Double Rainbow* ice cream), or coined terms (e.g., *Kodak*) are considered strong, and thus entitled to a relatively broad scope of offensive rights. On the other hand, marks which describe some function or characteristic of the product (e.g., "*Rapidcompute* computers" or "*Relieveit*" for an analgesic) are considered weaker and won't enjoy as broad a scope of offensive rights. Although the above differences may seem somewhat arbitrary, they really aren't. The courts give fanciful, coined, or other arbitrary marks a stronger and broader monopoly than descriptive marks because descriptive marks come close to words in common usage and the law protects everyone's right to use these.

In addition to the strong/weak mark dichotomy, trademarks may be denied offensive rights if they become commonly used to describe an entire class of products, i.e., they have become "generic." Thus, "aspirin," once a trademark which enjoyed strong offensive rights, became a generic word (no offensive rights) for any type of over-the-counter painkiller using a certain chemical. Why? Because its owner used it improperly as a noun (e.g., "Buy *Aspirin*") rather than as a proper adjective (e.g., "Buy *Aspirin* [brand] analgesic"), and the public therefore came to view it as synonymous with the product it described.

3. Relationship of Trademark Law to Patent Law

As indicated above, trademarks are very useful in conjunction with inventions, whether patentable or not. A clever TM can be used with even an unpatentable invention to provide it with a unique aspect in the marketplace so that purchasers will tend to buy the trademarked product over a generic one. For example, consider the *Crock Pot* slow cooker and the *Hula Hoop* exercise device. These trademarks helped make both of these unpatentable products successful. In short, a trademark provides brand name recognition to the product and a patent provides a tool to enforce a monopoly on its utilitarian function. Since trademark rights can be kept forever (as long as the TM continues to be used), a TM can be a powerful means of effectively extending a patent monopoly.

4. Overview of How Offensive Rights to Trademarks Are Acquired

Here's a list of steps you should take if you come up with a trademark and you want to acquire offensive rights to it and use it properly. Because this is a patent book, I haven't covered this topic in detail. Probably the best available source for learning how to understand your trademarks is the *Trademark Law Handbook*, U.S. Trademark Association, Clark Boardman.

a. Preserve Your Mark as a Trade Secret Until You Use It

As I explain in subsection d. below, you must take certain action before you can acquire offensive rights in a mark. This means that during the developmental stage you must treat your trademark as a trade secret so that others won't adopt your proposed mark and use it first. (see Section P below for an overview of acquiring offensive rights to trade secrets).

b. Make Sure the Mark Isn't Generic or Descriptive

Ask yourself if the mark is generic or descriptive. A generic mark is a word or other symbol which the public already uses to designate the goods or service on which you want to use the mark. Thus you can't acquire offensive rights on The Pill for a birth-control pill since it's already a generic term. A descriptive mark is similar to a generic mark in that it describes the goods, but hasn't yet gotten into widespread public use. Thus if you came up with a new electric fork, the mark Electric Fork wouldn't have offensive rights, since it merely describes the product.

c. Make Sure Your Mark Isn't Already In Use

It's not wise to select a mark that is in use by someone else. The good will you develop around the mark may go up in smoke in the event of a trademark infringement contest and you may be liable for damages as well. Even if your proposed mark isn't identical to the already-used mark, if it is similar, you're likely to be prevented from using it by the other mark's owner if, in the eyes of the law, there is a likelihood of customer confusion. To determine if your mark is already in use, you'll have to make a trademark search or hire someone to do it for you.

d. Use or Apply to Register Your Trademark [2]

The first to actually use or apply to register a trademark federally (with the PTO) owns the trademark, that is, acquires offensive rights against infringers. Briefly, this means once you (a) actually use the mark, or (b) apply to register it with an intent to use it, and then do use it, you can validly sue a later person who uses a similar mark for similar goods in a context that is likely to mislead the public. Contrary to popular belief, trademarks do not have to be registered for offensive rights to be acquired. However, as explained in Subsection e just below, registration can substantially add to these offensive rights.

e. Use and Register Your Trademark

If you apply to register your mark federally on the bases of your intent to use it, you will, as stated, eventually have to actually use it on your goods to get it registered. You must thus follow through by actually using it and proving such use as part of your registration application.

---

[2]Prior to 1989-11-16, you had to actually use your mark on the goods in order to apply for registration. After this date you can apply for registration merely on the basis of your expressed intent to use the mark.

If you do adopt and use a trademark on your goods before applying for registration, you should register it in your state trademark office if it's used exclusively in your state, and/or the PTO if it's used across a stateline. This is because once your mark is registered, it will be much easier to sue infringers. The registration will cause the court to presume that you have exclusive ownership of the mark and the exclusive right to use it. If you don't register your trademark and it's infringed, you'll have much more difficulty when you go to court. To register a trademark in your state, call or write to your Secretary of State in your state's capitol for a trademark application form and instructions; the cost will be about $10 to $50. To register it federally, write to the Patent and Trademark Office, Washington, DC 20231 for an application to register a trademark (specify whether you're an individual, corporation, or partnership and whether you want to apply on the bases of actual use or intent to use); the cost is high—see the PTO Fee Schedule in the Appendix.

f. Use Your Trademark Properly

The law considers it very important to use a trademark properly once you've adopted it as a brand name for your goods. Before it's registered, you should indicate it's a trademark by providing the superscript "™" or "Ⓣ" after the mark. Once the mark is federally registered, provide the superscript "®" or indicate that the mark is registered in the PTO (e.g., "Reg. U.S. Pat. & TM. Off.").

Word trademarks should always be used as brand names. That is, they should be used as adjective modifiers in association with the general name of the goods to which they apply, and shouldn't be used as a substitute for the name of the goods. For example, if you're making and selling can openers and have adopted the trademark *Ajax*, always use the words "can opener" after *Ajax* and never refer to an *Ajax* alone. Otherwise, the name can become generic and be lost, as happened to "cellophane" and "aspirin," and as could soon happen *to Xerox* (doesn't it

somehow feel more natural to use the word "xerox" than "photocopy"?).

5. What Doesn't Qualify as a Trademark (for the purpose of developing offensive rights)

The courts won't enforce trademark offensive rights, nor will the PTO or state TM offices grant TM registrations, on the following:

- lengthy written matter (copyright is the proper form of coverage here);
- slogans (unless they're used as a brand name);
- trade names (protect these using fictitious name registrations, corporation registrations, and the doctrine of unfair competition);
- immoral, deceptive, scandalous, or disparaging matter;
- governmental emblems, personal names, or likenesses without consent;
- marks which they consider close enough to existing marks as to be likely to cause confusion;
- pure surnames or purely geographical designations; or
- generic or descriptive words.

P. Copyright

A COPYRIGHT IS another offensive right given by law, this time to an author, artist or composer, to exclude others from publishing or copying literary, dramatic, musical, or artistic works. While a patent can effectively provide offensive rights on an idea per se, assuming it's reduced to hardware form, a copyright covers only the author's or artist's particular expression of an idea. Thus, while a copyright can provide offensive rights on the particular arrangement of words which constitute a book or play, it can't cover the book's subject matter, message, or teachings.

Some specific works that are covered by copyright are books, poetry, plays, songs, catalogs, photographs, computer programs, advertisements, labels, movies, maps, drawings, sculpture, prints and art reproductions, game boards and rules, and recordings. Certain materials such as titles, slogans, lettering, ideas, plans, forms, useful things and non-original material aren't covered through copyright. U.S. government publications, by law, aren't covered by copyright and may almost always be freely copied and sold by anyone, if desired.

1. What Is Copyright?

Now that we've seen what a copyright covers, what exactly is a copyright? As stated, a copyright is the offensive right that the government gives an author of any original work of expression (such as those mentioned above) to exclude others from copying or commercially using the work of expression without proper authorization.

The copyright springs into existence the instant the work of expression first assumes some tangible form, and lasts until it expires by law (the life of the author plus 50 years or 75-100 years for works made for hire, depending on when the work is first published).

2. Copyright Compared With Patent

The process involved in obtaining a patent differs significantly from that of registering a copyright. A copyright is deemed to exist automatically upon creation of the work, with no registration being necessary. On the other hand, to obtain patent rights, an application must be filed with the PTO, that office must review, approve, and issue a patent.

If a copyright is registered with the Copyright Office (which technically is part of the Library of Congress) on any copyrightable material, a certificate of registration will be granted without examination as to the work's novelty. The PTO (part of the U.S. Department of Commerce), on the other hand, makes a strict and thorough novelty and unobviousness examination on all patent applications and won't grant a patent unless it considers the invention novel and unobvious.

Finally, with some exceptions, the two forms of offensive rights cover types of creation that are mutually exclusive. Simply put, things that are entitled to a patent are generally not entitled to a copyright, and vice versa. However, it's important to understand that there is a small gray area where this generalization isn't necessarily true. In other words a few creations may be eligible for both types of coverage.

How do you protect a work by copyright?

While no longer necessary for works published after March 1,1989, it's still advisable first to place the familiar copyright notice (e.g., Copyright © 1988 David Pressman) on each published copy of the work. This tells anyone who sees the work that the copyright is being claimed, who is claiming it, and when the work was first published. (The year isn't needed on pictures, sculptures, or graphic works.)

Next you should register he work with the U.S. Copyright Office. If done in a timely manner, registration makes your case better if and when you prosecute a court action (for example, you can get minimum statutory damages and attorney fees). It's useful to distinguish between steps (a) and (b), placing the copyright notice on the work and actually getting a copyright registration. Thus I suggest that you don't say, "I copyrighted my program," but rather say, "I put a copyright notice on my program," or " I applied for a copyright registration on my program."

3. Areas Where Patent and Copyright Law Overlap

Let's look at these principal areas where you may be able to obtain offensive rights on intellectual property under either patent or copyright coverage, or both.

a. Computer Software

Computer programs are the best example of a type of creative work that may qualify for both a patent and copyright protection.

Viewed one way, computer programs are in fact nothing more than a series of numerical relationships (termed routines) and as such cannot qualify for a patent (although they can, of course, be covered under the copyright laws because they have been held to constitute a creative work of expression). However, viewed from another perspective, computer programs are a set of instructions that make a machine (the computer) operate in a certain way. And, in recent years patents have been issued on computer programs where the program was claimed (described in the patent application) in combination with some hardware. Simply put, a programmed machine, programmed system, or process using a program with some hardware may qualify for a patent, whereas the program as such couldn't. More on this in Chapter 5, Section C and Chapter 9, Section G.

Why patent a "program" as opposed to simply registering a copyright on it? Because the patent affords a 17-year, broad, hard-to-design-around scope of protection for the program; i.e., a patent effectively covers the idea (in this case the structure of the code comprising the program) underlying a machine, hardware, or idea. What is the drawback? It takes up to three years to obtain a patent, and because much software will probably become obsolete in a much shorter time, your software may well be a dinosaur by the time the patent issues. Thus, you often don't need the 17-year term of

coverage the patent offers, and if so, money spent on obtaining one may well be wasted. Accordingly, I advise that you apply for a patent on only those programs which utilize very valuable and different software concepts and which are likely to last a long time.

While copyrighting of programs is relatively inexpensive as well as easy to accomplish, the coverage gained isn't as broad as is offered by a patent. This is so because copyright covers only the particular way the program is written, not what it does. For instance, all major word processing programs accomplish pretty much the same tasks (e.g., cursor movement, screen and print formatting, search and find functions, and moving text from one location to another) but each does so through a differently expressed program and thus each is entitled to separate copyright status.

So when choosing whether to rely on copyright or patent offensive rights in respect to software (assuming the software can be "claimed" as part of a physical process) the software author must weigh the broader offensive rights that a patent brings against the expense and time in obtaining one. And the ease with which copyright is obtained must be counter-balanced by the narrow nature of its coverage.

b. Shapes and Designs

The inventor may also have a choice of patent or copyright protection in areas where an object's shape or design is both functional and aesthetic. Consider, for example, a new alphabet with letters that are attractive, yet which also provide more efficient, unambiguous spelling (such as the efficient alphabet which Shaw used to write Androcles and the Lion), or which are easier to read in subdued light. Patent or copyright can be used, the former will afford broader coverage to whatever principles can be identified and the latter will be cheaper, quicker, and easier to obtain., but limited to the specific shapes of the letters. Note that unlike design patents, copyright can be used to cover some aesthetic shapes even if they also have a significant function

In many areas both forms of coverage can be used together for different aspects of the creation. Thus in parlor games, the game apparatus, if sufficiently unique, can be patented, while the gameboard, rules, box, and design of the gamepieces can be covered by copyright. The artwork on the box or package for almost any invention can be covered by copyright, as can the instructions accompanying the product. Also the name of the game (e.g., *Dungeons and Dragons*) is a trademark and can be covered as such.

If the invention can also be considered a sculptural work, or if it's embodied or encased in a sculptural work, copyright is available for the sculpture. However, copyright can't be used for a utilitarian article, unless it has an aesthetic feature which can be separated from and can exist independently of the article.

Of course, to emphasize my earlier point, both copyrights and patents generally have their exclusive domains. Assuming they don't have any aesthetic components, patents are exclusive for machines, compositions, articles, processes, and new uses per se. On the other hand, copyrights are exclusive for writings, sculpture, movies, plays, recordings, artwork, etc., assuming they don't have any functional aspects.

c. *Copyright Compared to Design Patents*

There's considerable overlap here since aesthetics are the basis of both forms of coverage. Design patents are used mainly to protect industrial designs where the shape of the object has ornamental features and the shape is inseparable from the object. E.g., Dizzy Gillespie's trumpet with its upwardly-bent bell is perfect for a design patent, but a surface decal on his trumpet, which could be used elsewhere, is not. Copyright, on the other hand, can be used for almost any artistic or written creation, whether or not it's inseparable from an underlying object, so long as the

aspect of the work for which copyright is being sought is ornamental and not functional. Thus copyright can be used for pure surface ornamentation, such as the artwork on a can of beans, as well as sculptural works where the "art" and the object are integrated; such as a statue. For instance, the shape of a toy was held to be properly covered by copyright since the shape played no role in how the toy functioned and since a toy wasn't considered utilitarian. The same principle should apply to "adult toys," provided they are strictly for amusement and don't have a utilitarian function.

What are the differences in the coverage afforded by design patents and copyright? Design patents are relatively expensive to obtain (the filing fee is higher, an issue fee is required—see Fee Schedule in Appendix), a formal drawing is required, a novelty examination is required, and the rights last only 14 years. However, a design patent offers broader rights than a copyright in that it covers the aesthetic principles underlying the design. This means that someone else coming up with a similar, but somewhat changed design would probably be liable for patent infringement.

Copyright, on the other hand, provides relatively narrow offensive rights (minor changes in all of the artwork's features will usually avoid infringement), the government fee for registration is very small (see Fee Schedule), the term of protection is long (the life of the creator plus 50 years, or a flat 75 or 100 years for works classified as made for hire). And as no novelty examination is performed, you're virtually assured of obtaining a copyright registration certificate if you file.

Because the distinctions between design patents and copyrights are especially confusing, I've provided a comparison chart to summarize the distinctions between these two forms in Fig. 1-A.

4. When and How to Obtain Copyright Coverage

If you desire to copyright an invention, program, creation, or for instructions, packaging, or artwork that goes with your invention, you don't need to do anything until the item is distributed or published. This is because, as mentioned, your copyright rights arise when your work is first put into tangible form. And, although there is no requirement for a copyright notice on your work before it's generally distributed to the public, I advise you to put the proper copyright notice on any copyrightable material right away since this will give anyone who receives the material notice that you claim copyright in it and they shouldn't reproduce it without permission.

When your material is distributed to the public, it's even more desirable (though no longer mandatory for works published after March 1, 1989) that you place a copyright notice on it to notify others that you claim copyright and to prevent infringers from claiming they were "innocent" and thus entitled to reduced damages. This notice should

consist of the word "Copyright, followed by a "c" in a circle © [or a "p" in a circle for recordings and records], followed by the year the work is first published (widely distributed without restriction), followed by the name of the invention's owner, followed in turn by the phrase "All Rights Reserved."[3] Thus the copyright notice on this book appears as "Copyright © 1989 David Pressman. All Rights Reserved."

Strictly speaking, this is all you have to do to preserve your copyright. However, if anyone infringes your copyright (i.e. without your permission someone copies, markets, displays, or produces a derivative work based on your original work) and you want to go to court to prevent this from happening and collect damages, you first have to register your work with the U.S. Copyright Office. Moreover, if you register within three months of the time your item is distributed or published, or before the infringement occurs, you may be entitled to attorney fees, costs, and damages that don't have to be proved by you (called statutory damages). All things considered, I strongly advise you to register your work as soon as it's created if you think you're entitled to copyright coverage.

The registration forms, accompanying instructions, and circulars of general information can be obtained free by writing to the Copyright Office, Washington DC 20559, or by calling them at (202) 287-9100 (24-hour service). For detailed step-by-step guidance regarding copyright in general, ask for Circular 1 from the Copyright Office; for software, I recommend Salone, *How to Copyright Software,* Nolo Press.

---

[3]In the U.S. there are other and shorter acceptable forms of notice. For example, you can abbreviate the word "Copyright" to "Copr." or you can use just the circled "c." However, for the maximum protection worldwide, you should do it the way I suggest here.

Q. Trade Secrets

1. Definition

Thanks to the intensive coverage of the high-tech industry by the media, the term "trade secret" has become virtually a household word. You've probably heard of the IBM suit against Hitachi; here IBM claimed that Hitachi stole IBM's trade secrets by buying these from disloyal IBM employees. Similarly, Commodore Computer company has sued certain former employees to prevent them from revealing Commodore trade secrets to their new employer Atari.

What are these trade secrets and why are they valuable enough to warrant corporations paying millions of dollars to high-priced attorneys to protect them? In a sentence, a trade secret is any information, design, device, process, composition, technique or formula which is maintained as a secret and which affords its owner a competitive business advantage.

Among the items considered as trade secrets are chemical formulas, such as the formula for the paper used to make U.S. currency, manufacturing processes, such as the process used to form the eyes in sewing needles, and "magic-type" trade secrets, such as the techniques used to produce laser light shows and fireworks.

Obviously since these types of information and know how can go to the very heart of a business and its competitive position, businesses will often expend a great deal of time, energy, and money to guard their trade secrets.

When I refer to trade secrets in this book, I mean those which consist of technical information, such as in the examples given above. However, virtually every business also owns "business-information" type trade secrets, such as customer lists, names of suppliers, pricing data, etc. The law will enforce rights to both types of trade secrets, provided the information concerned was kept confidential and can be shown to be truly valuable.

More so than in any of the intellectual property categories, the primary idea underlying trade secrets is plain common sense. If a business knows or has some information which gives it an edge over competitors, the degree of offensive rights which the law will afford to a trade secret is proportional to the business value of the trade secret and how well the secret is actually kept. If a company is sloppy about its secrets, the courts will reject its request for relief. Conversely, a company that takes reasonable measures to maintain the secret will be afforded relief against those who wrongfully obtain the information. These central factors underlying trade secrets have profound implications for those who are seeking patents.

2. Relationship of Patents to Trade Secrets

When a patent issues, the public has complete access to the ideas, techniques, approaches, methods, etc. underlying the invention. This is because, as we'll see in Chapter 8, a patent application must include everything the inventor knows about how to make and use the invention. Since the application is printed verbatim when the patent issues, all of this "know-how" will become public. This public disclosure doesn't usually hurt the inventor, however, since the patent can be used to prevent anyone else from commercially exploiting the underlying information.

DESIGN PATENT   v.   COPYRIGHT

1. Permissible for All of the Following

The aesthetic aspects of original and ornamental articles of manufacture, jewelry, furniture, fabrics, vehicles, industrial equipment.

Literary and artistic content of written materials, lectures, periodicals, plays, musical compositions, maps, artworks, software, reproductions, photographs, prints, labels, translations, movies, sculpture.

2. Disadvantages

Must prepare a formal application with ink drawings, must prosecute before the PTO with legal briefs, large filing fee, and issue fees, lasts only 14 years, takes a long time (one to three years) to secure rights.

Gives a narrow scope of offensive rights, no doctrine of equivalents, no protection of concepts (only particular form of expression thereof), only good against proven actual copiers (not independent creators).

3. Advantages

Broader scope of offensive rights, including doctrine of equivalents (see Chapter 15), can cover concepts, good against independent creators.

Only need fill out a simple form with samples of the actual work, no formal drawings needed, no need for legal briefs, only small filing fee, no issue fee, lasts a very long time (life + 50 years or 75-100 years), instant offensive rights.

4. Can't Be Used For

Articles where the novel features have a utilitarian function (use utility patent); writings, flat artwork, photos, maps, drawings, programs, prints, labels, movies (use copyright).

Utilitarian articles, unless the aesthetic features are separable from and can exist independently of the article (toys aren't considered utilitarian), machines, processes, systems, concepts, principles, or discoveries.

5. Recommended For

The aesthetic shape or layout of utilitarian articles.

Articles of manufacture which aren't utilitarian, or if utilitarian, have aesthetic aspects which can be separated and exist independently, jewelry, furniture, fabrics, literary content of written materials, lectures, periodicals, plays, maps, musical compositions, artworks, software, reproductions, photographs, prints, labels, translations, movies, sculpture.

Fig.1-A—Design Patents Compared To Copyrights

What happens, however, if the patent isn't granted? Because patent applications are treated as confidential by the PTO, it is possible to apply for a patent and still maintain the underlying information as a trade secret (assuming it has been kept strictly secret) during the patent application process. Then, if the patent is later denied, the competition will still not know about the invention and any competitive advantage inherent in that fact can be maintained. Thus, even though the patent application must contain all the details of the invention, including any applicable trade secrets, if no patent is granted, the PTO won't publish anything and the trade secret will remain intact.

Even if a decision is later made by the PTO to grant a patent, you'll have an opportunity to reject it and continue relying on trade secret principles to enforce offensive rights on your invention. This means, in essence, that every patent applicant can both apply for a patent and maintain the invention as a trade secret for the full pendency of the patent application process, which often takes up to three years. Only later, in the event the PTO decides to award a patent, need a decision be made as to which path to follow.

The following material discusses the pros and cons of each form of protection.

3. Advantages of Trade Secret Protection

Often I advise people to choose trade secret rights over that afforded by a patent, assuming it's possible to protect the creation by either. Let's look at some of the reasons why.

- The main advantage of a trade secret is the possibility of perpetual protection. While a patent is limited by statute to 17 years and isn't renewable, a trade secret will last indefinitely if not discovered, e.g., some fireworks and sewing needle trade secrets have been maintained for decades.

- A trade secret can be maintained without the cost or effort involved in patenting.

- There is no need to disclose details of your invention to the public for trade secret rights (as you have to do with a patented invention).

- With a trade secret, you have definite, already existing rights and don't have to worry about whether your patent application will be allowed.

- Since a trade secret isn't distributed to the public as a patent is, no one can look at your trade secret and try to design around it, as they can with the claims of your patent.

- A trade secret can be established without naming any inventors, as must be done with a patent application. Thus no effort need be made to determine the proper inventor and a company needn't request its inventor-employee to assign (legally transfer) ownership of the trade secret to it, as is required with a patent application.

- A trade secret doesn't have to be a significant, important advance, as does a patented invention.

- A trade secret can cover more information, including many relatively minor details, whereas a patent generally covers but one broad principle and its ramifications. For example, a complicated manufacturing machine with many new designs and which incorporates several new techniques can be covered as a trade secret merely by keeping the whole machine secret. To cover it by patent, on the other hand, many expensive and time-consuming patent applications would be required, and even then the patent wouldn't cover many minor ideas in the machine.

- Trade secret rights are obtained immediately, whereas a patent takes several years to obtain, in which time rapidly evolving technology can bypass the patented invention.

4. Disadvantages of Trade Secret Versus Patenting

Before you stop reading this book, please understand that I spent three years writing it for a good reason. Or put more clearly, there are many circumstances in which the trade secret rights have important disadvantages. In these contexts, using the rights offered by a patent is essential.

The main reason that trade secrets are often a poor way to cover your work is that they can't be maintained when the public is able to discover the information by inspecting, dissecting, or analyzing the product (called "reverse engineering"). Thus mechanical and electronic devices that are sold to the public can't be kept as trade secrets. However, the essential information contained in certain chemical compositions sold to the public (e.g., cosmetics), and in computer programs (assuming they're distributed to the public in object code form) often can't be readily reverse engineered and thus can be maintained as trade secrets. However, because very sophisticated analytic tools are now available, such as chromatographs, Auger analyzers, spectroscopes, spectrophotometers, scanning electron microscopes, etc., most things can be analyzed and copied, no matter how sophisticated or small they are. And remember, the law generally allows anyone to copy and make anything freely, unless it is patented or subject to copyright coverage, or unless

its shape is its trademark, such as the shape of the Photomat huts.

Strict precautions must always be taken and continually enforced to maintain the confidentiality of a trade secret. If your trade secret is discovered legitimately, or by any other method, it's generally lost forever, although you do have rights against anyone who purloins your trade secret by illegal means. You can sue the thief and any conspirators for the economic loss you suffered as a result of the thief's actions. In practice this amount can be considerable since it will include the economic value of the trade secret.

Regardless of these offensive rights, individuals rarely will be able to respond adequately in damages; hence the individual's new employer or the purchaser of the trade secret (who usually has a deeper pocket) is usually sued. For example, IBM sued Hitachi as well as the individuals concerned and actually obtained millions of dollars in compensation from Hitachi. In addition, a trade secret is more difficult to sue on and enforce than a patent. A patent must be initially presumed valid by the court, but a trade secret must be proven to exist before the suit may proceed.

A trade secret can be patented by someone else who discovers it by legitimate means. Thus, suppose you invent a new formula, say for a hair treatment lotion, and keep it secret. Jane M., who is totally unconnected with you and who has never even heard of your lotion, comes up with the same formula and decides to patent it, which she does successfully. She can legitimately sue and hold you liable for infringing her patent with your own invention!

What conclusion should you draw from this discussion? Because offensive rights connected with trade secrets continue as long as the trade secret itself is maintained, and because infringement of patents on "trade-secretable" inventions is difficult to discover, if you have an invention which can be kept as a trade secret for approximately 20 years, you may be better off doing so than obtaining a patent on it.

5. Acquiring Trade Secret Rights

I've mentioned several times that trade secret rights are available only for information which its owner has taken all reasonable precautions to keep secret. Over the years the courts have devised a number of tests for determining what these reasonable precautions should be and whether a trade secret owner has taken them. If you're interested in further reading on the subject, I recommend Pooley, *Trade Secrets*, Osborne-McGraw Hill and Remer & Elias, *Legal Care for Software*, Nolo Press. Also, see the heading "Trade Secrets" in the Bibliography.

R. Unfair Competition

THE AREA OF "UNFAIR COMPETITION" is the most difficult to explain. Although anyone who is creative, or is in a competitive business, will encounter unfair competition problems or questions from time to time, any attempts to define this area are necessarily fraught with confusion. And no wonder! The scope of unfair competition law is nebulous in the first place and is regularly being changed by judges who make new and often contradictory rulings.

1. When Unfair Competition Principles Create Offensive Rights

Fortunately, this is a patent book rather than a law school course. And, for the purpose of this book all you really need to understand about unfair competition law can be summarized in several sentences.

- An unfair competition situation exists when one business represents its goods or services in such a way as to potentially cause the class of buyers who purchase the particular type of goods or services

to confuse them with goods or services offered by another business.

- Unfair competition law is usually only available as a source of offensive rights when no offensive rights are available under the trademark, copyright, or patent laws.

- Unfair competition can be used to cover advertising symbols, methods of packaging, slogans, business names, "trade dress" (i.e., anything distinctive used by a merchant to package or house its goods, such as the yellow container which has come to be identified with Kodak film), titles, etc. In other words, when the characteristics of a product or service aren't distinctive or defined enough to be considered a trademark, then unfair competition may be the appropriate way to cover it.

- If a business has engaged in unfair competition, a judge will issue an injunction (legal order) prohibiting the business from any further such activity and will possibly award compensation (monetary damages) to the injured business (i.e. the business who lost profits because of the public's confusion).

2. How does the law of unfair competition affect you?

- If you already have a product or service you find has been copied or pirated, and the traditional methods (patents, copyrights, trademarks, and trade secrets) are no help (perhaps because it's not patentable or it's too late to patent it, it doesn't qualify under the copyright or trademark laws, or it doesn't qualify as a trade secret), you still may be able to get relief under the doctrine of unfair competition.

- If you're contemplating coming out with a product or service, try to make it as distinctive as reasonably possible in as many ways as reasonably possible so that you'll be able to establish a secondary meaning easily. For example, you would be wise to use unique and distinctive packaging ("trade dress"), unique advertising slogans and symbols, a unique title, business name, advertising campaign, etc.

S. Acquisition of Offensive Rights In Intellectual Property—Summary Chart

THE FOLLOWING CHART Summarizes intellectual property law as it applies to various types of mental creations.

Underlying Mental Creation	How to Acquire Offensive Rights	Legal Remedy For Misappropriation
Invention	File a patent application as soon as possible, but within one year of sale or publication.	Patent infringement litigation.
Trademark	Use the TM as a proper adjective with the "™ superscript"; register the TM as soon as possible, with the state and/or the PTO.	TM infringement litigation, either before or after registration.
Writings, Music, Recordings, Art, Software, Etc.	Put a copyright notice on the work; also advisable to register copyright within three months of publication.	Copyright infringement litigation, after registration.
Confidential Technical Or Business Information	Keep it secret; keep; keep good records so you can prove you kept it secret.	Trade secret litigation.
Distinctive Trade Dress Slogans, Etc.	Advertising and frequent use.	Unfair competition litigation.

T. Selection Guide to Which Type of Intellectual Property Is Best For Your Creation

NOW THAT YOU'RE FAMILIAR with all of the types of intangible property, let's summarize how to select the appropriate form for any type of mental creation.

If Your Creation Relates To:

The functional aspect of any machine, article, composition, or process or new use of any of therefore going— e.g., circuits, computer programs associated with hardware, gadgets, apparatus, machinery, tools, devices, implements, chemical compositions and industrial or other processes or techniques which one could discover from final product, toys, game apparatus, semiconductor devices, buildings, receptacles, and vehicles, cloth and apparel, furniture (functional structure), personal care devices, scientific apparatus, abrasives, hardware, plumbing, parts, alloys, laminates, protective coatings, drugs,[4] sporting goods, kitchen implements, locks and safes, timekeeping apparatus, cleaning implements, filters, refrigeration apparatus, environmental control apparatus, medical apparatus, etc. Also, under recent decisions, new and non-obvious genetically altered animals.

Acquire Offensive Rights By:

Utility patent (use the rest of this book.).

Any new design for any tangible thing where the design is nonfunctional and is part of and not removable from the thing, e.g., a bottle, a computer case, jewelry, a type of material weave, a tire tread design, a building or other structure, any article, item of apparel, furniture, tool, etc.

Design patent (use Chapter 9).

Any sexually or asexually-reproduced plant.[5]

Plant patent (see PTO Rules 161- 167).

Any symbol, sign, word, sound, design, device, shape, mark, etc. used as a brand name (trademark), service mark, certification mark, or collective mark. and a common name, e.g., "Ajax™ tools"; later register the TM.(See Section N above.)

Using it as a trademark with "™" and then register it in state and/or federal TM offices.

Any book, poem, speech, recording, computer program, work of art (statue, painting, cartoon, label), musical work, dramatic work, pantomime and choreographic work, photograph, graphic work, motion picture, videotape, map, architectural drawing, artistic jewelry, gameboard, gameboard box, and game instructions, etc.

Place a correct copyright notice on the work, e.g, "© 1986 M. Smith"; later secure copyright registration (See Section O above.)

Any information whatever which isn't generally known which will give a business advantage or is commercially useful, e.g., formulae, ideas, techniques, know-how, designs, materials, processes, etc.

Identify it as proprietary information or a trade secret, e.g., "This document contains Ajax Co.-confidential information:",or put it on an invention disclosure-type form (see Chapter 3) and limit its dissemination using appropriate means. (See Section P above.)

---

[4]Orphan drugs (those useful in treating rare diseases) can be covered under the Orphan Drug Act, 21 USC 360—; write to the Food and Drug Administration for details.

[5]Sexually reproduced plants can also be under the Plant Variety Protection Act, 7 USC 2321—; write to Plant Variety Protection Office, National Agriculture Library, Room 500, 10301 Baltimore Blvd., Beltsville, MD 20705.

If Your Creation Relates To:

Any distinctive design, slogan, title, shape, color, trade dress, package, etc.

Note on Games: You can theoretically acquire offensive rights on games by three different intangible property enforcement tools. You can patent the game apparatus (assuming it's sufficiently different from the prior art) by a utility patent, treat the name of the game as a trademark, and cover the gameboard, rules, and box design by copyright.

Acquire Offensive Rights By:

Use it publicly as much as possible, in advertising, etc., so as to establish a "secondary meaning" to enable you to win an unfair competition lawsuit. (See above.)

Note on Computer Programs: These can also be covered by patents, copyright, or by trade secret. If the program is to be narrowly disseminated under a license agreement so that you have some control over its purchasers or users, keep it as a trade secret, having your purchasers or users sign non-disclosure agreements. If the program is to be widely disseminated, so that trade secret protection wouldn't be practicable, apply for a patent if the program involves one or more valuable or highly-novel algorithms that can be claimed as part of a machine or physical process and which you expect to be useful for more than a few years. Use copyright if the algorithms aren't that valuable or novel. Also, the name of the program is a trademark and should be treated accordingly. The instructions should be covered by copyright.

U. Invention Exploitation Flow Chart

TO GET YOU ORIENTED and make it easier to use this book, here's a chart which shows the overall steps to use to exploit your invention and where the details of these steps are found.

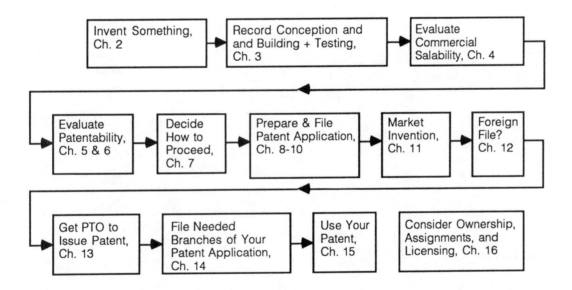

Chapter 2

THE SCIENCE AND MAGIC OF INVENTING

BEFORE WE GET TO PATENTS, the primary subject-matter of this book, I want briefly to talk about inventions and inventing. Why do this? To begin, you may be a first-time inventor and thus have no experience in the real world of protecting and patenting inventions. I believe that too many first-timers get discouraged without trying hard enough. To inspire you to hang in there, I include here some past success stories. Hopefully, when you see that many other small, independent inventors have found their pot of gold, you'll be stimulated to press on.

Inventing provides new creations which enhance our lives, making them more interesting, pleasurable, exciting and rewarding. As the noted Swiss psychologist, Piaget, once said, "We learn most when we have to invent." Remember that everything of significance, even the chair you're probably seated in now, started with an idea in someone's brain. If you come up with something, don't dismiss it; it could turn out to be something great!

▲

Common Misconception: *The day of the small inventor is over; an independent inventor no longer has any chance to make a killing with his or her invention.*

Fact: *As you'll see by the examples given later in this chapter, many small, independent inventors have done extremely well with their inventions. Billions of dollars in royalties and other compensation are paid each year to independent inventors for their creations. In fact 73% of all inventions which have started new industries have come from individual inventors. So, don't be a victim of the "no-use-going-on-with-it because-surely-someone-has-invented-it-already" syndrome. While I recommend that you don't rush blindly ahead to patent your work without making a sensible investigation of prior inventions, (in the ways I discuss later), I urge you not to quit without giving your invention a fair chance.*

Another reason for this chapter is that many inventors come up with valuable inventions, but they haven't developed them enough so that they can be readily sold. If their creations could be improved with further work, they'd have a far greater chance of

success. So here I'll also give some hints about improving your inventions and solving problems about workability, drawbacks, etc.

If you've already made an invention, or are even in the business of inventing, I believe the techniques in this chapter which increase your creativity and provide additional stimulation will help you to make more and better inventions. On the other hand, I also recognize that the information in this chapter may not be particularly helpful to the experienced inventor, the corporate inventor or the corporate patent department worker. After all, you're already firmly in the inventing business. If you would rather skip this information for now, skip to Chapter 3, where my discussion of record keeping should prove of value to even the most seasoned of inventors.

A. What I Mean By "Invention"

FOR THE PURPOSE OF THIS BOOK, an invention is any thing, process, or idea which isn't generally and currently known; which, without too much skill or ingenuity, can exist in or be reduced to tangible form or used in a tangible thing; which has some use or value to society; and which you or someone else has thought up or discovered.

Note that under this definition, an invention can be a process or even an idea, so long as it can be made tangible in some way, "without too much skill or ingenuity." On the other hand, the definition eliminates fantasies and wishes, such as time-travel or perpetual motion machines, since these obviously (at least to me) can't be made tangible.

An invention must have some use or value to society; otherwise what good is it, and how will you sell it? It must be generally unknown anywhere in the world (at the time you invent it), and it must have been thought up or discovered by you or someone else—otherwise it doesn't really have inventive value.

Why do I bother to define the term invention in such detail? So you'll begin to understand it and have a better feel for it, as well as to define the limits of its usage in this book. As you'll see, my primary concern is with inventions that qualify for a patent (i.e., patentable inventions). However, nonpatentable inventions can also be valuable as long as society finds them at least somewhat special and useful.

B. Inventing by Problem Recognition and Solution

NOW THAT YOU KNOW what an invention is, how do you make one? Most inventions are conceived by the following two-step procedure: (1) recognizing a problem, and (2) fashioning a solution.

Although it may seem like duck soup, recognizing a problem often amounts to about 90 percent of the act of conceiving the invention. "To be an inventor is to perceive need." In these situations, once the problem is recognized, conceiving the solution is easy. Consider some of the Salton products—the home peanut-butter maker, for instance, or the plug-in ice-cream maker for use in the freezer. In both cases, once the problem was defined (the need for an easy homemade version of a product normally purchased at the store) implementing the solution merely involved electrification and/or size reduction of an existing appliance. Once the problem was defined, any competent appliance designer could accomplish its solution. True, during the implementation of the idea, i.e., the design of the actual hardware, designers and engineers often contribute the very aspects of the invention that make it ingenious and patentable. Still, the main ingredient leading to a successful outcome for most inventions consists of recognizing and defining the problem that needs to be solved. Although Edison seemed to contradict this when he said that inventing is 10% inspiration and 90% perspiration, he was referring to the whole experience of inventing, including conception, making a practicable model, and licensing or selling the invention. Here, I'm referring just to the conception part of inventing—what Edison called "inspiration."

Of course, in some contexts, the recognition of a problem plays no part in the invention. Most improvement inventions fall into this category, such as, for example, the improvement of the mechanism of a ball-point pen to make it cheaper, more reliable, stronger, etc. But in general, you will find it most effective to go about inventing via the two-step process of identifying a problem and solving it.

Let's look at some simple inventions that were made using this two-step process and which have been commercially implemented. I delineate the problem [P] and solution [S] in each instance. Where I know an Independent Inventor was responsible, I add an "[II]".

1. Grasscrete. [P] Wide expanses of concrete or asphalt in a parking lot or driveway are ugly. [S] Make many cross-shaped holes in the paving and plant grass in the earth below so that the grass grows to the surface and makes the lot or driveway appear mostly green; grass is protected from the car's tires because of its subsurface position.

2. Stick-On Transparent Watch Calendar. [P] Calendar watches don't display a full month's calendar. [S] Print a set of transparent overlay stickers with the calendar for one month of the year on each; a new sticker can be placed on the watch crystal each month, yet the watch hands can still be seen. [Robert Merrick, II]

3. Buried Plastic Cable-Locator Strip. [P] Construction excavators often damage buried cables (or pipes) because surface warning signs often are removed or can't be placed over the entire buried cable. [S] Bury a brightly colored plastic strip parallel to and above the cable; it serves as a warning to excavators that a cable is buried below the spot where they're digging. (This is a "new-use" invention since the plastic strip per se was obviously already in existence.)

4. Magnetic Safety Lock for Police Pistols. [P] Police pistols are often fired by unauthorized persons. [S] A special safety lock inside the pistol releases only when the pistol is held by someone wearing a finger ring containing a high-coercive-force samarium-cobalt magnet.

5. Wiz-z-er™ Gyroscopic Top. [P] Gyroscopes are difficult to get running: they require the user to wind a string around a shaft surrounded by gimbals and then pull it steadily but forcefully to set the rotor in motion. [S] Provide an enclosed gyro in the shape of a top with an extending friction tip which can be easily spun at high speed by moving it across any surface. (Paul Brown, II. Mr. Brown came up with this great invention because, while at a party, he had repeated difficulty operating a friend's son's gyro. His first royalty check from Mattel was five times his annual salary!)

6. Dolby® Audio Tape Hiss Elimination. [P] Audio tapes played at low volume levels usually have an audible hiss. [S] Frequency-selective companding of the audio during recording and expanding of the audio during playback to eliminate hiss. (Ray Dolby, perhaps the most successful II of modern times.)

7. Xerography. [P] Copying documents required messy, slow, complicated photographic apparatus. [S] Xerography—the charging of a photosensitive surface in a pattern employing light reflected from the document to be copied and then using this charged surface to pick up and deposit black powder onto a blank sheet. (Chester Carlson, II. When Mr. Carlson, a patent attorney, brought his invention to Kodak, they said it could never be commercially implemented and rejected it. Undaunted, he brought it to The Haloid Co., which accepted it and changed its name to Xerox Corporation; the rest is history.)

8. Flip-Top Can. [P] Cans of beverage were difficult to open, requiring a church key or can opener. [S] Provide the familiar flip-top can. (Ermal Frase, II)

9. FM, CW, and AGC. [P] Information wasn't conveyable by radio. Noisy, limited frequency response and fade-out of AM reception. [S] Provide CW, FM, and AGC circuitry, familiar to all electronic engineers. (Edwin Howard Armstrong, II, the genius of high fidelity.)

10. Thermostatic Shower Head. [P] Shower takers sometimes get burned because they inadvertently turn on the hot water while standing under the shower. [S] Provide a thermostatic cut-off valve in the supply pipe. (II)

11. Typewriter With Acoustic Key Decoding. [P] Typewriters require awkward, heavy levers to couple the keys to the typing head, or unreliable key switches. [S] Provide an acoustic key-strike sound bar with two microphones at its respective ends for providing two pulses in response to each key strike; the timing and order of the pulses is decoded electronically to determine which key was struck. (Used in SCM's ingenious ultrasonic typewriter.)

12. Organic Production Of Acetone. [P] During WWI, the U.K. desperately needed acetone to make explosives since its normal source was cut off. [S] Use an anaerobic bacterium to produce acetone from locally-available corn mash. (Dr. Chaim Weizmann. This invention helped save one nation and start a new one: It was instrumental in helping the U.K. and the Allies survive WWI and defeat the Germans. The U.K. rewarded Weizmann with the Balfour Declaration which helped lead to the eventual formation of the State of Israel.)

13. Grocery Shopping Cart. [P] Shoppers in grocery stores used their own small, hand-carried wicker baskets and bought only the small amounts which they could carry in the baskets, thereby necessitating several trips to the grocery and causing sales to be relatively low per customer visit. [S] Provide a "grocery cart," i.e., a large wire basket in a frame on wheels so that it can be rolled about and carry a large amount of groceries. Sylvan Goldman, II. [When Mr. Goldman first introduced his carts (about 1925), shoppers wouldn't use them and stores wouldn't buy them despite his extensive efforts. He eventually found a way to get his carts accepted: he hired crews of men and women to wheel the carts about his store, pretending they were shoppers, and also hired a woman at the entrance to the store to offer the carts to entering shoppers. Goldman then made millions from patents on his cart and its improvements (nesting carts and airport carts). This illustrates the crucial value of perseverance and marketing genius.]

The inventors of these inventions necessarily went through the problem-solution process (though not necessarily in that order) to make their invention. Even if an inventor believes the invention came spontaneously, you'll usually find that problem-solution steps were somehow involved, even if they appear to coalesce.

So, if you either don't have an invention or want to make some new ones, you should begin by ferreting out problem or "need" areas. This can often be done by paying close attention to your daily activities. How do you or others perform tasks? What problems do you encounter and how do you solve them? What needs do you perceive, even if they're as simple as wanting a full month's calendar on your calendar watch? Ask yourself if something can't be done more easily, cheaply, simply, or reliably, if it

can't be made lighter, quicker, stronger, etc. Write the problems down and keep a list. Make sure you take the time to cogitate on the problems or needs you've discovered.

Sometimes the solution to the problem you identify will be a simple expedient, such as electrification or reduction in size. Generally, however, it will be more involved, as in some of the examples listed above. But you don't have to be a genius to come up with a solution. Draw on solutions from analogous or even nonanalogous fields. Experiment, meditate, look around. When a possible solution strikes you, write it down, even if it's in the middle of the night. History records a great number of important scientific and conceptual breakthroughs occurring during sleep or borderline-sleep states.

Also, remember that sometimes the "problem" may be the ordinary way something has been done for years, and which no one has ever recognized as a problem. Consider shower heads. Although essentially the same device operated satisfactorily for about 50 years, the inventor of water-massaging shower heads recognized the deficiency of an ordinary constant spray that didn't create any massage effect. He thus developed the water-massaging head that causes the water to come out in spurts from various head orifices, thereby creating the massaging effect.

Don't hesitate to go against the grain of custom or accepted practice if that's where your invention takes you. Many widespread erroneous beliefs have abounded in the past which were just waiting to be shattered. The medical field, in particular, had numerous nonsensical practices and beliefs, such as the use of "poudrage" (pouring talcum powder onto the heart to stimulate it to heal itself), bloodletting, and blistering, and the belief that insanity could be cured by drilling holes in the head to let the demons out.

You'll probably find the going easier if you invent in fields with which you're familiar. In this way you won't tend to "reinvent the wheel." Also, think about uncrowded fields or newly emerging ones where you will find ample room for innovation. But even if you work in an established area, you will find plenty of opportunity for new inventions. For example, more patents issue on bicycles than anything else. Still, you would make millions if you could invent an automatic, continuous bicycle transmission to replace the awkward derailleur. Or how about a truly-compactable bicycle (or wheelchair) which could easily be carried onto a train or into the office but worked as well as the standard variety?

The U.S. Government publishes a quarterly list of needed products requiring inventive effort. Write to the U.S. Small Business Administration, Office Of Innovation, Research And Technology, SBIR, 1441 L Street, N.W., Washington, DC 20416; ask to be put on the list to receive its Quarterly Solicitation Announcements.

One important principle to successful inventing is to remember the acronym KISS (Keep It Simple, Stupid!). If you can successfully eliminate just one part from any machine, its manufacturer (or a competing manufacturer!) will be overjoyed: the cost of the machine will be reduced, it will be lighter, and, of course, it will be more reliable. Another way to look at this is Sandra Bekele's (an inventor-friend) admonition to (figuratively) "eliminate the corners." Or, to quote jazz great Charlie Mingus, "Anybody can make the simple complicated. Creativity is making the complicated simple."

C. Inventing By Magic (Accident and Flash of Genius)

WHEN I DON'T UNDERSTAND how something is done, I sometimes call it "magic." Inventions made by "magic" don't involve the problem-solution technique which I just described; rather, they usually occur by "accident" or by "flash of genius." The PTO and the courts really don't care how you come up

with an invention, so long as they can see that it wasn't already accomplished and it looks substantially different from what's been done before. In the hopelessly-stilted language of the law, "Patentability shall not be negatived by the manner in which the invention was made" (35 USC 103).

Many famous inventions have resulted from accident or coincidence. For example, Goodyear invented rubber vulcanization when he accidentally added some sulphur to a rubber melt. A chemist accidentally left a crutcher (soap making machine) on too long, causing air to be dispersed into the soap mixture. He found that the soup floated when it hardened, thus giving birth to floatable soap bars, such as Ivory® brand. Another chemist accidentally mixed some chemicals together and spilled them, finding they hardened to a flexible, transparent sheet (later known as cellophane). When Alexander Flemming accidentally contaminated one of his bacterial cultures with a mold, he was sufficiently alert and scientifically-minded to notice that the mold killed the bacteria, so he carried this discovery forward and isolated the active ingredient in the mold, which later was named penicillin. (Unfortunately he didn't patent it so he got the fame, but not the fortune.)

The law considers the fact that these inventions came about by total accident, without the exercise of any creativity by their "inventors," legally irrelevant. All other things being equal, a patent on cellophane would be just as strong as one on nylon (another former trademark), the result of 12 years' intensive and brilliant work by duPont's now-deceased genius, Dr. Wallace Carothers, of Wilmington, Delaware.

Since I don't understand how the "magic" occurs, I can't tell you or even suggest how to invent by accident. Please remember, however, that in case you ever come up with an accidental invention, treat it like any other invention; the law will.

The other type of "magical" invention I'll refer to as the product of a "flash of genius." While "flash of genius" inventions inherently solve a need, the inventive act usually occurred spontaneously and not as a result of an attack on any problem. Some examples of this type are the electric knife and the previously-discussed Salton inventions which actually created their own need, the Pet Rock (not a real invention by traditional definitions, but rather, a clever trademark and marketing ploy, but highly profitable just the same), Bushnell's "Pong" game, the Cabbage Patch dolls, Ruth Handler's Barbie Dolls, and a client's Audochron® clock, which announces the time by a series of countable chimes for the hours, tens of minutes, and minutes. With these inventions, the inventor didn't solve any real problem or need, but rather came up with a very novel invention which provided a new type of amusement or a means for conspicuous consumption (showing off).

Although I don't understand how the creativity in these types of cases occurs, I suggest in Part E of this chapter several techniques for stimulating and unlocking such creativity. Using these techniques, many inventors have come up with valuable inventions and profitable ideas and marketing ploys.

manufacturers may not see it that way. So, it's best to have as many alternatives handy as possible.

4. When you apply for a patent, the more ramifications you have, the easier it will be to make your patent stronger (see Chapter 8).

5. Conversely, if the broad concept or initial embodiment of your invention is "knocked out" by a search of the "prior art" [see Chapter 5, Section B(3)] made by you, your searcher, or the examiner in the Patent and Trademark Office, you'll have something to fall back on, so you'll still be able to get a patent.

6. Ramifications often help you understand your basic invention better, see it in a new light, see new uses or new ways to do it, etc.

In some situations, you'll find that you won't be able to ramify beyond your basic conception. But give it a try anyway, and make sure you record in writing any ramifications you do come up with as soon as possible (see Chapter 3).

One way to make ramifications is to pretend that a part of your device can't be made due to a law or shortage and then try to come up with a replacement.

D. Making Ramifications of Your Invention

ONCE YOU'VE MADE AN INVENTION, write down the problem and solution involved. Then, try to ramify it—that is, to do it or make it in other ways so it will be cheaper, faster, better, bigger (or smaller), stronger, lighter (or heavier), longer- (or shorter) lasting, or even just different. Why ramify?

1. Most inventors usually find that their initial solution can be improved or made more workable.

2. By conceiving of such improvements first, you can foreclose future competitors from obtaining patents on them.

3. Even if you believe your first solution is the best and most workable, your potential producers or

E. Solving Creativity Problems

UNFORTUNATELY, hardly any invention ever works right or "flies" the first time it's built. You need to build and test it to be aware of the working problems. If you don't, the first builder, whoever it is, will inevitably face them. If this is a corporation to which you've sold or licensed your invention, it's sure to create problems. If your first construction doesn't work, don't be discouraged; expect problems and expect to solve them through perseverance. If you don't believe me, consider Edison's views on this subject:

"Genius? Nothing! Sticking to it is the genius! Any other bright-minded fellow can accomplish just as much if he will stick like hell and remember nothing that's any good works by itself. You've got to make the damn thing work! . . . I failed my way to success."

If you show your invention to someone and you get static in return, don't necessarily get discouraged; the history of invention abounds with quotes from naysayers who were proved to be disastrously wrong. The enlightening book *303 Of The World's Worst Predictions* by W. Coffey (see Bibliography), is full of amusing and insightful erroneous quotes. Here are a few teasers:

- *Everything that can be invented has been invented.* — U.S. Patent Office Director, urging President McKinley to abolish the Office (1899).

- *What, sir? You would make a ship sail against the wind and currents by lighting a bonfire under her decks? I pray you excuse me. I have no time to listen to such nonsense.* — Napoleon Bonaparte to Robert Fulton, after hearing Fulton's plans for a steam engine driven boat.

- *I think there is a world market for about five computers.* — Thomas J. Watson, IBM President. (1958)

- *Man won't fly for a thousand years.* — Wilbur Wright to Orville after a disappointing experiment in 1901.

To solve problems you have with your creation, such as workability, you can draw on any of the following techniques which are known to enhance creativity.

Frame It Differently: One of the most effective ways to solve a problem is to "frame" the problem properly. Framing is another way of describing the way in which one looks at a situation. A common example of framing a problem occurs when you try to move a bulky sofa through a small doorway. If the first way doesn't work, frame the problem differently by turning the sofa upside down and trying again. Or take another example: If you have an apparatus which includes a lever, and you can't find a design shape for the lever which the machine will accommodate, look at the situation another way; perhaps you can redesign the apparatus to eliminate the lever altogether!

Use Your Right Brain: In the course of trying to solve a problem with an invention, you may encounter a brick wall of resistance when you try to think your way logically through the problem. This is surprising, as thinking through a problem is a linear type of process (i.e., one step follows another) which utilizes our rational faculties usually located in the left side of our brain. This works fine when we're operating in the realm of what we know or have experienced. However, when we need to deal with new information, ideas, and perspectives, linear thinking will often come up short. Creativity, on the other hand, by definition involves the application of new information to old problems and the conception of new perspectives and ideas. For this you will be most effective if you learn to operate in a non-linear manner, i.e., use your right brain or creative faculties. Stated differently, if you're a linear thinker, you'll tend to be conservative and keep coming up with techniques which are already known. This, of course, is just what you don't want.

One way to engage your right-brain faculties in a search for a creative solution to your quandary is to pose the problem in clear terms and then forget about it and think of something completely different.

For example, if you can't fit that lever in your apparatus, think of a different activity, or just take a break (how about a nice boating trip or a hike in the woods). Your subconscious will work on the problem while you're "away." Then come back to the problem and force your different activity onto your problem. In other words, try to think of the apparatus and your boating trip or hike simultaneously. You may find that a solution appears by magic (e.g., you may realize a way to design the machine without the lever!).

Meditation: Another way to bring out your creativity is to meditate on the problem or meditate merely to get away from the problem. Either will help. As strange as it seems, some experts say that creativity can be enhanced during reverie by listening to a largo movement from a baroque symphony. At least you'll enjoy it! Also, the use of biofeedback machines can induce or teach deep relaxation with enhanced alpha, or even theta brain waves, a very effective stimulus to creativity.

Dreams: Most creative people find dreams the most effective way of all to solve problems. Or as Edison said: "I never invented anything; my dreams did."

Elias Howe solved the basic problem of his sewing machine in a dream. He saw some tribal warriors who ordered him to come up with a solution or they would kill him. He couldn't make a solution, so the warriors then threw their spears at him. When the spears came close, he saw that each had a hole near its tip. He awoke from the nightmare in terror, but soon realized the symbology: he put a hole near the tip of his bobbin needle and passed the thread through. Again, the rest is history.

Similarly, Mendelev came up with the periodic table of the elements in a dream.

To stimulate creative dreaming, first immerse yourself in the problem near bedtime. Then forget about it—do something completely different and go to sleep. Your subconscious will be able to work on the problem. You'll most likely have a dream with an inspiration or insight. Then remember the dream and evaluate the insight to find out if it's correct (sometimes it won't be!).

Note that you'll forget most dreams, so keep a dream diary or notebook handy, by your bedside. Also, you'll find a pen with a built-in flashlight is also helpful. Before you go to bed, repeat fifteen times, "I'll remember my dreams." Whenever you do dream, wake up (you'll find it possible to do this if you intend to do so beforehand) and write your dream down promptly. Once they are written down, forget about them, go back to sleep, and try to figure them out in the morning. Sometimes a week or more will pass before the meanings become clear. Or talk your dreams over with an equally inventive friend and see if he or she can get the meaning—sometimes talking about it helps.

Good luck. And pleasant dreams!

Computerized Creating: As strange as it may seem, computers can be used to enhance creativity, solve problems, and bust through conceptual roadblocks. Several programs for this purpose exist and I believe they can be of significant help in this area. The programs work by first asking you to enter lots of details of your problem or area and then it rearranges the details and suggests lots of modifications and permutations for you to consider. One good program is called *The Idea Generator* from Experience In Software, Inc., Berkeley, CA.

The Hot Tub Method: This has been used by many creative geniuses, starting with Archimedes who discovered the principle of volumetric measurement while in his tub. It works like this: When you relax in a hot tub for a long period, the heat on your body mellows you out and dilates your blood vessels so as to draw blood from your analytical brain, allowing your creative subconscious to come to the fore.

Unstructured Fanaticism: As "excellence guru" Tom Peters states, structured planners rarely come up with the really great innovations; monomaniacs who pursue a goal with unstructured fanaticism often do.

So let yourself go and become an unreasonable madman—it may do the trick!

Group Brainstorming: If all else fails, get a group of friends or trusted associates together and throw the problem to the group. For some unknown reason, a group of people working together often come up with more good ideas than the same individuals working separately. This synergistic method is often used in corporations with great success.

F. Contact Other Inventors

IN RECENT YEARS, many inventors' organizations have developed or sprung up in order to provide inventors with information and ideas, model makers, lists of searchers, speakers, and patent attorneys, etc., as well as to sponsor various seminars and trade fairs where inventions can be exhibited. One or more of these organizations may provide you with invaluable assistance in your inventing efforts. As far as I'm aware, all of these organizations are legitimate and honest, and provide reasonable value for the membership or other fees charged, but check yourself before investing a significant amount of your time or money. Some of these organizations are:

American Society of Inventors
Box 58426
Philadelphia, PA 19102

Sunshine Inventors' Assn.
3319 Maquire Blvd., Suite 155
Orlando, FL 32803

Inventors' Assistance League
345 West Cypress
Glendale, CA 91204

Inventors of California
3201 Corte Malpaso, Suite 304
Camarillo, CA 93010

Inventors' Workshop International
Box 6664
Woodland Hills, CA 91356
(has local branches)

Minnesota Inventors' Congress
Box 71
Redwood Falls, MN 56283

Mississippi Society of Scientists & Inventors
Box 2244
Jackson, MS 39225

Oklahoma Inventors' Congress
Box 53043
Oklahoma City, OK 73152

Technology Transfer Society
11720 W. Pico Blvd.
Los Angeles, CA 90064

IWI Education Group
340 Rosewood Ave.
San Jose, CA 95117

CIC Inventors' Group
869 Alameda De Las Pulgas
Redwood City, CA 94061

Ohio Inventors Association
146 South High Street, Suite 206
Akron, OH 44308

G. Beware of the Novice Inventor's "PGL Syndrome"

AS HIGHLY SUCCESSFUL inventor (Whiz-z-er top) Paul Brown has discovered, many novice inventors have a very different attitude from experienced inventors. This attitude can be summarized as the "PGL (Paranoia, Greed, Laziness) syndrome." Let's discuss the components of this syndrome in more detail since each usually is a significant hindrance for inexperienced inventors.

Paranoia: Extremely common with inexperienced inventors, paranoia (excessive suspicion of other people's motives) makes them afraid to discuss or show their invention to others— some even go as far as refusing to disclose it to a patent attorney. I do advise some measure of caution with unpatented inventions. However, once you record your invention properly (as discussed in Chapter 3), you can and should disclose it to selected persons, provided you take adequate measures to document whom you've disclosed it to and when. Don't be as paranoid as my friend Tom who invented a very valuable stereo movie invention but kept it totally to himself out of fear of theft, only to see it patented and commercialized by someone else.

Greed/Overestimation: Most people have heard fabulous stories of successful inventors who've collected millions in royalties. As a result, some novice inventors think that their invention is worth millions and demand an unreasonably large royalty or lump-sum payment for their creation. This is seldom wise. It is much better to set your sights at a reasonable level (see Ch. 16) so you won't miss out on commercial opportunities.

Laziness: Some novice inventors believe that all they need to do is show their invention to a company, sign a lucrative contract, and let the money roll in. Unfortunately it hardly ever happens so easily. To be successful, you usually have to record your invention properly (Chapter 3), build and test a working model (desirable but not always necessary), file a patent application, seek out suitable companies to produce and market the invention, and work like hell to sell the invention to one of these companies.

H. Don't Bury Your Invention

IF YOU BELIEVE that you have what will turn out to be a successful idea, but you have doubts because it's very different, or you get negative opinions from your friends, consider that Alexander Graham Bell was asked by an irate banker to remove "that toy" from his office. The "toy" was the telephone. Or if that doesn't convince you, ponder these words of Mark Twain and Albert Einstein:

The man with a new idea is a crank—until the idea succeeds.—Mark Twain

For an idea that does not at first seem insane, there is no hope.—Einstein

And as a recent successful inventor, Nolan Bushnell (*Pong*) said, "Everyone who's ever taken a shower has an idea. It's the person who does something about it who makes a difference."

I hope you've received my message in this chapter loud and clear. If you have a worthwhile invention, and you scrupulously follow all the advice and instructions given in this and the succeeding chapters, and persevere, I believe you'll have a very good chance of success.

Chapter 3

Documentation Can Be Vital

A. Introduction

IT'S TRUE IN LIFE generally that the better the documentation you keep, the easier it will be for you to retrieve important ideas, information, and, when necessary, proof that something happened. When it comes to inventing, good documentation is even more vital than in most other aspects of our lives. There are two distinct and important reasons why all inventors should document all of their work. The first has to do with the inventing process itself. The second involves the possibility that you will need to prove 1) that you are the inventor and 2) that you came up with the invention first. Let's examine these reasons in order.

Note: To help you properly document your invention, Nolo Press publishes the *Inventor's Notebook*. See the back of this book for more information.

B. Documents Are Vital to the Invention Process

1. Good Engineering Practice

It's good engineering practice to keep a "technical diary," i.e., accurate, detailed documents of your ideas, work done, and accomplishments. Good engineers and technicians record their developments in chronological order so that they can refer back to their engineering diary days, weeks, months or even years later. First, this enables them to avoid running up the same blind alley twice. Second, good records will shed light on subsequent developments, will allow the inventor to find needed data and details of past developments, and will provide a base for new paths of exploration and ramifications, especially if failures have occurred.

2. Psychological Stimulus

Many of us come up with great ideas, especially when we're engaged in some other activity (including dreaming), and we forget to write them down. Later, we may recall that we had a brilliant idea the night before, or during the office party, but because we went back to sleep or were too busy, we forgot it. If we could get into the habit of writing down our thoughts on a piece of paper, later on we'd find that piece of paper there to bug us, almost forcing us to do something about it. So, keep a small pencil or pen and some paper with you at all times, even by your bedside, and in your wallet, and write down your thoughts as soon as they occur. Later on, you'll be glad you did.

3. Analyzation Stimulus

A WWII admiral (Raborn) once said: "If you can't write it down, you don't really know what you are doing."

Have you ever had an idea, plan, or concept which you really didn't fully understand yourself? I'll bet you discovered that when you tried to write it down, you were forced to figure it out, and only then finally realized fully or exactly what you had. Writing your ideas down forces you to think about them and crystallize them into communicable form. Note that no matter how great your idea, and no matter how much of the work you do yourself, you'll never be able to make a nickel from it until you can communicate it to others, e.g., to get a patent, to license it, or to sell the product.

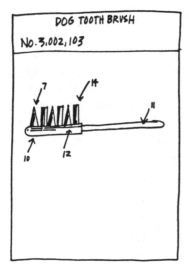

C. Documentation Is Vital for Proving Date and Fact of Invention

IF YOU KEEP CLEAR, signed, dated, and witnessed documents of your creations, this will prove to others that you are a methodical, diligent and reliable person. Who cares? While you may not be particularly interested in establishing such a reputation, you'll find it invaluable in case you ever get into any invention contest, such as an interference, theft situation, etc. Also, when you go to license the invention, or undertake any other activity with it, as well as taking any tax deductions for your expenditures (see below), you'll find that having such documents will greatly enhance your standing with anyone who sees, evaluates, or considers your invention, or any aspect of your inventive activity.

Good documents can be crucial when you may be asked to prove (hopefully this will never come!) that you were the inventor, or the first to invent a particular invention. Because many valuable inventions are independently and simultaneously conceived and brought to fruition, while others are misappropriated from the true inventor, it's

important that you embrace the very specific record keeping techniques suggested later in this chapter.

Let's start with the law: First, note generally that in patent law, dates are crucial. Thus you should date everything you receive or send; you can never tell when you'll need to rely on (prove) any date.

Second, and specifically, an inventor who's first to conceive of an invention will be entitled to a patent on it, as against the competing claims of other inventors, as long as he or she can show adequate (diligent) efforts were also made to build the invention and/or file the patent application. Because of this, the date of conception is extremely important in resolving disputes when two or more inventors claim a patent to the same underlying invention (called an "interference" and covered in Chapter 16). For this reason I have made the recording of the fact of an invention the first of my seventeen "Inventor's Commandments."

INVENTOR'S COMMANDMENT #1:
After conceiving of an invention, you shouldn't proceed to develop, build, or test it, or reveal it to outsiders, until you first 1) make a clear description of your conception (using ink), 2) sign and date the same, and 3) have this document signed and dated by two trustworthy people who have "witnessed and understood" your creation.

There are five reasons why it's legally important promptly and properly to record your conception of your invention:

1. In Case of an Interference

As mentioned above, the primary legal reason for recordation is to deal with the claim of others that they invented your invention first. A description of your invention will be an invaluable item of proof to show that you came up with the invention on the date given and that you [and your co-inventor(s), if any], are the actual and true inventor(s) of the creation.

2. Proof in Case of Theft

Similarly, if someone sees or hears about your invention and attempts to "steal" it by claiming it as his or her own invention (in actuality, a rare occurrence), there will probably be a lawsuit or other proceeding in which the true and first inventor must be ascertained. In such a proceeding, the side with the best and most convincing evidence will win. 'Nuff said!

3. Proof in Case of Confusion of Inventorship

There's also, commonly, confusion as to who is the actual, true, and first inventor of a particular invention. Often several engineers or friends will be working on the same problem, and if conception isn't promptly recorded, memories fade and there will be confusion as to who is (are) the actual inventor(s). Also, bosses and other supervisors have been known to claim inventorship, or joint inventorship, in an employee's invention. If all inventors promptly record their inventions and get them witnessed, preferably by co-workers (including bosses), there would be very few cases of such confusion of inventorship.

4. Antedate References

As we'll see later in Chapter 13, if the PTO examiner cites a "prior-art reference" against your application (i.e., finds a prior publication that casts doubt on the originality of your invention), you can eliminate that reference as prior-art (i.e., prevent the examiner from using it) if you're able to show:

a. you filed your application within the year after the reference's publication date; or

b. you built and tested the invention prior to such publication date; or

c. you conceived of the invention prior to such publication date and you were then diligent in building and testing it.

This process of antedating a cited reference is called "swearing behind" the reference. Naturally, to be effective and acceptable when swearing behind a reference, your records should be detailed, clear, and witnessed.

5. Supporting Tax Deductions

Once you make an invention and spend any money on your creation, the IRS considers that you are "in business," thus enabling you to file a "Schedule C" or "Schedule E" (Form 1040) with your tax return to deduct all expenditures you made on your invention from ordinary or invention income that you received. The IRS will be far more inclined to allow these deductions if you can support them with full, clear and accurate records of all of your invention activities, including, but not limited to, expenditure receipts.

D. Trade Secret Considerations

IN CHAPTER 1, Section P, you learned that an invention can qualify as a trade secret right up until the time a patent issues on it. Keeping an invention secret can provide its owner with certain obvious commercial advantages and the owner may have recourse in the courts against any person who improperly discloses the secret to others.

Making a witnessed record of your invention doesn't conflict in any way with this trade secret protection. Even if you show your invention to witnesses, this won't compromise the trade-secret status of your invention because of the implied understanding that witnesses to an invention should keep it confidential. However, I recommend that you don't merely rely on this implied understanding, but actually have your witnesses agree to keep your invention confidential. A verbal agreement is good, but a written agreement is far better and will really tie down the confidentiality of your invention. I've incorporated a keep-confidential agreement in the disclosure form (Form 3-2—discussed below), but you can also have your witnesses sign the "Proprietary Materials Agreement" (Form 3-1—discussed below) when you give them your lab notebook or disclosure to sign.

Whether your invention is to be patented or kept as a trade secret (you can decide later—see Chapter 7), you should first record it properly so as to be able to prove that you invented it and you did so as of a certain date. Since you can keep your notebook confidential, at least for the time being, no loss of any potential trade secret protection will result from your making a proper record of your conception.

Remember that while recording your invention can be vital in the situations outlined above, it provides only limited rights, since it won't give you any weapon to use if anyone independently comes up with your creation, or if anyone copies your invention once it's out on the market. To acquire full offensive rights in these situations, you need to obtain a patent on your invention. As discussed in Chapter 1, a patent will give you rights against independent creators of your invention and those who copy it once it's out on the market.

E. Record the Building and Testing of Your Invention

AFTER YOU CONCEIVE of your invention and prepare the proper record, you should follow my Inventor's Commandment #2.

INVENTOR'S COMMANDMENT #2:
(1) Try to build and test your invention (if at all possible) as soon as you can, (2) keep full and true written, signed, and dated records of all the efforts, correspondence and receipts concerning your invention, especially if you build and test it, and (3) have two others sign and date that they have "witnessed and understood" your building and testing.

If, as part of the testing of your invention, you have to order any special part or material, or if you have to reveal to or discuss your invention with anyone to get it built, be cautious about how and whom you contact. And when you do make any specific revelation, have the recipient of the information about your invention (which is a trade secret or proprietary information at this stage—see Chapter 1, Section P), sign a Proprietary Materials (keep confidential) Agreement (Form 3-1 in the Appendix). Model makers and machine shops are used to routinely signing these agreements.[1] The agreement is completed merely by identifying the materials (documents or hardware) in the first section, your name in the second section (you're the Lender), and the name of your recipient in the third section (he, she, or it is the Borrower). Have your Borrower sign and date the bottom of the agreement. I recommend that you give a copy of the signed agreement to your Borrower, as well as any extra copies that may be needed if any other persons in your Borrower's organization are to sign also.

Note that the Agreement calls the delivery of your proprietary materials to the recipient a "loan." This will give you maximum rights if the recipient makes unauthorized use of or refuses to return the materials.

This agreement will cover almost all situations where you need to deliver proprietary materials under a keep-confidential arrangement. However it isn't cast in stone: If, for example, you are making more than a loan of the materials, feel free to redraft the Agreement, e.g., by changing "loaned" to — deliver—and "Lender" to —Owner—.[2]

Why should you painstakingly record the activities involved in the building and testing of your invention? This is an easy question to answer. All of the reasons discussed for recording the facts of your invention in the first place are applicable here, in spades. This is because the building and testing of an invention can be as (or even more) important than its conception, especially as proof of your invention in case of theft, confusion of inventors, interferences, the need to swear behind references, and the need to

establish tax deductions. However, recordation of your efforts to build and test your invention isn't necessary to obtain a patent, unless an interference or other special situation occurs that requires you to prove your development efforts.

To illustrate the value of recordation, recently I prepared a patent application for a client. As she was reviewing it, I got a flier in the mail from a store listing for sale an item almost identical to that which my client wished to patent! Since the item was being sold and was published before she had filed the application, the flyer theoretically could preclude my client's invention from being considered as novel and thus lead to the rejection of her application. But fortunately, my client had built and tested the invention, had made records of her building and testing, and had signed and dated these and had gotten them witnessed months before. She could thus go ahead and file without fear of the fact that the flier was published before her filing date. This is because she could use her records to "swear behind" the flyer. Simply put, by following Inventor Commandments #1 and #2, my client was still able to obtain a patent on her invention. On the other hand, had she failed to properly record her conception and efforts, her application would have been totally barred and she would have lost all rights to her invention!

F. How to Record Your Invention

1. The Lab Notebook

Hopefully I've managed to sell you on the need to carefully and accurately record your thoughts and activities that normally occur in the course of inventing. The best, most reliable and most useful way to record your invention is to use the *Inventor's Notebook*, by Grissom and Pressman, Nolo Press.

---

[1] If you make an appointment with someone whom you wish to sign the agreement prior to showing your invention, it's only courteous and proper business practice to advise your counterpart in advance that you'll be asking to have a keep confidential agreement signed; don't spring the agreement in a surprise manner.

[2] The agreement will also cover oral disclosures, but for reasons of difficulty of later proof (if needed) you should make disclosures only by actually delivering written materials.

Specifically designed for use with this book, the *Inventor's Notebook* provides organized guidance for properly documenting your invention. More information about the *Inventor's Notebook* and how to order it can be found at the end of this book.

If you choose not to use the *Inventor's Notebook*, what type of notebook should be used? Preferably one with a closed spiral binding or with a stiff cover, with the pages bound in permanently, e.g., by sewing or gluing. Also, the pages should be consecutively numbered. Lab notebooks of this type are available at engineering and laboratory supply stores, and generally have crosshatched, prenumbered pages with special lines at the bottom of each page for signatures and signature dates of the inventor and the witnesses. As should be apparent, the use of a bound, paginated notebook that's faithfully kept up provides a formidable piece of evidence if your inventorship or date of invention is ever called into question, for instance in an interference proceeding or lawsuit. A bound notebook with consecutively dated, signed and witnessed entries on sequential pages establishes almost irrefutably that you are the inventor (i.e., the first to conceive the invention) on the date indicated in the notebook.

If you don't have or can't get a formal lab notebook like this, a standard bound letter-paper size crackle-finish school copybook will serve. Just number all of the pages consecutively yourself, and don't forget the frequent dating, signing and witnessing, even though there won't be special spaces for this. You should date each entry in the notebook as of the date you and your co-inventor(s), if any, make the entries and sign your name(s). If you made the entries over a day or two before you sign and date them, add a brief candid comment to this effect, e.g., "I wrote the above on July 17, but forgot to sign and date it until now." Similarly, if you made and/or built the invention some time ago, but haven't made any records until now, again state the full and truthful facts and date the entry as of the date you write the entry and sign it. E.g., "I thought of the above invention while trying to open a can of truffles at my sister's wedding reception last July 23 (1984), but didn't write any description of it until now when I read *Patent It Yourself*."

EYE PROTECTOR FOR FOWL
NO. 620,832

Fig 2.

2. How to Enter Technical Information in the Notebook

Fig. 3-A is an example of a properly-completed notebook page showing the recordation of conception, and Fig. 3-B shows recordation of building and testing.

The sketches and diagrams should be clearly written (preferably double-spaced) in ink to preclude erasure and later-substituted entries. Your writing doesn't have to be beautiful and shouldn't be in legalese. Just make it clear enough for someone else to understand without having to read your mind. Use sketches where possible. Many inventors have told me they put off writing up their invention in a notebook or invention disclosure because they didn't know the proper "legal" terms to use, or had writer's block. However as indicated, legalese isn't necessary or desirable and you can bypass writer's block by relying mostly on sketches with brief labels which explain the parts and their functions.

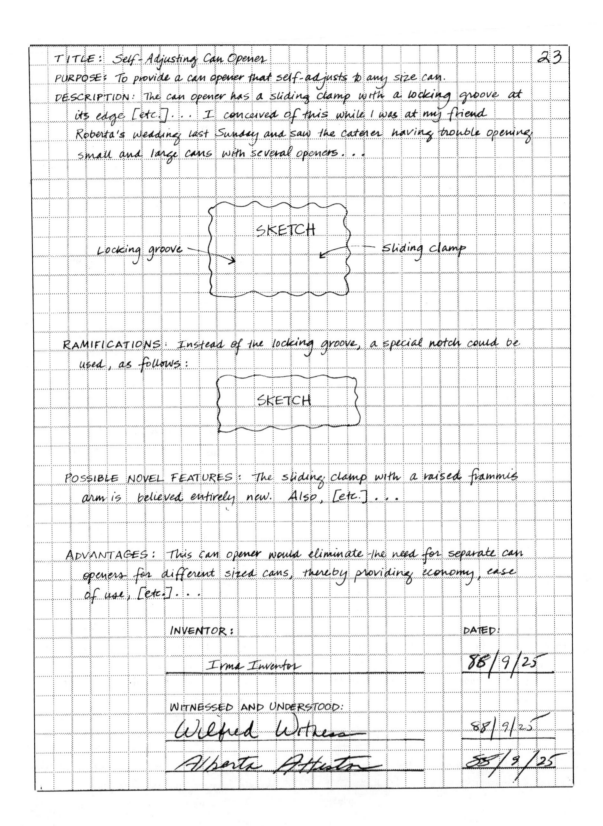

Fig. 3-A—Properly Completed Notebook Page Showing Conception

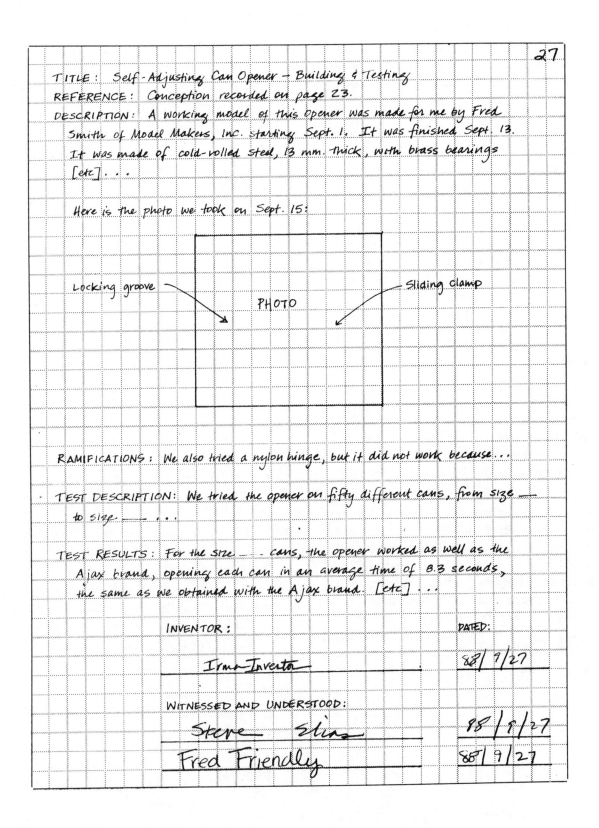

Fig. 3-B—Properly Completed Notebook Page Showing Building and Testing

No large blank spaces should be left on a page—it should be filled from top to bottom. If you do need to leave space to separate entries, or at the bottom on a page where you have insufficient space to start a new entry, draw a large cross over the blank space to preclude any subsequent entries, or, more accurately, to make it clear that no subsequent entries could have been made in your notebook.

If you make a mistake in an entry, don't attempt to erase it; merely line it out neatly and make a dated note why it was incorrect. The notation of error can be made in the margin adjacent to the correct entry, or it can be made several pages later, provided the error is referred to by page and date. Don't make cumulative changes to a single entry. If more than one change is required, enter them later with all necessary cross-references to the earlier material they supplement. Refer back to earlier material by page and date.

If possible, all entries should be made directly in the notebook, or transferred there from rough notes on the day the notes were made. If this isn't possible, make them as soon as practicable with a notation explaining when the actual work was done, when the entries were made and why the delay occurred.

If you've made an invention several months ago, and are now going to record it because you've just read this book, you should date the entries in the notebook when you actually write them, but you should also write when you actually made the invention and explain the delay with honesty and candor! Since the notebook is bound, you will have to handwrite the entries in it. Again, don't worry about the quality of your prose—your goal is only to make it clear enough for someone else to understand; use labeled sketches if writer's block occurs.

3. What Should Be Entered in the Notebook

Your notebook should be used as a "technical diary," so you should record anything you work on of technical significance, not just inventions. The front of the notebook should have your name and address and the date you started the notebook. When you record the conception of your invention, you and anyone who later sees the notebook will find it most meaningful if you use the following headings:

- Title (what your invention is called);
- Purpose (what purpose the invention is intended to serve);
- Description (a functional and structural description of the invention);
- Sketch (an informal sketch of the invention);
- Ramifications (all ramifications of the invention that you have conceptualized);
- Novel features (all possible novel features of the invention);
- Closest known prior art (the closest known existing approach of which you're aware);
- Advantages (of the invention over previous developments and/or knowledge—see the example in Fig. 3-A).

Don't forget to sign and date your conception and have two witnesses also sign and date the record of conception. See Section 5 below.

To record the subsequent building and testing of your invention at a later page of the notebook, you will find it most useful to record the following items:

1. Title and Back Reference;
2. Technical Description;
3. Photos and/or Sketches;
4. Ramifications;
5. Test Description;
6. Test Results;
7. Conclusion.

Fig. 3-B (above) shows a properly-done lab notebook record of the building and testing of an invention. Don't forget to sign, date, and have your witnesses also sign and date the building and testing record, as well as the conception record. See Section 5 below.

If you're skilled enough to conceive, build and test your invention all at once, just combine all of the items of Figs. 3-A and 3-B as one entry in your notebook.

I strongly recommend that you record as much factual data as possible; keep conclusions to a minimum and provide them only if they are supported by factual data. Thus, if a mousetrap operated successfully, describe its operation in enough detail to convince the reader that it works. Only then should you put in a conclusion, and it should be kept brief and non-opinionated. For example, "Thus this mousetrap works faster and more reliably than the Ajax brand." Sweeping, opinionated, laudatory statements tend to give an impartial reader a negative opinion of you or your invention. However, it's useful to include the circumstances of conception, such as how you thought it up, where you were, etc. This makes your account believable and helps refresh your memory later.

The entries should be worded to be complete and clear in themselves so that anyone can duplicate your work without further explanation. While the lab notebook shouldn't be used as a scratch pad to record every calculation and stray concept or note you make or think about, it shouldn't be so brief as to be of no value should the need for using it as proof later arise. If you're in doubt as to whether to make an entry, make it; it's better to have too much than too little.

Also, you'll find it very helpful to save all of your "other paperwork" involved with the conception, building, and testing of an invention. Such paperwork includes correspondence, purchase receipts, etc. These papers are highly probative (trustworthy and useful as evidence) since they are very difficult to falsify. For example, if you buy a thermometer or have a machine shop make a part for you in January 1977, you should save receipts from these expenditures since they'll tie in directly with your notebook work.

4. How to Handle Large or Formal Sketches, Photos, Charts or Graphs Drawn on Special Paper

If possible, items that by their nature can't be entered directly in the notebook by hand should be made on separate sheets. These, too, should be signed, dated and witnessed and then pasted or affixed in the notebook in proper chronological order. The inserted sheet should be referred to by entries made directly in the notebook, thus tying them in to the other material. Photos or other entries which cannot be signed or written should be pasted in the notebook and referenced by legends (descriptive words, such as "photo taken of machine in operation") made directly in the notebook, preferably with lead lines which extend from the notebook page over onto the photo, so as to preclude a charge of substituting subsequently made photos (see Fig. 3-B). The page the photo is pasted on should be signed dated, and witnessed in the usual manner.

If an item covers an entire page, it can be referred to on an adjacent page. It's important to affix the items to the notebook page with a permanent adhesive, such as white glue or non-yellowing transparent tape.

If you have to draw a sketch in pencil and want to make a permanent record of it (to put in your notebook) without redrawing the sketch in ink, simply make a photocopy of the penciled sketch: voila´—a permanent copy!

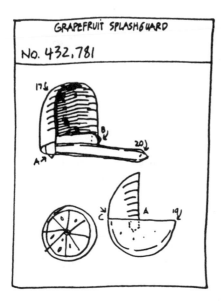

5. Witnessing the Notebook

As I've repeatedly stressed earlier in this chapter, it's important that the notebook entries be witnessed. This is because an inventor's own testimony, even if supported by a properly completed notebook, will often not be adequate for proving an entry date. The witnesses chosen should be as impartial and competent as possible, which means they shouldn't be close relatives or people who have been working so closely with you as to be possible co-inventors. A knowledgeable friend, business associate, or professional will make an excellent witness, given the necessary technical ability or background to understand the invention. The witness should also be someone who's likely to be available later. Obviously, a person who's seriously ill, or of very advanced age, wouldn't be a good choice. Don't ask your patent attorney (if you are using one) to perform this function since the courts and the PTO won't allow an attorney to represent someone and also be that person's witness.

If the invention is a very simple mechanical device, practically anyone will have the technical qualifications to be a witness. But if it involves advanced chemical or electronic concepts, obviously a person with an adequate background in the field will have to be used. If called upon later, the witnesses should be able to testify to their own knowledge that the physical and/or chemical facts of the entry are correct. Thus they shouldn't just be witnesses to your signature (such as when you write a will and have it witnessed), but to the actual technical subject material in the notebook, including the actual tests if they are witnessing the building and testing (Fig. 3-B). Obviously, then, you should call in your witnesses to observe your final tests and measurements so that they can later testify that they did witness them.

While one witness may be sufficient, two are preferred, since this will enhance the likelihood of at least one of them being available to testify at a later date. If both are available, your case will be very strong. Also, if a dispute occurs between two inventors, the one with the greater number of witnesses will prevail, assuming all other considerations are substantially a wash..

Some notebooks already contain, on each page, a line for the inventor's signature and date, together with the words "Witnessed And Understood" with lines for two signatures and dates. If your notebook doesn't already contain these words and signature lines, merely write them in as indicated in Figs. 3-B to 3-D. To really tie down the trade secret status of your invention, you should add the words "The above confidential information is" just before the words "Witnessed And Understood" as has been done on Form 3-2 and on Figs. 3-B, 3-C, and 3-D. You and the witnesses should sign and enter the date on the appropriate lines at the end of your description of the conception of your invention and at the end of your description of your building and testing.

G. Another Way to Keep Records— The Invention Disclosure

SUPPOSE YOU CONCLUDE that for some good reason it's too difficult or inconvenient for you to keep a notebook or technical diary. There's a second, albeit somewhat inferior, way for you to record the conception of your invention. This is by using a document called an "Invention Disclosure."

Despite its formidable name, an Invention Disclosure is hardly different from a properly completed notebook entry of an invention. It should be a complete record of your invention, including a title, its purposes(s), advantages, a detailed description of it, possible novel features, ramifications, details of its construction if you built it, and results obtained, if any. These entries should be made on a separate sheet of paper which has no other information on it except details of your invention and your name and address. For your convenience, Form 3-2 in the Appendix provides an Invention Disclosure form, and Fig. 3-C illustrates how the form should be completed to record conception.

Since an Invention Disclosure isn't bound, the writing on it can, and preferably should, be typed. But if you do write rather than type, just make sure your handwriting is legible. A sheet of professional or personal letterhead (if you have it) is suitable for an Invention Disclosure document. Otherwise print your name, address, and telephone number at the top (or bottom, after your signature). Business letterhead is okay if the invention is to be owned by your business. If the disclosure runs to more than one page, you should write the title of your invention on the second and each succeeding page, followed by the word "continued," numbering each page and indicating the total number of pages of the entire disclosure—for example, "Page 1 of 3."

As before, the description of your invention should be signed and dated by you, marked "The above confidential information is Witnessed and Understood," and signed and dated, preferably by two witnesses, who, as before, are technically competent to understand your invention and who actually do understand and have witnessed the subject matter you have entered on your Invention Disclosure. See Section F(5) above. If you use more than one page, each should be signed and dated by both the inventor and the witnesses.

As with the notebook, if you conceive of an invention on one date, and build and test the invention later, you should make two separate invention disclosures, one to record conception and the second to record the building and testing. The second should refer to the first, and both should be signed and dated by you and the witnesses. I haven't provided an example of an invention disclosure completed to show building and testing, but it would be similar to the notebook entry to show same, i.e., Fig. 3-B set out in Section F above.

H. The Disclosure Document Program (DDP)—or How to Make the PTO Your Witness

SEVERAL YEARS AGO the Patent and Trademark Office (PTO) started a program under which it accepts Invention Disclosures and preserves them for two years, or longer if a patent application is filed which refers to the Invention Disclosure. The purpose of this service, for which the PTO charges a small fee (see Fee Schedule in the Appendix), is to provide credible evidence of the conception date and inventorship for inventors who, for some reason, cannot or don't wish to rely on witnesses. There's no doubt that, in case of an interference or other proceeding where the date of invention or inventorship itself is at issue, the PTO will regard a copy of a PTO-filed Disclosure Document as excellent of evidence of conception.

Invention Disclosure

Page__ of__

Inventor(s): _____ Irma Inventor _____

Address(es): _____ 1919 Chestnut St., Philadelphia, PA 19103 _____

Title of Invention: Self-Adjusting Can Opener _____

To record **Conception**, describe: 1. Circumstances of conception, 2. Purposes and advantages of invention, 3. Description, 4. Sketches, 5. Ramifications, 6. Possible novel features, and 7. Closest known prior art. To record **Building and Testing**, describe: 1. Any previous disclosure of conception, 2. Construction, 3. Ramifications, 4. Tests, and 5. Test results. Include sketches and photos, where possible. Continue on additional identical sheets if necessary.

1. I thought of this can opener while at my friend Roberta's wedding last Sunday. I saw the caterer having trouble opening small and large cans with several openers. Thinking there was a better way, I recalled my Majestic KY3 sewing machine clamp and how it was adjustable and thought to modify the left arm to accommodate a can opener head.

2. My can opener will work with all sizes of can and is actually cheaper than the most common existing one, the UR4 made by Ideal Co. of Racine, WI.

3. My can opener comprises a sliding clamp 10, a pincer groove 12, [etc.] as shown in the following sketch:

4. Sketch:

5. Instead of sliding clamp 10, I can use a special notch, as follows:

6. I believe that the combination of sliding clamp 10 and pincer groove 12 is a new one for can openers. Also I believe that it may be novel to provide a frammis head with my whatsit.

7. The Acme KZ12 can opener, mfgd. by Acme Kitchenware of Berleley, CA and p. 417 of "Kitchen Tools And Their Uses" (Reddy Publishers, Phila. 1981) show the closest can openers to my invention, in addition to the devices already mentioned.

Inventor(s): _Irma Inventor_____

Dates of Signatures: 1988 July 6 _____

The above confidential information is Witnessed and Understood:

_____Grisilda Himmelfarb_____ 88/ Jul / 7

_____Leonore Zimala_____ 88/ Jul / 10

Fig. 3-C—Invention Disclosure

Despite these advantages, I generally recommend that most inventors *not* use the DDP. Although the PTO's small fee per invention may not seem like much, the methods I've described earlier for recording your invention are free and in fact will give you (a) equally good or better evidence of conception and, more importantly, (b) evidence of building and testing. This is because live witnesses can testify to additional facts surrounding conception, and can also testify that they actually saw the building, testing, and operation of your invention. Also, a DDP record isn't as useable in court, as a notebook or invention disclosure is since you won't have witnesses to validate the DDP record. Simply put, I hate to see you unnecessarily spend money for an inferior "product."

Warning: In the event you do choose to utilize the PTO's DDP, remember several things which I've put into Form 3-3 to remind you: the DDP is not a substitute for filing a patent application or for building and testing the invention. Also, it won't provide you with any "grace period" or any other justification for delaying the filing of a patent application. It's merely an alternative way to get your conception disclosure witnessed. Moreover, even if you use the DDP to record your conception, you should still use a lab notebook or separate sheets with proper witnessing to record all the pertinent facts if you build and test your invention. Finally, filing a Disclosure Document with the PTO doesn't allow you to refer to the invention as "Patent Pending" or "Patent Applied For." To say either of these things you must file an actual, complete patent application, as described in Chapters 7 through 9. (It's actually a criminal offense, punishable by a $500 fine, to refer to an invention as "Patent Pending" where no patent application has been filed.)

If you still wish to file a disclosure with the PTO under its Disclosure Document Program, simply send the following five items to the Commissioner of Patents and Trademarks, Washington, DC 20231:

1. A letter requesting that the attached disclosure be accepted under the Disclosure Document Program (complete Form 3-3 in the Appendix);

2. A photocopy of the request letter of Item 1;

3. A check for the specified fee, payable to the Commissioner of Patents and Trademarks (regrettably I can't provide a form for this; you'll have to use your own check!);

4. A copy of your notebook entry (see F above) or Invention Disclosure (completed Form 3-2—see Section G above);

5. A stamped envelope addressed to you.

The disclosure sheets can be any size, but if they're large, they should be folded so as not exceed 21.5 x 33 cm (8 1/2" x 13"). Each sheet should be numbered. I suggest that you submit a photocopy of your original signed and witnessed disclosure and keep your original. (You don't have to have an original ink signature, or any signature at all, on the copy of the disclosure which you send to the PTO, but I recommend that you do sign and date your notebook or disclosure to make them more believable in case you ever have to go to court.)

The PTO will stamp all of the papers with the date of receipt and an identifying number and will return the duplicate of your request letter (Form 3-3) in your envelope. The date and number on the returned duplicate is important and should be carefully preserved.

If you file a Disclosure Document with the PTO and then do nothing else, the PTO will keep the original of your request letter and your disclosure for two years and then destroy them. However, if you later file a patent application on the invention described in your disclosure, you should do so within two years of filing the disclosure. You should also file a separate reference letter in the application, referring to the disclosure (use Form 3-4 in the Appendix). The PTO will then retain the disclosure indefinitely in case you ever need to rely on it in connection with your patent application. Be sure to file the reference letter within two years of filing the disclosure document; otherwise the disclosure will have already been destroyed by the PTO, even if you have filed a patent application. (This is another serious disadvantage of the DDP.)

Beware of two scams involving the DDP which I recently encountered. In one, an inventor was charged $200 by an "invention developer" to file his DDP. In another, a company advertised as follows:

PATENT PROTECTION FOR ONLY $10

A **government sponsored** program requires: Only a $10 registration fee to legally secure your "first priority" filing status for a 2-year period. Conserve your capital. Market NOW—patent later. For application kit send $10 to: xxxxxxxxxxx. 100% guaranteed.

Both firms indicated that the inventor could "secure priority," "reserve rights," or take advantage of a "grace period" for two years. However, as stated, the DDP is merely a substitute for witnesses to your *conception*; nothing more!

I.   Don't Use the So-called "Post Office Patent" to Document Your Invention

THERE'S A MYTH that you can document the date you conceived of your invention (or even protect your invention) by mailing a description of your invention to yourself by certified (or registered) mail and keeping the sealed envelope. In fact law regards the use of these "Post Office Patents" as tantamount to worthless and no substitute for the signatures of live witnesses on a description of your invention, or even for the PTO's Document Disclosure Program. The PTO's Board of Appeals and Patent Interferences, which has great power in these matters, has specifically said that it gives a sealed envelope little evidentiary value.

Chapter 4

WILL YOUR INVENTION SELL?

A. Why Evaluate Your Invention for Salability?

NOW THAT YOU'VE MADE an invention (Chapter 2) and recorded it properly (Chapter 3), it's time to do two things before proceeding further: evaluate it for commercial potential and make a patentability search. While you can do these in any order, I recommend that you do the easiest and/or cheapest one first. Since, for most people, it's the commercial evaluation, I put this chapter first. However, if you live across the street from the PTO, then go to Chapters 5 and 6 first. Also, if you're a corporate inventor, the decision as to whether a particular invention is sufficiently marketable to justify being patented may not be yours. In any event I recommend that you at least skim through this material for new ideas that might help you assess your work in a different light before proceeding to Chapters 5 and 6, where I discuss patentability and searching.

Why is a commercial evaluation so important? Because the next steps you take will involve the expenditure of significant money and effort. Specifically, your next step, in addition to searching the invention, is to build and test it for feasibility and cost (if possible), and then to file a patent application on the invention. Naturally, you won't want to do all of this unless you feel you have some reasonable chance that your efforts and expenditures will be justified.

Many people believe that if they get a patent, they'll be assured of fame and fortune. However, the fact that you can get a patent doesn't mean that you will make any money from the invention. In fact, less than one out of ten patented inventions make any money for their owners, mainly because the commercial prospects of the inventions were not adequately assessed at the outset.

The purpose of this chapter, then, is to help you reduce the risk of a "patented failure" by assisting you in checking your invention out for salability. In fact, before you proceed with a search, or the actual filing of a patent application, I recommend that you be reasonably confident that your invention is likely to make you at least $50,000, or at least twenty times the cost of what you plan to spend for searching, building a model, and patenting. Of course if you can do the search easily, or if you're into inventing for the sheer fun of it, you can disregard these financial requirements.

If after reading this chapter, you're still not sure about the commercial prospects of your invention, you may want to test market it. This can legally be done for up to one year since you can file a patent application up to one year after the invention is first sold or offered for sale. A test marketing is feasible if you're able to make (or have made) reasonable quantities of your invention cheaply. However, you must be willing to sacrifice your foreign rights.[1] Obviously, a field or use test of a working model of an invention will tell you much more than a theoretical "paper" evaluation discussed in this chapter.

Warning: If you do decide to test market the invention before filing, you must keep in mind the "one-year rule" which I'll also discuss in the next chapter. This rule, contained in Section 102 of the patent statutes (35 USC 102) requires that in order to be valid, a U.S. patent application must be filed

---

[1]As will be explained in the next chapter and in Chapter 12, you'll lose most of your foreign rights if you sell or otherwise release your invention to public scrutiny before you file for a patent in the U.S.

within one year after you first sell your invention (this includes test marketing), offer for sale or publish it—See Chapter 5, Section B.

B. Start Small But Ultimately Do It Completely

WHEN YOU EVALUATE your invention for commercial potential, try to do it on a small scale at first in order to avoid a large, wasted expenditure. For example, if you make metal parts as part of building a prototype to test operability, try to have them made by the economical electric discharge machining (EDM) technique rather than with molds. Similarly, prior to conducting extensive interviews, try to consult with a single expert to be sure you're not way out in left field. If your initial, small-scale investigation looks favorable and you don't run into any serious impediments, I advise that you then do it carefully, completely, and objectively, using the techniques of this chapter.

If after you do the full evaluation your idea looks like it has great commercial potential, but some other factor such as patentability or operability doesn't look too promising, don't make any hasty decision to drop it. Continue to explore the negative areas. On the other hand, if after a careful evaluation you are truly convinced that your invention won't be successful, don't waste any further time on it. Move on.

C. You Can't Be 100% Sure of Any Invention's Commercial Prospects

THERE'S ONLY ONE QUESTION you need to answer in commercially evaluating your invention: If my invention is manufactured and sold, or otherwise commercially implemented (for example, as a process that is put into use), will it be profitable? Unfortunately, no one can ever answer this question with complete certainty. The answer will always depend on how the invention is promoted, how well it's designed, how well it's packaged, the mood of the market, the timing of its commercial debut, and dozens of other intangible factors. Most marketing experts say that five "P" factors must all be "right" for a new product to make it: Production, Price, Position (its place in the market), Promotion, and Perseverance.

As stated, the method by which an invention is marketed can be crucial to its success. Consider a relatively recent invention, the Audochron® clock which indicates the time by a series of countable chimes. Given this technical feature only, the clock probably wouldn't have sold too well. But a talented designer put the works in a futuristic case shaped like a flattened gold sphere on a pedestal in which a plastic band at the center of the sphere lit with each chime. As a result, it became a status symbol and sold relatively large quantities at a high price; it even appeared in Architectural Digest, shown in a photo of a U.S. President's desk!

The trademark chosen to sell your invention under can also make a big difference as to whether it's a commercial success. If you doubt this, consider Vaseline's hand lotion. The lotion would very likely have been just another member of the bunch, consigned to mediocre sales, had not some clever marketing person come up with the trademark *Intensive Care*. This helped make it a sales leader. Ditto for the *Hula-Hoop* exercise device and the *Crock Pot* slow cooker, both of which certainly weren't hurt by evocative names. Even something as dull as roach traps were blasted into marketing stardom by the trademark *Roach Motel* and its brilliant ad campaign ("Roaches check in, but they don't check out"). Even something as prosaic as raisins were given a mighty boost recently with the "dancing raisins" TV campaign thought up by a marketing genius.

D. Take Time to Do a Commercial Feasibility Evaluation

DESPITE THE MARKETING UNCERTAINTIES, most experts believe that you can make a useful evaluation of the commercial possibilities of an untested invention if you will take the time to do some scientific and objective work in three areas:

- The positive and negative marketing factors attached to your invention;

- Consultation with experts, potential users of the invention, marketing people, and others; and

- Research into prior developments in the same area as your invention.

    Let's take a look at each.

The Positive and Negative Factors Test

Every invention, no matter how many positive factors it seems to have at first glance, inevitably has one or more significant negative ones. To evaluate the positive and negative factors objectively, carefully consider each on the list below. Using Form

4-1, the Positive and Negative Factors Evaluation Sheet, (a copy is in the Appendix), assign a commercial value or disadvantage weight to each factor on a scale of 1 to 100, according to your best, carefully-considered estimate.

For example, if an invention provides overwhelming cost savings in relation to its existing counter–parts, assign an 80 or higher to the "Cost" factor (#1) in the positive column. If it requires only a moderate capital expenditure (tooling) to build, a 50 would be appropriate for this factor (#43), in the negative column and so on.

The following balance scale analogy will help to understand the positive and negative factors evaluation: Pretend the positive factors are stacked on one side of a balance scale and the negative factors are stacked on the other side, as indicated in Fig. 4-A.

Cheaper Cost	Legal Problems
Lighter Weight	Operability Problems
Smaller Size	Difficult to Develop
Safer	Hard to Sell At a Profit
Etc.	Etc.
Positive Factors	**Negative Factors**

▲

Yes   Potential Salability?   No

Fig. 4-A—Conceptual Weighing of Positive v. Negative Factors

If the positive factors strongly outweigh the negative, you can regard this as a "go" indication, i.e., the invention is commercially viable. Obviously this balance scale is just an analogy. It can't be used quantitatively because no one has yet come up with a way to assign accurate and valid weights to the factors. Nevertheless, you'll find it of great help in evaluating the commercial prospects of your invention.

Before you actually take pen (or word processor) in hand and begin your evaluation, read through the following summary of positive and negative factors.

Positive Factors Affecting the Marketability of Your Invention

1. **Cost.** Is your invention cheaper to build or use than current counterparts?

2. **Weight.** Is your invention lighter (or heavier) in weight than what is already known, and is such change in weight a benefit? For example, if you've invented a new automobile or airplane engine, a reduction in weight is a great benefit. But if you've invented a new ballast material, obviously an increase in weight (provided it doesn't come at too great a cost in money or bulk) is a benefit.

3. **Size.** Is your invention smaller or larger in size or capacity than what is already known, and is such change in size a benefit?

4. **Safety/Health Factors.** Is your invention safer or healthier to use than what is already known? Clearly there's a strong trend in government and industry to improve the safety and reduce the possible chances for injury or harm in most products and processes, and this trend has given birth to many new inventions. Often a greater increase in cost and weight can be tolerated if certain safety and health benefits accrue.

5. **Speed.** Is your invention able to do a job faster (or slower) than its previous counterpart, and is such change in speed a benefit?

6. **Ease of Use.** Is your invention easier (or harder) to use (the current buzzword is "ergonomic") or learn to use than its previously known counterpart? An example of a product where an increase in difficulty of use would be a benefit is the child-proof drug container cap.

7. **Ease of Production.** Is your invention easier or cheaper (or harder or more expensive) to manufacture than previously known counterparts? Or can it be mass-produced, whereas previously known counterparts had to be made by hand? An example of something which is more difficult to manufacture yet which is highly desirable are the new credit cards with holographic images: they're far more difficult to forge.

8. **Durability.** Does your invention last longer (or wear out sooner) than previously known counterparts? While built-in obsolescence is nothing to be admired, the stark economic reality is that many products, such as disposable razors, have earned their manufacturers millions by lasting for a shorter time than previously known counterparts.

9. **Repairability.** Is it easier to repair than previously known counterparts?

10. **Novelty.** Is your invention at all different from all previously known counterparts? Merely making an invention different may not appear to be an advantage per se, but it's usually a great advantage: It provides an alternative method or device for doing the job in case the first method or device ever encounters difficulties, (e.g., from government regulation), in case the first device or method infringes a patent that you want to avoid infringing, and also provides something for ad people to crow about.

11. **Convenience/Social Benefit.** Does your invention make living easier or more convenient? Many inventions with a new function provide this advantage. Although you may question the ultimate wisdom and value of such gadgets as the electric knife, the remote-control TV, and the digital-readout clock, the reality remains that, in our relatively affluent society, millions of dollars have and are being made from devices that save labor and time (even though the time required to earn the after-tax money to buy the gadget is often greater than the time saved by using it). Then too, many new industries have been started by making an existing

invention easier and convenient to use. Henry Ford didn't invent the automobile; he just produced it in volume and made it convenient for the masses to use. Ditto for George Eastman with his camera. And in modern times, the two Steves (Jobs and Wozniak) did much the same for the computer.

12.   Reliability. Is your invention apt to fail less or need repair less often than previously known devices?

13.   Ecology. Does your invention either make use of what previously were thought to be waste products? Does it reduce the use of limited natural resources? Does it produce less waste products, such as smoke, waste water, etc.? If so, you have an advantage which is very important nowadays and which should be emphasized strongly.

14.   Salability. Is your invention easier to sell or market than existing counterparts?

15.   Appearance. Does your invention provide a better-appearing design than existing counterparts?

16.   Viewability. If your invention relates to eye use, does it present a brighter, clearer, or more viewable image? For example, a color TV with a brighter picture, or photochromic eyeglasses which automatically darken in sunlight were valuable inventions.

17.   Precision. Does your invention operate or provide greater precision or more accuracy than existing counterparts?

18.   Noise. Does your invention operate more quietly? Does it turn unpleasant noise into a more acceptable sound?

19.   Odor. Does your invention emanate less or more unpleasant fumes or odors?

20.   Taste. If your invention is edible or comes into contact with the taste buds (for example, a pill or a pipe stem), does it taste better? A foul taste (or smell) can also be an advantage, e.g., for poisons to prevent ingestion by children, and for telephone cables to deter chewing by rodents.

21.   Market Size. Is there a larger market for your invention than for previously known devices? Because of climatic or legal restrictions, for example, certain inventions are only usable in small geographical areas. And because of economic factors, certain inventions may be limited to the relatively affluent. If your invention can obviate these restrictions, your potential market may be greatly increased, and this can be a significant advantage.

22.   Trend of Demand. Is the trend of demand for your device increasing? Of course you should distinguish, if possible, between a trend and a fad. The first will provide a market for your invention while the second is likely to leave you high and dry unless you catch it in the beginning stages.

23.   Seasonal Demand. Is your invention useful no matter what the season of the year? If so, it will have greater demand than a seasonal invention, such as a sailboat.

24.   Difficulty of Market Penetration. Is your device an improvement of a previously accepted device? If so, it will have an easier time penetrating the market than a device which provides a completely new function.

25.   Potential Competition. Is your invention so simple, popular, or easy to manufacture that many imitators and copiers are likely to attempt to design around it, or break your patent as soon as it's brought out? Or is it a relatively complex, less popular, hard-to-manufacture device, which others wouldn't be likely to produce in competition with you because of the large capital outlay required for tooling and production, etc?

26.   Quality. Does your invention produce or provide a higher quality output or result than existing counterparts? For example, laser disks provide a much better audio quality than do phonorecords or magnetic tape.

27.   Excitement. (The Neophile and the Conspicuous Consumer/Status Seeker). Almost all humans need some form of excitement in their lives: some obtain it by watching or participating in sports, others by the purchase of a new car or travel, and still others by the purchase of new products, such as a 50-inch T.V., a laser disk player, or a friendly household robot. Such purchasers can be called "neophiles" (lovers of the new); their excitement comes from having and showing off their new "toy." Purchasers of expensive products, like the Mercedes Benz or a Rolex watch, are commonly motivated by what Thorsten Veblen has called "conspicuous consumption," and what we now call "status seeking." They enjoy showing off an expensive or unique item which they've acquired. Thus, if your invention can provide consumer excitement, either through sheer newness or through evidence of a costly purchase, it has a decided advantage.

28. Mark-Up. If your invention is in an excitement category (i.e., if it's very different, novel, innovative or luxurious), it can command a very high mark-up, a distinct selling advantage.

29. Inferior Performance. Yes, I'm serious! If your invention performs worse than comparable things which are already available, this can be a great advantage, if put to the proper use. Consider the 3M Company's fabulously successful Scotch® Post-It® note pads: Their novelty is simply that they have a strip of stickum which is *inferior* to known adhesives, thus providing removable self-stick notes. Here the invention may not be so much the discovery of an inferior adhesive as the discovery of a new use for it.

30. "Sexy" Packaging. If your invention is or comes in a "sexy" package, or is adaptable to being sold in such a package, this can be a great advantage. Consider the Hanes l'Eggs® stockings where the package (shaped like an egg) made the product!

31. Miscellaneous/Obviation of Specific Disadvantages of Existing Devices. This is a catchall to cover anything I may have missed in the previous categories. Often the specific disadvantages which your invention overcomes will be quite obvious; they should be included here, nonetheless.

32. Long Life Cycle. If your invention has a potentially long life cycle, i.e., it can be made and

sold for many years before it becomes obsolete, this is an obvious strong advantage which will justify a capital expenditure for tooling, a big ad campaign, etc.

33. Related Product Addability. If your invention will usher in a new product line, as did the computer, where many related products, such as disc drives, printers, can be added, this will be an important advantage with potentially enhanced profits.

34. Satisfies Existing Need. If your invention will satisfy an existing, recognized need, such as preventing drug abuse, or avoiding auto collisions, your marketing difficulties will be greatly reduced.

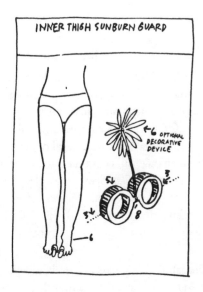

Negative Factors Likely to Affect the Marketability of Your Invention

Alas, every invention, no matter how great and disadvantage-free it seems, has one or more negative factors, even if the negative factor is merely the need to change or design and produce new production equipment. I've seen inventions and developments which were better in every way than what already existed, but which weren't used solely because the improvement didn't justify the cost of replacing existing production equipment, or the cost associated with manufacturing and promoting the device.

The negative factors of your invention are generally more important and require more consideration than the positive factors, since if your invention fails, it will obviously be one or more of the negative factors that causes it. Since all the positive factors listed above can be disadvantages when viewed in reverse, they should be carefully considered, but won't be reproduced here. For example, consider Factor #23., Seasonal Demand. This will be a negative, rather than a positive factor if the invention is something like skis or a holiday decoration, which does have a seasonal demand, rather than an all-year-around one. Also each negative factor can be used as a positive factor if applicable.

1-34—Reverse of Positive Factors Listed Above

35. Legality. Does your invention fail to comply with, or will its use fail to comply with, existing laws, regulations, and product and manufacturing requirements? Or, are administrative approvals required? If your invention carries legal difficulties with it, its acceptance will be problematic no matter how great its positive advantages are. And if ecological or safety approvals are required (for example, for drugs and automobiles), this will be viewed as a distinct disadvantage by prospective buyers.

36. Operability. Is it likely to work, or will significant additional design or technical development be required to make it practicable and workable? Usually problems of operability will become abundantly clear when you try to build a working model, which you should try to do as soon as possible, if at all practicable. (Don't forget to fill out another copy of Form 3-2 after you build and test it.)

37. Development. Is the product already designed for the market, or will additional engineering, material selection, appearance work, etc., be required?

38. Profitability. Because of possible requirements for exotic materials, difficult machining steps, great size, etc., is your invention likely to be difficult to sell at a profit?

39. Obsolescence. Is the field in which your invention is used likely to die out soon? If so, most manufacturers won't be willing to invest money in production facilities.

40. Incompatibility. Is your invention likely to be incompatible with existing patterns of use, customs, etc.?

41. Product Liability Risk. Is your invention in an area (such as drugs, firearms, contact sports, automobiles, etc.) where the risks of lawsuits against the manufacturer, due to product malfunction or injury from use, are likely to be greater than average?

42. Market Dependence. Is the sale of your invention dependent on a market for other goods, or is it useful in its own right? For example, an improved television tuner depends on the sale of televisions for its success, so that if the television market goes into a slump, the sales of your tuner certainly will fall also.

43. Difficulty of Distribution. Is your invention so large, fragile, perishable, etc., that it will be difficult or costly to distribute?

44. Service Requirements. Does your invention require frequent servicing and adjustment? If so, this is a distinct disadvantage. But consider the first commercial color TVs which, by any reasonable standard, were a service nightmare, but which made millions for their manufacturers.

45. New Production Facilities Required. Almost all inventions have this disadvantage. This is because the manufacture of anything new requires new tooling and production techniques.

46. Inertia Must Be Overcome. An example of a great invention that so far has failed because of user inertia is the Dvorak typewriter, which, although much faster and easier to use, was unable to overcome the awkward but entrenched Qwerty keyboard. There's a risk in introducing *any* new product, and when any invention is radically different, potential manufacturers, users and sellers will manifest tremendous inertia, regardless of the invention's value.

47. Too Advanced Technically. In the 60's, I got a client a very broad patent on a laser pumped by a chemical reaction explosion; we were very pleased with this patent. However it was so advanced at the time that the technology behind it is just now being implemented in connection with the "Star Wars" defense effort. Unfortunately, the patent expired in the meantime. The moral? Even if you have a great invention, make sure it can be commercially implemented within about 17 years.

48. Substantial Learning Required. If consumers will have to undergo substantial learning in order to use your invention, this is an obvious negative. An example: the early personal computers. On the other hand, some inventions, such as the automatically-talking clock make a task even easier to do and thus have an obvious strong advantage.

49. Difficult To Promote. If it will be difficult to promote your invention, e.g., because it's technically complex, has subtle advantages, or is very expensive, large, awkward, etc., you've got an obvious disadvantage.

Complete Form 4-1 by assigning a weight to each listed factor, either in the positive or negative column. Also list and assign weights to any other factors you can think of which I've omitted. Then compute the sum of your positive and negative factors and determine the difference to come up with a rough idea of a net value for your invention. I suggest that you continue to pursue inventions with net values of 50 and up, that you direct your efforts elsewhere if your invention has a net value of less than 0, and that you make further critical evaluation of inventions with net values between 0 and 50.

The list has many other valuable uses:

- Using the list may cause you to focus on one or more drawbacks which are serious enough to kill your invention outright;

- The list can be used to provide a way of comparing two different inventions for relative value so that you'll know which to concentrate more effort on.

You now should extract all factors on the list of Form 4-1 which have any value other than 0 and write these factors and their weights on Form 4-2, the Positive and Negative Factors Summary Sheet. (a copy is in the Appendix). This sheet, when completed, will provide you with a concise summary of the advantages and disadvantages of your invention. You can use it in at least four valuable ways:

1. To provide you with a capsule summary of your invention for commercial evaluation purposes (this chapter);

2. To help you prepare your patent application (see Chapter 8);

3. To help you to sell or license your invention to a manufacturer (see Chapter 11); and

4. To help you to get the PTO to grant you a patent (see Chapter 13).

Don't hesitate to update or re-do Forms 4-1 and 4-2 if more information comes to mind.

E. Check Your Marketability Conclusions Using the Techniques of Consultation and Research

1. How to Go About It

If your evaluation of the above positive and negative factors affecting the marketability of your invention gives the positive side the edge, I recommend that you extend your investigation by doing some consultation and research. If you continue to get positive signs, extend your search still further until you've learned all you can about the field of your invention. This knowledge will also be of great benefit when you make your patentability search, prepare your application, market your invention, and deal with the PTO.

Confidentiality Note: In Part 2, below, I suggest a number of procedures to use when you're disclosing your ideas to others so that they won't be stolen and so their trade secret (TS) status will be maintained. Here, I simply warn you at the outset that you shouldn't disclose ideas and information without utilizing appropriate safeguards; otherwise you may lose them to others.

The areas of consultation and research which you should investigate include asking both nonprofessionals and experts in the particular field for an opinion, and researching the relevant literature. As you do this, keep in mind and ask about all of the positive and negative factors listed above. Your consultation efforts and research will almost surely give you more information useful in assessing many of them. If so, again don't hesitate to redo your Forms 4-1 and 4-2.)

As indicated, nonprofessionals can often be an excellent source of information and advice, especially if your invention is a consumer item that they are likely to have an opportunity to purchase if it's ever mass-produced. Consult your lay friends and associates, i.e., those who have no special expertise in the field in which you are interested, but whose opinion you trust and feel will be objective. Often you may find it valuable not to tell them that you are the inventor so you will get a more objective evaluation. You may also want to inquire as to what price they'd be willing to pay. It's especially helpful if you've built a working model (see Part F below) so you can show it to them and ask if they'd buy it and for what price.

Experts to be consulted in the particular field of your invention include any and all of the following who can supply you with relevant feedback:

- Salespeople and buyers in stores which sell devices similar to yours;
- Engineers, managers, or technicians in companies in the field of your invention; and
- Scholars, educators, or professors who do research in the area of your invention; and
- Friends who are "in the business."

Naturally you may not know all of these experts. Getting to them will require the creative use of the contacts you do have so as to arrange the proper introductions. Once you do, however, most people will be flattered that you've asked for their advice and hence pleased to help you.

After you show your invention (preferably a working model) note the person's initial reaction. If you hear a "Well, I'll be damned!" or "Why didn't I think of that!", you know you're on the right track.

For your literature search, I suggest that you start by locating a research librarian who's familiar with the area of your concern. Large technical and business libraries, and those associated major universities are obvious places to start. The library literature which you should investigate includes product directories, how-to-do-it books, catalogs, general reference books, patents (if available), etc. (see Chapter 6).

Remember that the purpose of the literature search isn't to determine whether your invention is new or patentable, but rather to give you additional background in the field so you can evaluate the positive and negative factors listed above. However, while you're doing your literature search, you may find that your invention was publicly known before you invented it. This is especially likely to occur if you search the patent literature. If so, you'll either have to drop the invention, since you'll know you aren't the first inventor, or try to make a new invention by improving your first effort. You'll be surprised how much better a feel you'll have for your invention once you've done some research and become familiar with the field.

If you work for or have access to a large company, visit its purchasing department and ask for permission to look through its product catalogs. Most companies have an extensive library of such catalogs and you'll often find much relevant and valuable information there that you won't find in even the biggest and best public libraries.

Note: This search isn't the equivalent of the "patent search" which occurs before you apply for your patent. Covered in the next chapter is the more formal "patent search" which obviously will provide you with considerably more background in the area of your invention.

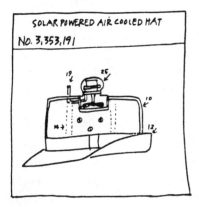

2. Precautions to Take During Consultation

If you do show your invention to others or discuss it with them to any extent, a degree of care is mandatory to preserve the trade secret status of your invention and to prevent theft of your ideas, or to prove it in case it occurs (See Chapter 1, Section P). Remember that any of the agreements discussed below are only as good as the parties who have signed them. Thus you shouldn't disclose your invention to anyone you don't trust. Suing someone for breaching a non-disclosure agreement is no substitute for picking a trustworthy person in the first place.

Here are some good alternatives that can be used to protect your invention from being misappropriated by others:

- Have disclosees sign a receipt or log book entry indicating that they have seen your invention. The log book entry can be simply a page in your inventors notebook which says at the top, "The undersigned have seen and understood Tom Brown's confidential [name of invention] as described on pages ___ of this book, on the dates indicated." You may also want to add a "Comments" column to your book to indicate that you value their opinion. Doing this also makes it easier to ask your consultants to sign your receipt page or log notebook.

- Ask those to whom you show your invention to sign and date your disclosure as witnesses. Witnesses can hardly ever claim that they invented independently of you if they're on record as having witnessed your invention. If there are more than two or three witnesses, however, this method won't work as there won't be room in your book for more.

- Get your consultants to sign the Proprietary Materials Agreement (Form 3-1). However, it may be difficult for you to ask someone who's doing you a favor to sign this agreement.

- Although inferior to the other devices listed above, send a confirming or thank-you letter before and after your consultation so you'll have written, uncontradicted records that you showed your invention to the person on a specific date and that you asked it to be kept confidential. Your confirmatory letter can simply say, "Thanks very much for looking at my [name of invention] at your office last Wednesday, July 3. This letter is to confirm that you agreed that the details of my [name of invention] should be maintained in strictest confidence. Thanks for your cooperation. Sincerely, [your name]."

While care in disclosing your invention is necessary to prevent loss of its TS status and theft, don't go overboard with precautions. Many new inventors get a such a severe case of "inventor's paranoia" that they're afraid to disclose their brainchild to anyone, or they're willing to disclose it only with such stringent safeguards that no one will want to look at it! In practice, most stolen inventions are taken only after they're out on the market and proven successful. This is because thieves are generally after a sure thing. While I don't totally approve, highly-successful inventor Paul Brown usually shows his inventions freely: he says, "Let them steal it—they don't know how much work they're in for!"

F. Now's the Time to Build and Test It (If Possible)

1. Why Do It

As stated under #36 in Part D above, if you haven't already done so, it's very desirable to build and test a working model (prototype) of your invention, if at all possible. The reasons: a working model will give you something real to show your marketing consultants, plus valuable information about operability, cost, technical problems, and most of the other factors on the positive and negative factors list. If it's impractical to build a working model, often a non-working model, or scale model, will give you almost as much valuable data. As stated, don't forget to fill out another copy of Form 3-2 (Inventor's Disclosure) after you build and test it.

2. If You Use a Model Maker, Use a Consultant's Agreement

If you can't build and test it yourself, many model makers, engineers, technicians, teachers, etc. are available who will be delighted to do the job for your for a fee, or for a percentage of the action. If you do use a model maker (consultant), you should take precautions to protect the confidentiality and proprietary status of your invention. There's no substitute for checking out your consultant carefully by asking for references (assuming you don't already know the consultant by reputation or referral).

In addition, have your consultant sign a copy of the Consultant's Work Agreement (Form 4-3 in the Appendix). Note that this Agreement includes fill-in blanks to describe the names and addresses of the inventor and consultant, the name of the project or invention (e.g., "New Sweater-Drying Form"), detailed description of the work to be done (e.g., "build a wire-frame, plastic-covered, sweater-drying collapsible form in accordance with plans in attached Exhibit A — finished form to operate smoothly and collapse to 14" x 23" x 2" (or less) size."), and manner of payment(usually 1/3 at start, 1/3 upon construction, and 1/3 on acceptance by you, the Contractor) and which state's law should govern (pick the state where you reside if the Consultant's out of state.)

Note that the agreement requires the Consultant to sign any patent applications which you choose to file on the Consultant's Inventions, and also to assign such inventions to you. Note also that the inventions which belong to you, the Contractor, are those which arise out of the Consultant's work under the agreement, even if conceived on the Consultant's own time. This is a customary clause in employment agreements (see Ch. 16) and is provided so that the Consultant won't be able to claim that a valuable invention made under the agreement isn't yours because it was made on the Consultant's time. Generally the Consultant will be a sole inventor (who should be the only one named in the patent application if the Consultant's invention can exist independently of yours), and a joint inventor with you if the invention is closely related to or improves on yours. More on inventorship in Ch. 10 Sec. F.) This is because all of the true inventor(s) must be named as inventor(s) in all patent applications. I provide an assignment form and a Joint Owners' Agreement in the Appendix; see Chapter 16.

G. Summary

ONCE YOU'VE COMMERCIALLY EVALUATED your invention (i.e. garnered all your input, filled out your evaluation and summary sheets with the positive and negative factors, etc.), you're in a better position to decide whether or not to go ahead. If you decide to, your next step is to decide whether the invention will qualify for a patent under the patent laws. To do this you should make a formal patent search (Chapter 6) to determine whether your invention meets the legal requirements for a getting a patent (Chapter 5).

If, on the other hand, your commercial evaluation leaves you uncertain, though you feel there's good potential, wait a while before proceeding. The passage of time may give you a new perspective that can make your decision easier. If after a couple of weeks you still can't make up your mind, it's probably best to proceed to the next step (the search). If the search discloses that your invention is already known, that's the end of the road. But if the search shows that you have a new invention, you should probably attempt to patent and market it rather than let a potentially valuable and profitable idea die without being given its day in the sun.

Chapter 5

WHAT IS PATENTABLE?

HERE WE DEAL with the specific subject of what's legally patentable and what's not. Over many decades, both Congress and the courts have hammered out a series of laws and accompanying rules of interpretation which the PTO and the courts (and hence you) must use to separate the patentable wheat from the unpatentable chaff. All of these laws and rules are introduced in this chapter and then referred to repeatedly in later chapters where I take you through the ins and outs of obtaining a patent.

Because an understanding of the material in this chapter is crucial to the rest of the book, and to an understanding of patents in general, I urge you to relax and read it carefully. If later you become confused by terminology or about the criteria for determining the patentability of an invention, return for a refresher.

A. Patentability Compared to Commercial Viability

IF YOU READ CHAPTER 4 and assessed the commerciability of your invention, and your invention received a passing grade on the commercial scorecard, your next question probably is, "Can I get a patent on it?" The answer to this question can be crucial since you're likely to have a difficult time commercially exploiting an invention that isn't patentable, despite its commercial feasibility. Although you may be able to realize value from an invention by selling it to a manufacturer as a trade secret (a difficult sale to make!), or by selling it yourself and using a clever trademark, or (in some cases) by relying on copyright protection and unfair competition laws (as explained in Chapter 1), such approaches are usually inferior to the broad offensive rights which a patent offers. Concisely put, if your invention fails to pass the tests of this chapter, reconsider its commercial prospects and whether other intellectual property devices will provide adequate protection.

B. Legal Requirements for Getting a Utility Patent

AS YOU CAN SEE from Fig. 5-A, the legal requirements for getting a patent can be represented by a mountain having four upwardly-sloping sections, each of which represents a separate test which every invention must pass to be awarded a patent. The PTO is required by statute to examine every utility patent application to be sure it passes each of these tests. If it does, the PTO must award the inventor(s) a patent.

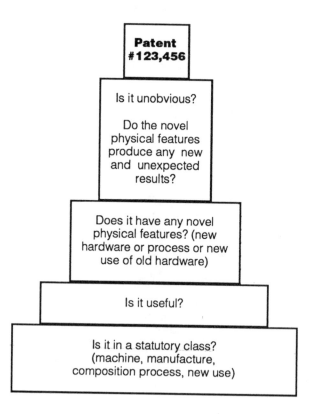

Fig. 5-A—Patentability Mountain

Design and Plant Patents

Design patent applications must cover a new, original, and *ornamental* design for an article of manufacture and are examined in the same way and must pass the same obviousness test as utility patent applications, except that the "better functioning" tests which are used to evaluate unobviousness (see Sec. F below) are now used since only the aesthetics of a design invention are relevant.

Plant patent applications are subject to the same legal requirements as utility patent applications, except that the statutory class requirement (first test) is obviously not relevant: plants provide their own statutory class. Since plant patents are relatively rare and are of very specialized interest, I won't go into detail except to set forth the additional legal requirements for getting one. They are: 1) the plant must be asexually or sexually reproduced; and 2) the plant must be a new variety. These may include cultivated sports, mutants, hybrids, and newly-found seedlings, but should not be a tuber propagated plant or a plant found in an uncultivated state

The four requirements and the pertinent respective statutes are:

1. Statutory Class: Does the invention fit into one of five classes established by Congress? Or put concretely, can it reasonably be called a:

- process,
- machine,
- manufacture,
- composition, or

- a "new use" of one of the first four. Title 35 of the US Code, Sec. 101, or in legal terminology, 35 USC 101.[1]

2. Utility: Can the invention properly be regarded as a useful one (or ornamental in the case of designs)? 35 USC 101.

3. Novelty: Can the invention properly be regarded as novel, i.e., does it have an aspect which is different in any way from all previous inventions and knowledge, i.e., the relevant prior art?[2] 35 USC 102.

4. Unobviousness: Can the invention be properly regarded as unobvious from the standpoint of someone who has ordinary skill in the specific technology involved in the invention, i.e., does it provide one or more new and unexpected results? (When dealing with designs, the question becomes, are the novel features of the design unobvious in an ornamental sense?) 35 USC 103.

As Fig. 5-A shows, the first three tests are represented by relatively short boxes. The last one, unobviousness, is relatively high. This is a real-life reflection of what commonly happens to patent applications before the PTO (or to patents when they're challenged in court). In other words, most inventions are found to 1) fit within at least one statutory class, 2) have utility (or ornamentality for designs), and 3) possess novelty. However, most of the patent applications which fail to reach the patent summit (almost half of all patent applications which are filed) are rejected because the PTO isn't convinced that the invention is unobvious.

---

[1]Congress derives its power to make the patent statutes from the US Constitution (Art. 1, Sec. 8). In turn, the statutes authorize the PTO to issue its Rules of Practice (which are relatively broad) and its Manual of Patent Examining Procedure (which is relatively specific)—see "Books of Use and Interest" in Appendix.

[2]While you might think that the definition of "prior art" would be simply anything known before your filing date, unfortunately (as in many other areas) the law has become far more complex, as we'll see in Sec. E of this chapter.

Let's now look at each of these requirements in more detail.

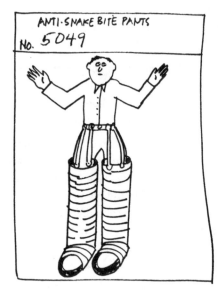

ANTI-SNAKE BITE PANTS
No. 5049

C. Requirement #1: The Statutory Classes

TO BE PATENTABLE, your invention must fall into one of the five statutory classes. If it does, it's "within a statutory class or category." I.e., it's one of the five types of subject matter on which the law authorizes the PTO to grant a patent monopoly, assuming the other requirements for a patent are met.

Fortunately, the statutory categories established by the patent laws, although only five, are very comprehensive. Accordingly, you'll usually be able to squeeze most inventions into at least one of them. In many instances an invention will fit into more than one category, since they overlap to some extent. This isn't a problem since you don't have to specify the one to which your invention belongs when you file your patent application. But you should be fairly sure it fits into at least one category. Otherwise, the PTO may reject it under Section 101 as "nonstatutory subject matter."

Again, the five statutory classes are:

1. Processes: Sometimes termed "methods," processes are ways of doing or making things which involve more than purely manual or mental manipulations, or more than manual manipulations with very simple implements. Processes always have one or more steps, each of which expresses some activity and manipulates or treats some physical thing. Examples are heat treatments, chemical reactions for making or changing something, ways of making products or chemicals, ways of changing things, etc. To give you an example, I represented one side in a patent lawsuit which involved a patent on a process of attaching a hairpiece to a bald person's scalp by putting suture anchors in the scalp and sewing the piece to the suture anchors.

2. Machines: Machines are devices or things for accomplishing a task. Like processes, they usually involve some activity or motion that's performed by working parts, but in machines the emphasis is on the parts or hardware, rather than the activity per se. Put differently, while a process involves the actual steps of manipulation of an item or workpiece (the machine which does the manipulation is of secondary import), a machine is the thing which does the manipulating and the steps or manner of its operation, and the process itself, or material worked upon, are of lesser import. Examples of machines are cigarette lighters, robots, sewage treatment plants, clocks, all electronic circuits, automobiles, boats, rockets, telephones, TVs, computers, VCRs, disk drives, robots, printers, lasers, photocopiers, etc. Many inventions can be claimed as a process and/or as a machine: e.g., an electric circuit or a weaving machine can be claimed in terms of its actual hardware and/or as a process for manipulating an electrical signal or weaving fabrics.

Note: As I've said, there are no clear lines between the five statutory classes. The important thing to realize is that it doesn't matter as long as your invention fits into at least one of them. Put differently, you needn't be able to tell a machine from a process to qualify for a patent.

3. Manufactures: Manufactures, sometimes termed "articles of manufacture," are items that have been made by human hands or by machines. This excludes naturally-occurring things, like rocks, gold, shrimp, and wood, or slightly-modified naturally occurring things, like a shrimp with its head and vein removed. But if you discover a new and unobvious use for a naturally-occurring thing, such as a way to use the molecules in a piece of gold as part of a computer memory, you can patent the invention as a new use (see below), or as a machine (the gold with the necessary hardware to make it function as a memory).

Manufactures are relatively simple things that don't have working or moving parts as prime features. Clearly you will see some overlap between the machine and the manufacture categories. Many devices, such as mechanical pencils, cigarette lighters, and electronic circuits can be classified in either. Examples of manufactures are erasers, desks, houses, wires, tires, books, cloth, chairs, containers, transistors, dolls, hairpieces, ladders, envelopes, buildings, floppy discs, knives, hand tools, boxes, etc.

4. Compositions of Matter: Compositions of matter are items such as chemical compositions, conglomerates, aggregates, or other chemically-significant substances which are usually supplied in bulk (solid or particulate), liquid, or gaseous form. Examples are road-building compositions, all chemicals, gasoline, fuel gas, glue, paper, soap, plastics, etc.

Although, as stated, naturally-occurring things like wood, rocks, etc. can't be patented, purified forms of naturally-occurring things, such as medicinals extracted from herbs, can be patented. One inventor even obtained a composition of matter patent on a new element he discovered. And recently, genetically-altered plants, microbes, and non-human animals have been allowed under this category. Compositions are generally homogeneous chemical compositions or aggregates whose chemical natures are of primary importance and whose shape is of secondary import, while manufactures are items whose physical shapes are significant, but whose chemical compositions are of lesser import.

5. New Uses of Any of the Above: A new-use invention is actually a new and unobvious process or method for using an old and known invention, whether it be an old and known process, composition, machine, or article. The inventive act here isn't the creation of a new thing or process per se, but the discovery of a new use for something which in itself is old.

If you discover a new and unobvious (unrelated) use of any old invention or thing, you can get a patent on your discovery. For example, suppose you discover that your venetian-blind cleaner can also be used as a seed planter: You obviously can't get a patent on the venetian-blind cleaner per se, since you didn't invent it—someone already patented, invented, and/or designed it first—but you can get a patent on the specific new use (seed planting) you've invented. In another example, one inventor obtained a patent on a new use for aspirin: feeding it to swine to increase their rate of growth.

New use inventions are relatively rare and technically are a form of, and must be claimed as, a process. 35 USC 100[b]. However, most patent experts treat them as a distinct category. See Chapter 9 for a discussion of patent claims.

Examples of Inventions That Don't Fit Within a Statutory Class

As mentioned, if you've invented something and you can't reasonably assign it to one of the five classes above, your invention can't be patented, no matter how new it is. Examples of types of inventions which aren't considered within a statutory class, and hence can't be patented, are:

- Processes performed solely by hand (such as a method of meditating), or by hand with only simple implements, such as a method of styling hair with a comb and brush or a method of speed reading;

- Naturally-occurring articles, even if modified somewhat;

- Abstract scientific or mathematical principles (e.g., John Napier's invention of logarithms in 1614 was immensely innovative and valuable, but it would never get past the bottom level of the patentability mountain);

- An arrangement of printed matter without some accompanying instrumentality; and

- Methods of doing business, such as sales plans, ad schemes, ideas for a quiz show, a method of teaching firewalking, etc.

- Computer programs per se, that is, naked computer instructions or algorithms[3] not tied in with any machine or hardware, are still not considered within a statutory class. However, most patent lawyers will agree that it's possible effectively to patent an algorithm (and thus software) if it can be described and claimed (see Chapter 9) a machine or as a process that utilizes a machine or some hardware or physical element (even if only a computer memory or display). The important thing is to recite some hardware in the claims (formalistic statements in your patent application that precisely define the scope of your invention) and to not wave a red flag in your patent examiner's face by using the words "program" or "algorithm."

Ideas per se, that is, thoughts or goals not expressed in concrete form or usage, are obviously not assignable to any of the five categories above. If you have an idea, you must show how it can be made and used in tangible form so as to be useful in the real world, even if only on paper, before the PTO will accept it. Often the act of figuring out how an idea can be made and used will be more significant than the idea. For example, if you have an idea for time-travel or telepathic transportation, or more realistically, a bicycle with a continuously-variable torque/speed control, your idea wouldn't be patentable, even if you were the first to think of it. But if you could come up with a way to make and use it, i.e., a tangible apparatus or workable process for accomplishing either of the above fantasies or the bicycle control, you'd be able to get a patent on your apparatus or process.

D. Requirement #2: Utility

TO BE PATENTABLE your invention must be useful. Problems are seldom encountered with the literal utility requirement; *any* usefulness will suffice.[4] In fact, it's hard for me to think of an invention which couldn't be used for some purpose. However utility is occasionally an issue in the chemical area when an inventor tries to patent a new chemical for which a use hasn't yet been found, but for which its inventor will likely find a use later. If the inventor can't state (and prove, if challenged) a realistic use, the PTO won't grant a patent on the chemical. Of course, a chemical intermediate which can be used to produce another useful chemical is itself regarded as useful.

Notwithstanding the fact that virtually all inventions are useful in the literal sense of the word, some types of inventions are deemed "not useful" as a matter of law and patents on them are accordingly denied by the PTO. These include:

---

[3]Generally speaking, an algorithm is any computational step, or series of steps, leading to the solution of a problem; all computer programs fit within this definition.

[4]However, See Chapter 4 (Will Your Invention Sell) where I recommend that the usefulness of your invention be relatively great in order to pass the "commercial viability" test.

DOG TRAINING PANTS
NO. 702,391,862

1. Unsafe New Drugs

The PTO won't grant a patent on any new drug unless the applicant can show that it isn't only useful in treating some condition, but also that it's relatively safe for its intended purpose. Put another way, the PTO considers an unsafe drug useless. Most drug patent applications won't be allowed unless the Federal Drug Administration has approved tests of the drug for efficacy and safety, but drugs which are generally recognized as safe, or are closely related to known safe drugs, don't need prior FDA approval to be patentable

2. Whimsical Inventions

Occasionally, the PTO will reject an application for a patent when it finds the invention to be totally whimsical, even though "useful" in some bizarre sense. Nevertheless in 1937 the PTO issued a patent on a rear windshield (with tail-operated wiper) for a horse (patent 2,079,053). They regarded this as having utility as an amusement or gag.

Most patent attorneys have collections of humorous patents. I could easily fill the rest of this book with my collection, but I'll restrain myself and briefly describe just a few.

- a male chastity device (patent 587,994—1897);
- a figure-eight shaped device to hold your big toes together to prevent sunburned inner thighs (patent 3,712,271—1973);
- dentures with individual teeth shaped like the wearer's head (patent 3,049,804—1962); and
- a dress hanger with breasts (design patent D226,943—1973).

Also, even though the PTO issued patent 2,632,266 in 1953 for a fur-encircled keyhole, the censor wouldn't let me show this on a TV show.

3. Inventions Useful Only for Illegal Purpose

An important requirement for obtaining a patent, which Congress hasn't mentioned, but which the PTO and courts have brought in on their own initiative (by stretching the definition of "useful"), is legality. For example, inventions useful solely for illegal purposes such as disabling burglar alarms, safecracking, copying currency, and defrauding the public might be incredibly useful to some elements in our society, but the PTO won't issue patents on them. However most inventions in this category can be described or claimed in a "legal" way: e.g., a police radar detector would qualify for a patent if it's described as a tester to see if a radar is working or as a device for reminding drivers to watch their speed.

4. Immoral Inventions

In the past, the PTO has—again on its own initiative—included morality in its requirements. But, in recent years, with increased sexual liberality, the requirement no longer exists. Thus the PTO now regularly issues patents on sexual aids, gags, and stimulants.

5. Non-operable Inventions, Including Perpetual Motion Machines

Another facet of the useful requirement is operability. The invention must appear to the PTO to be workable before they will allow it. Thus, if your invention is a perpetual-motion machine, or a metaphysical-energy converter, or, more realistically, a very esoteric invention which looks technically questionable (it looks like it just plain won't work or violates some well-accepted physical law), your examiner will reject it as lacking utility because of inoperability. In this case you would either have to produce a logical, technical argument refuting the examiner's reasons (you can include affidavits of witnesses and experts and test results), or bring the invention in for a demonstration to prove its operability.

Operability is rarely questioned since most patent applications cover inventions which employ known principles or hardware and will obviously work as described. If the examiner questions operability, however, you have the burden of proof. And note that all patent examiners have technical degrees (some even have Ph.D.s), so expect a very stringent test if the operability of your invention is ever questioned.

Despite the foregoing, the PTO occasionally issues a patent on what appears to be a perpetual-motion-like machine as they did in 1979 (patent 4,151,431). This raises an important point. The fact that a patent is granted doesn't mean that the underlying invention will work. It only means that the invention appears to work on paper (or that the PTO can't figure out why it won't work).

The PTO, however, has recently become more careful about perpetual energy or perpetual motion machines as you may have noted from a recently

publicized case where it denied an inventor a patent on an energy machine. The inventor took the case to the courts, but lost after the National Bureau of Standards, acting as a court-expert, found the machine didn't convert matter into energy.

It's a common misconception that the PTO won't "accept" patent applications on perpetual motion machines: the PTO *will* accept the application for filing (see Ch. 13), but will almost certainly reject it later as inoperative, after it makes a formal examination.

6. Nuclear Weapons

The invention must not be a nuclear weapon.; such inventions aren't patentable because of a special statute. However, if you've invented that doomsday machine, don't be discouraged: you can be rewarded directly by making an application with the DOE (Department of Energy), formerly the Atomic Energy Commission (AEC).

7. Theoretical Phenomena

Theoretical phenomena per se, such as the phenomenon of superconductivity or the transistor effect, aren't patentable. You must describe and claim (see Ch. 9) a practical, realistic, hardware-based version of your invention for the PTO to consider it useful.

BURPING BABY DOLL
No. 7,283,406

E. Requirement #3: Novelty

1. Prior Art

Your invention must also be novel in order to qualify for a patent. In order for your invention to meet this novelty test it must differ physically in some way from all given publicly-known or available prior developments or concepts. In the world of patent law, these prior developments and concepts are collectively referred to as "prior art." Accordingly, before I tell you how to determine whether your invention is novel, it's vital to understand what your invention must differ from.

a. When Is Art Considered Prior Art?

According to Section 102 of the patent laws, the term "prior art" means generally the state of knowledge existing or publicly available either before the date of your invention or more than one year prior to your filing your patent application.

b. Date of Your Invention

Clearly, in order to decide what prior art is with respect to any given invention, it's first necessary to determine the "date of your invention." Most inventors think it's the date on which one files a patent application. While this date is important, and you can always use it if you have nothing better, you can usually go back earlier than your filing date if you can prove the date you conceived the invention or the date you built and tested it (see Chapter 3).

Your date of invention is the earliest of:

1. the date you filed your patent application;

2. the date you can prove you built and tested your invention in the U.S.; or

3. the date you can prove you conceived of your invention, provided you were diligent thereafter in building and testing it or filing a patent application on it.

So, from now on, when I refer to "your earliest provable date of invention," this will mean the earliest of the above three dates (filing, building and testing, or conception accompanied by diligence) which you can prove.[5]

The kind of proof that the PTO and the courts typically rely on are the witnessed records of the type I described in Chapter 3. If you follow my recommendations in Chapter 3 about making proper records of conception, and building and testing, you'll be able to go back to your date of conception, which usually will be at least several months before your filing date. More on this in Chapters 13 and 16.

Now that you know what your earliest date of invention is, you also know that the relevant "prior art" is the knowledge that existed prior to that date. More precisely, prior art comprises all of the items in

---

[5]In the law, the building and testing of an invention is called a "reduction to practice." The filing of a patent application, while not an actual reduction to practice, is termed a "constructive" reduction to practice because the law will construe it in the same way it does an actual reduction to practice.

the categories discussed below. Any item in any of these categories can be used against your invention at any time, either by the PTO to reject your patent application, or later on (if the PTO didn't find it or didn't give it adequate weight) to invalidate your patent in court or in preliminary-to-litigation negotiations.

c. *Publicly Known More than One Year Prior to Application—The One-Year Rule*

Prior art is also knowledge about your invention that has become publicly known more than one year prior to the date you file your patent application. Known as the "one-year rule," the patent laws state that you must file a patent application within one year after you sell, offer for sale, or commercially or publicly use or describe your invention. If you fail to file within one year of such sale, offer for sale, public or commercial disclosure or use, the law bars you from obtaining a valid patent on the invention. Another way to put this, since we're talking about novelty, is that after a year following a sale, offer of sale, public or commercial use, or knowledge about your invention, it will no longer be considered novel by the PTO.

INVENTOR'S COMMANDMENT #4
One-Year Rule: You should treat the "one-year rule" as holy. You must file your patent application within one year of the date on which you first publish, publicly use, sell, or offer your invention, or any product which embodies same, for sale. Moreover if you wish to preserve your foreign rights and frustrate pirates of your creation, you should actually file your patent application before you publish or sell your creation.

Foreign Filing and the One Year Rule

While you have a year after publication or use to file in the U.S., most foreign countries aren't so lenient. So if you think you'll want to foreign file, you shouldn't offer for sale, sell, publicly use, or publish before you file in the U.S. For instance, suppose it's 1988 November 16[6] and you've just invented a new type of paint. If you have no intention of filing in another country, you can use, publish, or sell your invention now and still file your U.S. patent application any time up to 1989 November 16. However, if you think you may eventually want to foreign file on your invention, you should file in the U.S. before publicizing your invention. Then you can publish or sell the invention freely without loss of any foreign rights in the major industrial or "convention" countries, provided you file there within one year after your U.S. filing date. This is because, under an international convention, (agreement or treaty) you'll be entitled to your U.S. filing date in such countries. In "non-convention" countries (such as India and the Republic of China) you must file before you publicize the invention. See Chapter 12.

d. *Specifics of Prior Art*

Now that we've broadly defined prior art, let's take a closer look at what it typically consists of.

(1) **Prior Printed Publications Anywhere.** Any printed publication, written by anyone, and from anywhere in the world, in any language, is considered valid prior art if it was published either (a) before your earliest provable date of invention (see above), or (b) over one year before you file your patent

---

[6]This is the International Standards Organization (ISO) format for dates, which is also commonly used in computer applications. I use it since it provides a logical descending order which facilitates calculating the one-year rule and other periods.

application. The term "printed publication" thus includes patents (U.S. and foreign), books, magazines (including trade and professional journals), Russian inventor's certificates, publicly-available technical papers and abstracts, and even photocopied theses, provided they were made publicly available. The PTO has even used old Dick Tracy comic strips showing a wrist-watch radio as prior art!

Computer Note: While the statute speaks of "printed" publications, I'm sure that information on computer-information utilities or networks would be considered a printed publication, provided it was publicly available.

The "prior printed publications" category is the most important category of prior art and will generally constitute most of the prior art that you'll encounter. And most of the prior printed publications which the PTO refers to (cites) when it's processing your application, and which you will encounter in your search, will be patents, mainly U.S. patents.

(2) U.S. Patents Filed By Others Prior to Your Conception. Any U.S. patent which has a filing date earlier than your earliest provable date of invention (see above) is considered valid prior art. This is so even if such patent issues after you file your application. For example, suppose you conceive of your invention on 1989 June 9 and you file your patent application on 1989 August 9, two months later. Then, six months after your filing date, on 1990 February 9, a patent to Goldberger issues which shows all or part of your invention. If Goldberger's patent were any other type of publication, it wouldn't be prior art to your application since it was published after your filing date. However assume that Goldberger's patent application were filed on 1989 June 8, one day earlier than your date of conception. Under Section 102(e) of the patent laws, the PTO must consider the Goldberger patent as prior art to your application since Goldberger's application was filed prior to your date of conception.

Note: A common misconception is that only in-force patents (i.e., patents issued during the previous 17 years) count as prior art. This isn't true. Any earlier patent, even if it was issued 150 years ago, will constitute prior art against an invention.

(3) Prior Publicly-Available Knowledge or Use in the U.S. Even if there's no written record of it, any public knowledge of the invention, or use of it by you or others in the U.S., which existed or occurred either (a) before your earliest provable date of invention, or (b) over one year before you file your patent application, is valid prior art. For example, an earlier heat-treating process used openly by a blacksmith in a small town, although never published or widely known, can defeat your right to a patent on a similar process. Similarly a talk at a publicly-accessible technical society will constitute prior public knowledge.

For another example, suppose that you invented the new type of paint mentioned above and you use it to paint your building in downtown Philadelphia. You forget to file a patent application and leave the paint on for 13 months: it's now too late to file a valid patent application since you've used your invention publicly for over a year. Put another way, your own invention would now be prior art against any patent application you file.

This public use category of prior art is almost never used by the PTO since they have no way of uncovering it; they search only patents and other publications. Occasionally, however, defendants (infringers) in patent lawsuits happen to uncover (e.g., through depositions, interrogatories, and interviews), a prior public use that they then rely on to invalidate the patent.

Experimental Exception: If the prior public use was for bona fide (good faith) experimental purposes, it doesn't count as prior art. Thus suppose, in the "painted Philadelphia building" example above, that you painted your building to test the durability of your new paint: each month you photographed it, kept records on its reflectivity, wear resistance, adhesion, etc. In this case your one-year period wouldn't be initiated (begin to run) until your bona fide experimentation stopped.

(4) Your Prior Foreign Patents. Any foreign patent (this includes Russian inventor's certificates) of yours or your legal representatives which issued before your U.S. filing date and which was filed over a year before your U.S. filing date is valid prior art. This category is generally pertinent to non-U.S. residents who start the patenting process in a foreign country. If you're in this class, you must file your U.S. application either within one year after you file in the foreign country or before your foreign patent issues. However, if you want to get the benefit of a foreign filing date for use in the U.S., you should file within the one year after your foreign filing date (see Chapter 12).

(5) Prior U.S. Inventor. If anyone else in the U.S. invented substantially the same invention as yours before your date of conception, and he or she didn't abandon, suppress, or conceal it, then this other person's invention (even though no written record was made) can be used to defeat your right to a patent.[7] This uncommon situation usually occurs when two (or more) inventors each file a patent application on the same invention. The PTO will declare an "interference" between the two competing applications. See Chapter 16.

(6) Prior Sale or On-Sale Status in the U.S. Another instance where the law considers certain actions by humans to be "prior art," even when no paper records exist, is the "sale" or "on sale" category. Suppose you (or anyone else) offer to sell, actually sell, or commercially use your invention, or any product embodying your invention, in the U.S. You must file your U.S. patent application within one year after this offer, sale, or commercial use. This is another part of the "one-year rule." This means that

you can make sales to test the commercial feasibility of your invention for up to a year before filing in the U.S. Again, however, I advise you not to do so, since this will defeat your right to a patent in most foreign countries, as mentioned above, and as explained in more detail in Chapter 12.

(7) Abandonment. While I'm on the subject of Section 102 of the patent laws, I'll mention, for the sake of completeness, one other rarely-used facet of it which you may find pertinent even although it isn't strictly concerned with "prior art." If you "abandon" your invention by finally giving up on it in some way, and this comes to the attention of the PTO or any court charged with ruling on your patent, your application or patent will be rejected or ruled invalid. I've never personally had a case where this happened, but it has occurred.

Example: You make a model of your invention, test it, fail to get it to work, or fail to sell it, and then consciously drop all efforts on it. Later you change your mind and try to patent it. If your abandonment becomes known, you would lose your right to a patent. But if you merely stop work on it for a number of years because of health, finances, lack of a crucial part, etc., but intend to pursue it again when possible, the law would excuse your inaction and hold that you didn't abandon.

(8) Summary on Prior Art. If all of the above prior art rules seem complicated and difficult to you, you're not alone. Very few patent attorneys understand them fully either! Perhaps Congress will simplify Section 102 someday (write to your Congressperson!) but in the meantime, don't worry about it if you can't understand all of the rules. All that you really need to remember is that relevant prior art usually consists of:

- Any published writing (including any patent) which was made publicly available either (1) before your earliest provable date of invention (see above), or (2) over one year before you can get your patent application on file;

---

[7]Under a new statute, if your invention clears Section 102 (i.e., it's novel) and the prior inventor worked in the same organization as you, then the prior inventor's work won't be considered prior art under Section 103.

- Any U.S. patent whose issue date isn't early enough to stop you but which has a filing date earlier than your earliest provable date of invention;

- Any relevant invention or development (whether described in writing or not) existing prior to when your invention was conceived; or

- Any public use, sale or knowledge of the invention more than one year prior your application filing date.

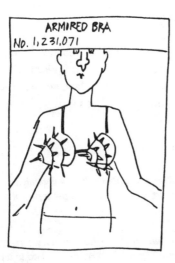

ARMIRED BRA
No. 1,231,071

2. Any Physical Difference Whatever Will Satisfy the Novelty Requirement

Any physical difference at all, no matter how trivial, will satisfy the novelty requirement. For example, suppose you've "invented" a bicycle which is painted yellow with green polka dots, each of which has a blue triangle in the center. Assume (this is easy to do) that no bicycle has never been so painted before. Your bicycle would thus clearly satisfy the requirement of novelty.

Rarely will an investigation into your invention's patentability (called a "patentability search") reveal any single prior invention or reference that could be considered a dead ringer. Of course, if your search does produce a dead-ringer reference for your invention—that is, an actual device or published description showing all the features of your invention and operating in the same way for the same purpose—obviously your patentability decision can be made immediately. Your invention lacks novelty over the "prior art." Another way of saying this is that your invention has been "anticipated" by a prior invention or conception and is thus definitely unpatentable. The concepts of anticipation and prior art are discussed in more detail in Requirement #4—Unobviousness.

Combinations: The PTO will consider your invention novel even if two or more prior-art references (actual devices or published descriptions) together account for all of your invention's physical characteristics. I.e., if your invention is a new combination of two old features, the law will consider it novel. For your invention to be considered as lacking novelty, and thus subject to rejection under Section 102 of the patent laws, *all* of its physical characteristics must exist in a *single* prior-art reference. For example, getting back to your bicycle, suppose you now "invent" a bicycle made of one of the recently-discovered, super-strength carbon fiber alloys. The bicycle per se is old, as is the alloy, but you're the first to "combine" the two old concepts. Your bicycle would clearly be considered to be novel since it has a new physical feature: a frame which is made, for the first time, of a carbon fiber alloy. But, remember, just because it's novel, useful, and fits within a statutory class, doesn't mean the bicycle is patentable. It still must surpass the high box of non-obviousness (covered in the following section).

Physical Differences: Also note that to satisfy the novelty requirement, the new feature(s) of your invention must involve some *physical* difference over the prior art. This means that if the invention is a machine, composition, or article, it must be or have one or more parts which have a different shape, value, size, color, or composition than what's already known. A physical difference can also be subtle or less apparent in the hardware sense so that it's manifested primarily by a different mode of operation. E.g., an electronic circuit which looks the same, but which operates in a different mode, or a pump which looks the same, but which operates at a higher pressure and hence in a different mode, will be considered novel.

If it's a process or new use, it must be or have a new method step (way of manipulating or using something).

It's often difficult for inventors to distinguish between a physical difference and a new result. When I ask clients, what's physically different about your invention?", they usually reply that theirs is lighter, faster, safer, cheaper to make, or use, portable, etc. However these factors are new *results* or *advantages*, not physical differences, and are primarily relevant to unobviousness (see next section(F)), not to novelty. That is, they won't per se help your invention satisfy the novelty requirement. Again, the new features must be *physical* (including operational) differences.

Processes: If your invention is a new process (including a new-use invention—see part C5. above), you don't need any novel hardware; your novelty is basically your new way of manipulating old hardware. Any novel step or steps whatever in this regard will satisfy the novelty requirement. If your process does involve novel physical hardware, technically it can't be a new-use invention.

If you're the type of person who thinks ahead, you're probably asking yourself, "Why is he bothering with novelty—isn't this quality inherent in unobviousness, i.e., if the invention is found to be unobvious won't it also be found to be novel"? Well, you're 100% correct. If an invention is unobvious, *a fortiori* (by better reason) it must be novel. However, the law makes the determination in two steps (Secs.

102 and 103) and most patent professionals have also found it far easier to first determine whether and how an invention satisfies the novelty requirement and then determine if it can be considered unobvious.

F. Requirement #4: Unobviousness

1. Overview

We're now entering what's probably the most misunderstood and difficult-to-understand (yet most important) aspect of patent law, i.e., whether your invention is unobvious. Let's start with a "common misconception."

▲

Common Misconception: *If your invention is different from the prior art, you're entitled to get a patent on it.*

Fact: *Under Section 103 of the patent laws, no matter how different your invention is, you're not entitled to a patent on it unless its difference(s) over the prior art can be considered "unobvious" by the PTO or the courts.*

Most of the time a patentability search will produce one or more prior-art references that show devices similar to your invention, or which show several, but not all, of the physical features of your invention. That is, you will find that your invention has one or more features or differences that aren't shown in any one prior art reference. However, even though your invention is physically different from such prior art, this isn't enough to qualify for a patent. To obtain a patent, the physical (or use) differences must be substantial and significant. The legal term for such a difference is "unobvious" or, commonly, "non–obvious" That is, the differences between your invention and the prior art must not be obvious to one with ordinary skill in the field. Because this concept is so important, let's examine it in detail.

2. Unobvious to Whom?

It doesn't tell anyone much to say an invention must be unobvious. The big question is, unobvious to whom? Under Section 103, you can't get a patent if a person having ordinary skill in the field of your invention would consider the idea of the invention "obvious" at the time you came up with it.

The law considers "a person having ordinary skill in the art to which said subject matter pertains" to be a mythical worker in the field of the invention who has 1) ordinary skill, but who 2) is totally omniscient about all the prior art in his or her field. This is a pure fantasy since no such person ever lived, or ever will, but realistically there's no other way to come even close to any objective standard for determining non-obviousness.

Let's take some examples. Assume that your invention has to do with electronics—say an improved flip-flop circuit. A person having ordinary skill in the art would be an ordinary, average logic-circuit engineer who's intimately familiar with all the prior-art logic circuits. If your invention has to do with chemistry, say a new photochemical process, a typical photochemical engineer with total knowledge of all photochemical processes would be your imaginary skilled artisan. If your invention is mechanical, such as an improved cigarette lighter or belt buckle, the PTO would try to postulate a hypothetical cigarette-lighter engineer or belt-buckle designer with ordinary skill and comprehensive knowledge. If your invention is a design, say for a computer case, the PTO would invent a hypothetical computer case designer of ordinary skill and full knowledge of all existing designs.

3. What Does "Obvious" Mean?

Most people have trouble interpreting Section 103 because of the word "obvious." If after reading my explanation you still don't understand it, don't be dismayed. Most patent attorneys, patent examiners, and judges can't agree on the meaning of the term. Many tests for obviousness have been used and rejected by the courts over the years. The courts have often referred to "a flash of genius," "a synergistic effect (the whole is greater than the sum of its parts)," or some other colorful term. One influential court said that unobviousness is manifested if the invention produces "unusual and surprising results."

Technically (for reasons mentioned below, I stress the term technically), none of these tests is used any longer. This is because the U.S. Supreme Court, which has final say in such matters, decreed in the famous 1966 case of *Graham v. John Deere* that Section 103 is to be interpreted by taking the following steps:

1. Determine the scope and content of the prior art;

2. Determine the novelty of the invention;

3. Determine the level of skill of artisans in the pertinent art;

4. Against this background, determine the obviousness or unobviousness of the inventive subject matter.

5. Also consider secondary and objective factors such as commercial success, long-felt, but unsolved need, and failure of others.

Unfortunately, while the Supreme Court has the last word in theory, in practice they added nothing to our understanding of the terms "obviousness" and "unobviousness" since in the crucial step (#4), they merely repeated the very terms (obvious and unobvious) they were seeking to define. Therefore, most attorneys and patent examiners continue to look for new and unexpected results which flow from the novel features when seeking to determine if an invention is obvious.

Despite its failure to define the term "obvious," the Supreme Court did add an important step to the process by which "obviousness" is to be determined. In Step #5, the court made clear that objective circumstances must be taken into account by the PTO or courts when deciding whether an invention is or isn't obvious. The court specifically mentioned three such circumstances: commercial success, long-felt but unsolved need, and failure of others.

So, although your invention might not, strictly speaking, produce "new and unexpected results" from the standpoint of one with "ordinary skill in the art," it still may be considered unobvious if, for instance, you can show that the invention has enjoyed commercial success.

Under the reasoning of the *John Deere* case, then, to decide whether or not your invention is obvious, you first should ask whether it produces "new and unexpected results" from the standpoint of one skilled in the relevant art. If it does, you've met the test for patentability. However, if there's still some doubt on this question, external circumstances may be used to bolster your position.

Note of Sanity: If you feel your head spinning, don't worry. It's natural. Because these concepts are so abstract, there's no real way to get a complete and comfortable grasp on them. However, if you take it slow (and take a few breaks from your reading), you should have a pretty good idea of when an invention is and isn't considered "unobvious." In Section 4 directly following, I discuss examples of "unobviousness" and "obviousness." Then, in Section 5 I cover the types of arguments based on external circumstances (called secondary factors) that can be made to bolster your contention that your invention is unobvious.

NECK SHOWER.
No. 821.716

4. Examples of Obviousness and Unobviousness

First, for some examples of unobvious inventions, consider all of the inventions listed in Chapter 2: the magnetic pistol guard, the buried plastic cable, the watch calendar sticker, "Grasscrete," the Wiz-z-er top, the shopping cart, etc. These all had physically novel features which produced new, unexpected results, i.e., results which weren't suggested or shown in the prior art.

Although generally you must make a significant physical change for your invention to be considered unobvious, often a very slight change in the shape, slope, size, or material can produce a patentable invention that operates entirely differently and produces totally unexpected results.

Example: Consider the original centrifugal vegetable juicer composed of a spinning perforated basket with a vertical side wall and a non-perforated grater bottom. When vegetables, such a carrots, were pushed into the grater bottom, they were grated into fine pieces and juice which was thrown against the cylindrical, vertical side wall of the basket. The juice passed through the perforations and was recovered in a container but the pieces clung to the side walls, adding weight to the basket and closing the perforations, making the machine impossible to run and operate after a relatively small amount of vegetable was juiced. Someone conceived of making the side of the basket slope outwardly so that while the juice was still centrifugally extracted through the perforated side of the basket, the pulp, instead of adhering to the old vertical side of the basket, was centrifugally forced up the new sloped side of the basket where it would go over the top and be diverted to a separate receptacle. Thus the juicer could be operated continuously without the pulp having to be cleaned out. Obviously, despite the fact that the physical novelty was slight, i.e., it involved merely changing the slope of a basket's side wall, the result was entirely new and unexpected, and deserved patent protection.

In general such a relatively small physical difference (changing the slope of the wall of a basket in a juicer) will require a relatively great new result (ability to run the juicer continuously) to satisfy the unobviousness requirement. On the other hand, a relatively large physical difference will need only minor new results for the PTO to consider it unobvious. I.e., in Fig 5-A (the patentability mountain) the height of the fourth box can be shortened if the height of the third box is increased.

As indicated, new-use inventions don't involve any physical change at all in the old hardware. However the new use must be 1) a different use of some known hardware or process, and 2) the different use must produce new, unexpected results.

Example: Again consider the venetian blind cleaner used as a seed planter, and aspirin used as a growth stimulant, discussed in Part C5 above. In both instances, the new use was very different and provided a totally unexpected result: thus they would be patentable. Also, in another interesting new use case, the patent court in Washington held that removing the core of an ear of corn to speed freezing and thawing was unobvious over core drilling to

speed drying. The court reasoned that one skilled in the art of corn processing could know that core removal speeds drying without realizing that core removal could also be used to speed freezing and thawing. Accordingly, the court held that the new result (faster freezing and thawing) was unexpected since it wasn't described or suggested in the prior art.

The courts have held that the substitution of a different, but similarly-functioning, element for one of the elements in a known combination, although creating a "novel" invention, won't produce a patentable one. For example, consider the substitution, in the 1950's, after transistors had appeared, of a transistor for a vacuum tube in an old amplifier circuit. At first blush this new combination of old elements would seem to the uninitiated to be a patentable substitution since it provided tremendous new result (decreased power consumption and weight). However, you'll soon realize that the result, although new, would have been entirely foreseeable since, just as in the carbon-fiber/bicycle case, the power-reduction and reduced weight advantages of transistors would have been already known as soon as a transistor made its appearance. Thus, substituting them for tubes wouldn't provide the old amplifier circuit with any unexpected new results. Accordingly, the PTO's Board of Appeals held the new combination to be obvious to an artisan of ordinary skill at the time.

If you're still a bit misty about all this, put yourself in the shoes of an electronic engineer who at the time of the replacement of the vacuum tube with the transistor, was skilled in designing vacuum tube circuits and was currently designing a flip-flop circuit. Along comes this newfangled "transistor" that uses no heater and weighs 1/10th that of a comparable tube, but which provides the same degree of amplification and control as the tube did. Do you think that it wouldn't be obvious to the engineer to try substituting a transistor for the tube in that flip-flop circuit? Similarly, the PTO would consider obvious the substitution of an integrated circuit for a group of transistors in a known logic circuit, or the use of a known radio mounting bracket to hold a loudspeaker enclosure instead of a radio.

The PTO will also consider as obvious the mere carrying forward of an old concept, or a change in form and degree, without a new result. For instance, when one inventor provided notches on the inner rim of a steering wheel to provide a better grip, the idea was held to be obvious because of medieval sword handles which had similar notches for the same purpose. And the use of a large pulley for a logging rig was held nonpatentable over the use of a small pulley for clotheslines.

On the other hand, one inventor merely changed the slope of a part in a paper-making (Fourdriner) machine; as a result the machine's output increased by 25%—a dramatic, new, and unexpected result which was held patentable.

In sum, the PTO will usually hold that substitution of a different material is obvious. But if the substitution provides *unexpected* new results, the law will hold it unobvious.

The courts and the PTO will also usually consider the duplication of a part obvious unless it can see new results. For instance, in an automobile, the substitution of two banks of three cylinders with two carburetors was held obvious over a six-cylinder, single carburetor engine since the new arrangement had no unexpected advantages. However, the use of two water turbines to provide cross flow to eliminate axial thrust on bearings was held unobvious over a single turbine; again, an *unexpected*, new result.

Similarly, making devices portable, making parts smaller or larger, faster or slower, effecting a substitution of equivalents (a roller bearing for a ball bearing), making elements adjustable, making parts integral or separated, and other known techniques with their known advantages, will be held obvious, unless new, unexpected results can be shown.

Design Patent Note: In design cases, the design must have novel features, and the PTO must be able to regard these as unobvious to a designer of ordinary skill. If the design involves the use of known techniques which together don't produce any new and unexpected visual effect, then the PTO will consider it obvious. But if they produce a startling or unique new appearance, then the PTO will hold it to be unobvious. Since only the ornamental appearance and not the function of a design is relevant, the degree of novelty of the design will be the main determinant of unobviousness: a high degree of novelty will always be patentable, while a low degree of novelty will encounter rough sledding unless you can set forth reasons why it has a very different appearance or visual effect.

5. Secondary Factors In Determining Unobviousness

As mentioned, if the new and unexpected results of your invention are marginal, you *may* still be able to get a patent if you can show that your invention possesses one or more secondary factors that establish unobviousness. While the Supreme Court listed only three in the *John Deere* case, I've compiled a list of eleven which the PTO and the courts actually consider. Also, in Section 6 below, I've listed eight additional factors that bear on obviousness when the invention combines elements from two or more prior-art references. In the real world, these secondary factors must generally only be dealt with if the PTO makes a preliminary finding of obviousness or if your invention is attacked as being obvious. However, when deciding whether your invention is legally entitled to a patent, you'll have a much better idea of how easy or difficult it will be to obtain if you apply these secondary factors to your invention.

Express Track Note: If you're sure that your invention is unobvious, feel free to skip this section, and Section 6, and proceed directly to Section 7.

Although some of these secondary factors may appear similar, try to consider each independently since the courts have recognized subtle differences between them. As part of doing this remember that lawyers like to chop large arguments into little ones so that it will appear that there are a multitude of reasons for their position rather than just one or two. While this approach may seem silly, it's nevertheless a fact (however sad) that the PTO and courts are used to hearing almost exclusively from lawyers (and, in the case of the PTO, from highly-specialized patent agents). Accordingly, the general rule is the more arguments you can use to claim unobviousness, the better your chances will be of getting a patent.

Now let's look at the eleven secondary factors in detail.

Factor 1. Previously unworkable techniques. If the invention goes against the grain of prior related inventions (prior art) by successfully employing techniques previously thought to be or found unworkable, this will be of great help to your application. For instance, suppose several prior-art references stated that electrostatic methods won't work for making photocopies and Chester Carlson (a patent attorney himself) came along and successfully used an electrostatic process to make copies. This would have enhanced his case for the patentability of his dry (xerographic) photocopying process.

Factor 2. Solves an unrecognized problem. Here the essence of your invention is probably the recognition of the problem, rather than its solution. Consider the shower head which automatically shuts off in case of excess water temperature discussed in Chapter 2. As the problem was probably never recognized in the prior art, the solution would therefore probably be patentable.

Factor 3. Solves an insoluble problem. Suppose that for years those skilled in the art had tried and failed to solve a problem and the art and literature were full of unsuccessful "solutions." Along you come and finally find a workable solution, e.g., a cure for the common cold: you'd probably get a patent.

Example: Consider an invention made by a client of mine—a circuit that, when connected across a light switch, holds the light on for about twenty seconds after the light is switched off, and can be used repeatedly and always operates in the same manner. The prior art showed "delayed-off" circuits, but all of these could only be used once every several minutes. By incorporating special discharging and reset circuits that had never before been used or suggested, my client's invention successfully solved a problem (lack of instant resettability) which was either not recognized before, or if recognized, wasn't before soluble. Thus the circuit was patentable over the prior art.

Factor 4. Commercial success. If your invention has attained commercial success by the time the crucial patentability decision is made, this militates strongly in favor of patentability. Nothing succeeds like success, right?

Factor 5. Crowded art. If your invention is in a crowded field (art), that is, a field which is mature and which contains many patents, such as electrical connectors or bicycles, a small advance will go further towards qualifying the invention for a patent than it will in a new, blossoming art, such as monoclonal antibodies.

Factor 6. Omission of element. If you can omit an element in a prior invention without loss of capability , this will count a lot since parts are expensive, unreliable, heavy, and labor-intensive (an example would be eliminating an inductor in an oscillator circuit).

Factor 7. Unsuggested modification. if you can modify a prior invention in a manner not suggested before, e.g., by increasing the slope in a paper making machine, or by making the basket slope in a centrifugal juice extractor, this act in itself counts for patentability.

Factor 8. Unappreciated advantage. If your invention provides an advantage which was never before appreciated, it can make a difference. In a recent case, a gas cap that was impossible to insert in a skewed manner was held to be patentable since it provided an advantage which was never appreciated previously.

Factor 9. Solves Prior Inoperability. If your invention provides an operative result where before only inoperability existed, then it has a good chance for a patent. For instance, suppose you come up with a gasoline additive which really prevents huge fires in case of a plane crash, you've got it made since all previous fire suppressant additives have been largely unsuccessful.

Factor 10. Successful implementation of ancient idea where others failed. The best example I can think of is the Wright Brothers' airplane. For millennia humans had wanted to fly and had tried many schemes unsuccessfully. The successful implementation of such an ancient desire carries great weight when it comes to getting a patent.

Factor 11. Solution of long-felt need. Suppose you find a way to prevent tailgate-type automobile crashes. Obviously you've solved a powerful need and your solution will be a heavy weight in your favor on the scales of patentability.

6. Secondary Factors In Determining Unobviousness for Combination Inventions

Inventions which combine two or more elements known in the prior art can still be held patentable, provided that the combination can be considered unobvious, i.e., it's a new combination and it produces new and unexpected results. In fact, most patents are granted on such combinations since very few truly new things are ever discovered. So let's examine some of the factors used especially to determine the patentability of "combination inventions" (i.e., inventions which have two or more features which are shown in two or more prior-art references).

Express Track Note: The following material is conceptually quite abstract and difficult to understand, even for patent attorneys. I'm presenting it in the interests of completeness. However, if you

wish, you can safely skip it for now and proceed directly to Section 7. If the PTO or anyone else suggests that two or more prior-art references, taken together, teach that your invention is obvious, come back and read it then.

a. Synergism (2 + 2 = 5)

If the results achieved by your combination are greater than the sum of the separate results of its parts, this can indicate unobviousness. Consider the pistol trigger release (Chapter 2) where a magnetic ring must be worn to fire the pistol. The results (increased police safety) are far in excess of what magnets, rings, and pistols could provide separately.

Example: For another example, suppose that a chemist combines, through experimentation, several metals which cooperate in a new way to provide added strength without added density. If this synergistic result wasn't reasonably foreseeable by a metallurgist, the new alloy would almost certainly be patentable.

Generally, if your invention is a chemical mixture, the mixture must do more than the sum of its components. For this reason, food recipes are difficult to patent unless an ingredient does more than its usual function. Similarly, if you combine various mechanical or electrical components, the courts and the PTO will usually consider the combination patentable if it provides more than the functions of its individual components.

As an example of an unpatentable combination without synergism, consider the combination of a radio, waffle iron, and blender in one housing. While novel, and useful, this combination would be considered obvious since there's no synergism or new cooperation: the combination merely provides the sum of the results of its components and each component works individually and doesn't enhance the working of any other component. On the other hand, the combination of an eraser and a pencil would be patentable (had it not already been invented) because the two elements cooperate to increase overall writing speed, a synergistic effect.

b. Combination Unsuggested

If the prior art contains no suggestion, either expressed or implied, that the references should be combined, this militates in favor of patentability. Examiners in the PTO frequently are assigned to pass on patent applications for combination inventions. To find the elements of the combination claimed, they'll make a search, often using a computer, to gather enough references to show the respective elements of the combination. While the examiners frequently use such references in combination to reject the claims of the patent application on unobvious grounds, the law says clearly that it's not proper to do so unless *the references themselves*, rather than an applicant's patent application, suggest the combination.

Example: Arthur B. files a patent application on a pastry molding machine. The examiner cites (or your search reveals) one patent on a foot mold and another on a pastry mold to show the two elements of the invention. It wouldn't be proper to "combine" these disparate references since they're from unconnected fields and thus it wouldn't be obvious to use them together against your invention.

An example of where the law would consider it obvious to combine several references is the case where, as discussed, you make a bicycle out of the lightweight carbon fiber alloy and, as a result, your bicycle is lighter than ever before. Is your invention "unobvious"? The answer is "No," because the prior art implicitly suggests the combination by mentioning the problem of the need for lighter bikes and the lightness of the new alloy. Moreover the result achieved by the combination would be expected from a review of existing bicycles and the new lightweight alloy. In other words, if a skilled bicycle engineer were to be shown the new, lightweight alloy, it would obviously occur to the engineer to make a bicycle out of it since bicycle engineers are always seeking to make lighter bicycles.

c. Impossible to Combine

This is the situation where prior-art references show the separate elements of the inventive combination, but in a way that makes it seem they would be physically impossible to combine. Stated differently, if you can find a way to do what appears to be physically impossible, then you can get a patent. For example, suppose you've invented the magnetic pistol release. The prior art shows a huge magnetic cannon firing release attached to a personnel shield. Since the step from a cannon to a small handgun is a large one, physical incompatibility might get you a patent, i.e., it would be physically impossible to use a huge cannon shield magnet on a small and very-differently-shaped trigger finger. Note, however, that sometimes by analogy the large can properly be used on the small if a mere change in size is all that's required.

d. Different Combination

Here your combination is A, B, and C, and the prior art references show a different, albeit possibly confusingly-similar combination, say A', B, and C. Since your combination hadn't been previously created, you've got a good case for patentability even though your creation is similar to an existing one. Again the last analogy holds: a personnel shield for a cannon, even though it has a magnetic firing release, is so far different from a finger ring, that the prior-art combination must be regarded as different from that of the invention.

e. *Prior-art References Would Not Operate In Combination*

Here the prior-art references, even if combined, wouldn't operate properly, e.g., due to some incompatibility. Suppose you've invented a radio receiver comprising a combiner tuner-amplifier and a speaker, and the prior art consists of one patent showing a crystal tuner and an advertisement showing a large loudspeaker. The prior art elements wouldn't operate if combined because the weak crystal tuner wouldn't be able to drive the speaker adequately; thus a combination of the prior-art elements would be inoperative. This would militate strongly in favor of patentability.

f. *Over Three Prior Art References Necessary to Show Your Invention*

While not a very strong argument, if it takes more than three references to meet your inventive combination, this militates in your favor.

g. *References Teach Away from Combining*

If the references themselves show or teach that they shouldn't be combined, and you're able to combine them, this militates in favor of patentability. For example, suppose you make that bike out of the carbon-fiber alloy and a reference says that the new carbon fiber alloy should only be used in structural members which aren't subject to sudden shocks. If you're able to use it successfully to make a bike frame, which is subject to sudden shocks, you should be able to get a patent.

h. *Awkward, Involved Combination*

Suppose that to make your inventive combination, it takes the structures of three prior-art patents, one of which must be made smaller, another of which must be modified in shape, and the third of which must be made of a different material. These factors can only help you.

7. How Does a Patent Examiner Determine "Unobviousness"

Because it's usually helpful to understand how a bureaucracy operates when you're dealing with it over significant issues, let's take a minute to examine how a patent examiner proceeds when deciding whether or not your invention is obvious. When patent examiners turn to the question of whether an invention is unobvious, they first make a search and gather all of the patents that they feel are relevant or close to your invention. Then they sit down with these patents (and any prior-art references you've provided with your patent application) and see whether your invention, as described in your claims (see Chapter 9), contains any novel physical features which aren't shown in any reference. If so, your invention satisfies Section 102: i.e., it's novel.

Next they see whether your novel physical features produce any unexpected or surprising results. If so, they'll find that the invention is unobvious and grant you a patent. If not (this usually occurs the first time they act on your case), they'll reject your application (sometimes termed a "shotgun" or "shoot-from-the-hip" rejection) and leave it to you to show that your new features do indeed produce new, unexpected results. To do this, you can use as many of the reasons listed above which you feel are relevant. If you can convince the examiner, you'll get your patent.

If a dispute over unobviousness actually finds its way into court (a common occurrence), however, both sides will present the testimony of experts who fit, or most closely fit, the hypothetical job descriptions called for by the particular case. These experts will testify for or against obviousness by arguing that the invention is (or isn't) new and/or that it does (or doesn't) produce unexpected results.

Because the question of whether an invention is unobvious is obviously crucial to whether a patent will issue and because Secs. 102 and 103 are widely confused, I have made the two-step evaluation of unobviousness the subject of another commandment:

INVENTOR'S COMMANDMENT #5:
To evaluate or argue the patentability (unobviousness) of any invention, use a two-step process:
a. First determine what novel physical (hardware) features the invention has over the closest separate prior-art references (a novel physical feature can be the combination for the first time, of two separate old and known features which are shown in different references); and
b. Then determine if the novel physical features produce any unexpected results, or otherwise "go against the grain of previous knowledge," or have achieved success so as to indicate unobviousness to one of ordinary skill in the art.

8. Weak Versus Strong Patents

Although in this section I've covered the basic legal requirements for obtaining a patent on an invention, there is, in reality, an additional practical requirement. If the claims in your patent are easy to design around or are so narrow as to virtually preclude you from realizing commercial gain, it's virtually the same as if a patent had been denied you in the first place. I'll come back to this point when I

cover how to conduct a patent search (Chapter 6) and how to draft your claims (Chapter 9).

9. Anyone May Apply for a Patent

You may have noticed that in discussing the requirements for obtaining a patent, I didn't mention any eligibility requirements or personal qualifications (e.g., the applicant should be over 21, etc.). That's because there are *no personal qualifications whatever*: So long as a person qualifies as a true inventor of the invention (discussed in Chapter 10), that person (or persons) may apply for a patent regardless of age, sex, citizenship, mental competence, health, physical disabilities, nationality, race, creed, religion, state of incarceration, etc. Even a dead or insane person can apply (through his legal representative, of course).

The manner of making the invention is also irrelevant, as we'll see by the next Common Misconception.

Common Misconception: *If a complete moron discovers something by accident, the law won't consider it to be as good an invention as if a genius had come up with it through years of hard, brilliant work.*

Fact: *The manner of making an invention is totally irrelevant to the law. The invention is looked at in its own right as to whether or not it would be obvious to one skilled in the art; the way it was made is never considered by the PTO.*

G. The Patentability Flowchart

TO GET A BETTER GRASP of the admittedly slippery concept of unobviousness and the role it plays in the patent application process, consider Fig. 5-B, the Patentability Flowchart. This flowchart is like a computer programmer's flowchart, except that

In addition to presenting all of the criteria used by the PTO and the courts for determining whether an invention is unobvious, the chart also incorporates the first three tests (statutory class, usefulness, and novelty) of Fig. 5-A. I strongly advise that you study this chart and the following description of it well since it sums up the essence of this crucial chapter. Also, you'll want to use this chart when making your search (next chapter) and when prosecuting your patent application (Chapter 13). This chart has been designed to cover and apply to anything you might come up with, so you can and should use it to determine the patentability of any creation whatever.[8]

Box A: Assuming that you've made an invention, first determine, using the criteria discussed above, whether you can reasonably classify your invention in one of the five statutory classes. If not, take the "N" (No) output of Box A to the "(X)" (Box X) on the right side near the bottom of the chart.

As indicated in Box X, the PTO will probably refuse to grant you a patent, so see if you can gainfully use another form of coverage (e.g., trade secret, copyright, design patent, trademark, unfair competition as discussed in Chapter 1). If this possibility also fails, you'll have to give up on the creation and invent something else. If the invention can be classified within a statutory class ("Y" or Yes output of Box A), move on to Box B.

Box B: Now determine, again using the criteria above, whether the invention has utility, including amusement. If not, move to Box X. If so, move on to Box C.

Box C: Here's the important novelty determination. If an invention has any physical features which aren't present in any single prior-art reference, it will clear Section 102, i.e., it has novelty: take the Yes output to Box E. If not, it lacks novelty, so take the No output and go to Box D to see if it's a new use invention.

Box D: Here determine if your invention is a new use of an old invention, i.e., your invention is any new and different use of any old machine, composition, article, or process. Remember the examples above—the use of aspirin to speed the growth of swine and the use of an old venetian blind cleaner as a seed planter. Any significantly new and different use of any old apparatus or process will get you to the Yes output of Box D and on to Box E. However, if your invention has no novelty and it isn't a new use, you'll have to take the No output of Box D and go to Box X where you'll have to forego the possibility of getting a patent.

Box E: This is the heart of the chart. You should now determine whether the novel feature(s) provide any new and unexpected results. Use the criteria and examples presented above in Sections F(1)-(4). If your answer is a clear "yes," i.e., the invention definitely provides new and unexpected results, take the Yes output to Box F. If you definitively feel your invention provides no new and unexpected results, take the No output from Box E to Box X. Of course you may not, at this point, be able to come up with a clear Yes or No. Instead, your invention will fall somewhere in between these two extremes.

Fortunately, as mentioned, marginally obvious inventions can still qualify for patents if they have one or more "secondary factors" tending to establish their underlying patentability. Follow the "possibly" output to Box F to determine whether your invention qualifies for a patent even though it doesn't produce especially new or surprising results.

---

[8]The chart is applicable only to utility inventions; not to designs or plants.

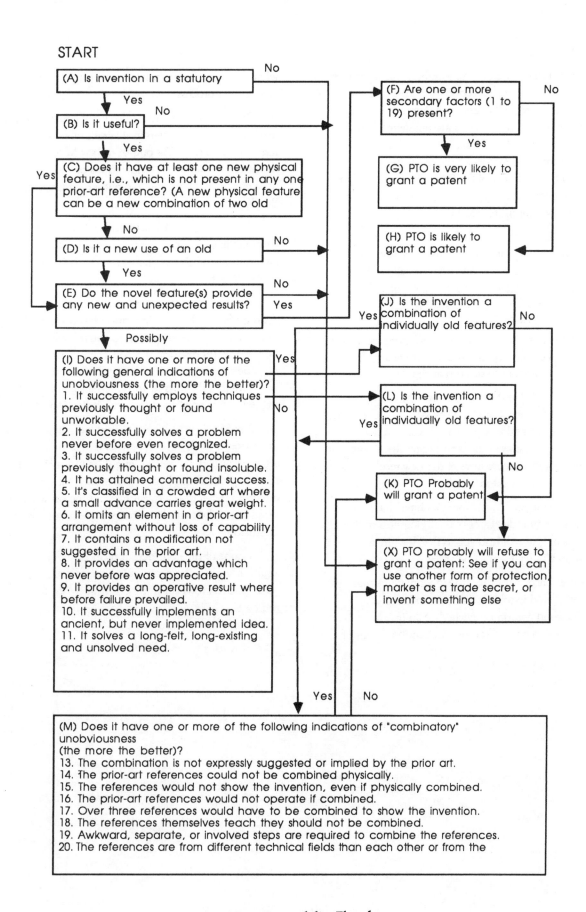

Fig. 5-B — Patentability Flowchart

Box F: Here determine whether your invention has one or more of the secondary factors that the PTO and courts will consider in determining whether it's unobvious. These factors are numbered 1 to 19 and are listed in Boxes I(1-11) and M (12-19). They're the same factors we discussed in Section F(5) and (6) above. If any of the items are applicable to your invention, take the Yes output to Box G. If not, take the No output to Box H.

Box G: Here you'll find that the PTO is very likely to grant you a patent. This is because your invention provides new and unexpected results and also has one or more secondary factors of unobviousness, an almost irresistible double whammy! Note that the stronger and more unexpected the new results, and the more secondary features you have and the stronger these are, the more likely you are to succeed at the PTO.

Box H: Even if your invention lacks any secondary factors of unobviousness, the PTO is likely to grant you a patent since your invention does provide some new and unexpected results. Again, the strength or unexpectedness of the new results is important.

Box I: If your invention does have one or more of the general indications of unobviousness of Box H, the Yes output will lead you to Box I, while a No decision will lead you to Box K.

Box I lists 12 of the 21 secondary items that can be considered in addition to new and surprising results when determining whether an invention is unobvious. See the discussion in Section F above for what these factors mean in the overall patent prosecution process.

Box J: If the invention also has a combination of features which are shown in several prior-art references, take the Yes output to Box M to consider the other "combinatory" indications of unobviousness. If not, the No output leads you to Box K.

Box K: Here, as indicated, the PTO may hold it patentable; the more secondary considerations that are present, and the more powerful each is, the better your chances.

Box L: This box is identical to Box J. If the decision here is No, the invention has no secondary indications of unobviousness, so you'll have to go to Box X. If a Yes decision is made, the invention has no general secondary indications, but several prior-art references show its features, so let's go to Box M to see if it has any "combinatory" indications.

Box M: Here you want to find out whether your invention is unobvious in light of two or more prior-art references that together might be interpreted as teaching (describing) your invention. This point was discussed in Section 6.h. earlier, and the nine factors listed in this box are the same as those set out in that discussion.

If none of the considerations in Box M is applicable, then the decision is No and you'll have to go to Box X. However if the decision is Yes, then go to Box K, where, as stated, you have a fighting chance of getting a patent.

Chapter 6

SEARCH AND YOU MAY FIND

SINCE YOU'VE LEARNED how to determine patentability from Chapter 5, you can now make a patentability search. The Patent and Trademark Office doesn't require a search, but I strongly believe that all inventors should make (or have made) a search prior to deciding whether to file a patent application. Thus I've made the "pre-ex" search (pre-examination patentability search) the sixth Inventor's Commandment.

INVENTOR'S COMMANDMENT #6:
You should make (or have made) a thorough patentability search of your invention before you decide whether to file a patent application.

A. Why Make a Patentability Search?

I'VE COME UP with twelve reasons for making a patentability search. These are:

1. To Determine Whether You Can Get a Patent

The main reason for making a patentability search of your invention is to discover if the Patent and Trademark Office (PTO) will be likely to grant you a patent on your invention. You may wonder why this should make any difference. After all, why worry about what the PTO will do before it does it? Simply because, if your search indicates that your invention is likely to qualify for a patent, you can go ahead with your development, marketing, and other work on the invention with far more assurance that your efforts will eventually produce positive results. Obviously, if a patent is ultimately granted, you will have a monopoly in the field of the invention for a number of years. Assuming, of course, that your invention has

economic value, this will allow you to sell or license it for a reasonable amount since you'll have at least some assurance that a right to exclude copiers will go with the invention.

If, on the other hand, your patentability search indicates that a patent isn't likely to be granted on your invention, you'll have to think long and hard about whether to proceed. The hard truth is that most manufacturers won't want to invest the money in tooling, producing, and marketing something that their competition can freely copy, and perhaps even sell, at a lower cost. As we'll see in Chapter 7, however, this isn't always true. While it's somewhat unusual, fortunes have sometimes been made manufacturing and selling unpatentable inventions.

2. To Avoid Needless Expenditures and Work

Another reason to make a patentability search has to do with time and money. It's a lot easier (and cheaper) to make a patentability search than to prepare a patent application which must contain specification, drawings, claims, filing fee, forms, etc. It makes sense to do a relatively small amount of work entailing a modest expenditure in order to gain useful information that may well allow you to avoid wasting considerable time and/or spending a relatively large amount of money.

3. To Provide Background to Facilitate Preparation of Your Patent Application

You'll find it far easier to prepare a patent application on your invention if you make a patentability search first. This is because a search will bring out prior art references (prior publications including patents and literature) in the field of your invention. After reading these, you're almost sure to learn much valuable background information that will make the task of writing your patent application far easier. Even patent attorneys routinely review some sample patents from the field of an invention before they begin preparation of a patent application in order to give them a "feel" for the "art" involved.

4. To Know Whether to Describe and Draw Components

This reason is closely allied with Reason #3. As we'll see in Ch. 8, a patent application must contain a detailed description of your invention, in sufficient detail to enable a person with ordinary skill in the "art" involved to make and use it. If your invention has certain components with which you aren't familiar, you won't have to take the trouble to draw and describe these in detail if you find them already described in prior-art publications, including patents.

5. To Provide More Information About Operability and Design

When you make a search, you almost always find patents in the field of your invention, possibly on inventions similar to yours. A reading of these patents will give you valuable technical information about your invention, possibly suggesting ways to make it work better and improve its design.

6. To Learn Commercial Information

The patents and other references which you uncover in your search will give you valuable commercial information about your invention, including possible additional advantages and disadvantages, possible new uses, past commercial failures, etc. For instance, suppose you see many patents on inventions that produce the same result as yours, and you know from your familiarity with the field (as a result of your commercial evaluation in Chapter 4 and your preliminary look—see Section 3 below) that none of these has attained commercial success. In this event, you might want to reconsider the wisdom of pushing ahead with your own invention. Or you might conclude that you can do better because the prior inventions were not commercially exploited properly or that they did not operate properly due to lack of proper materials, etc.

7. To Obtain Possible Proof of Patentability

Sometimes a search will uncover references which actually "teach away" from your invention, e.g., by suggesting that your approach won't work. You can cite such a reference to the PTO to help convince the examiner to regard your invention as unobvious (see secondary reason #1 in Fig. 5-B, Chapter 5, Section B).

For instance, suppose you've invented a bicycle frame made of a new carbon-fiber alloy which makes your bike far lighter and stronger than any previously made. Ordinarily, as discussed in Chapter 5, Section F, the substitution of a new material (here a carbon-fiber alloy for steel) would not be patentable since the substitution would not provide any unexpected results. But suppose during your search you find a prior art reference (e.g., an article in "Metallurgic Times") which states that the author has tried to use carbon-fiber alloys for bicycle frames without success. If you've found a way to use such alloys successfully, you can cite this reference to the PTO to show if that you've turned a past failure into success. Thus you'll have positive proof that your invention provides unexpected results.

8. To Facilitate Prosecution

By familiarizing yourself with the prior art, you'll be able to tailor the general thrust and the specific terms of your patent application around such art, thus saving yourself work and arming you with the proper terminology which you may need later in the "prosecution" stage (i.e., the stage where you actually try to obtain a patent from the PTO). More about this in Chapter 13. Also, an international application, discussed in detail in Chapter 12, requires that an invention be defined in a way which distinguishes it from the prior art. Your search will be of great help here.

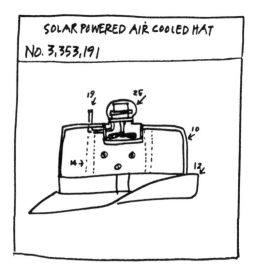

9. To Facilitate Licensing or Sale of Your Invention

When you attempt to sell or license your invention rights, your potential licensees will want to know if your patent application will be likely to get through the PTO. You can answer their concern, at least partially, by showing them your search results. This will give them confidence in your invention and will save them from having to do their own search, thereby speeding up and facilitating negotiations.

10. To Find Out What You've Really Invented

Yes, I'm serious! From over 25 years' experience I've found that many inventors don't realize or understand exactly what they've invented until they see a search report. Indeed, many inventors get a severe case of "search shock" when their "major advance" turns out to be relatively minor. On the other hand, occasionally an inventor, believing that the invention is a relatively small advance and that the basic broad idea of the invention must have already been invented, is very pleased to learn from the search results that the invention's a gold mine instead of a nugget!

11. To Get a Stronger Patent

The PTO itself will usually make a better search than you or a professional searcher will be able to do. Nevertheless, some examiners, at certain times, may miss a highly relevant reference. If anyone uncovers such a reference later, after you get your patent, and brings this reference to the attention of the PTO or any court, it may cast a cloud over, or even invalidate, your patent. However, if you find such a reference in your search, you can make a record of it in the PTO's file of your patent application, tailor your claims around it (see Chapter 9), and avoid any potential harm it may cause you later, thus making your patent stronger and less vulnerable.

12. To Get Your Patent Application Examined Ahead of Turn

For reasons explained in Chapter 10, Section N, I don't recommend that you get your patent application issued sooner, but if you really need to speed things up, you'll be entitled to get it examined ahead of its turn if you've made a pre-examination search.

B. When Not to Search

DESPITE MY INVENTOR'S COMMANDMENT about doing a patent search prior to filing, there are at least two situations where you can "skip the search."

If you are dealing in a very new or arcane field with which you're very familiar, obviously a search is highly unlikely to be profitable. For example, if you're a biotech engineer who's familiar with the state of the art, the newness of your field makes it highly unlikely you will find any early "prior art." Or, if you make semiconductors and have up-to-the-minute knowledge of all known transistor-diffusion processes, and you come up with a breakthrough transistor-diffusion process, a search will probably not produce any reference showing your idea. Before deciding not to search, however, you should be reasonably certain that you or someone else with whom you are in contact knows all there is to know about the field in question, and that you are fairly confident there is no obscure reference which shows your invention.

In addition, if you've made an improvement to an earlier invention which you've already searched, and you feel the search also covered your improvement, there's obviously no need to make a second search.

Designs: Generally I recommend not searching design inventions since the cost and time required to make the search is greater than the time and cost to prepare a design patent application. However, if you believe that reasons 6, 7, 9, 10, 11, and 12 of Sec. A above may be particularly relevant to your situation, you should make a search of your design.

C. The Two Ways to Make a Patentability Search

BASICALLY, THERE ARE only two ways in which you can get your search done: have someone do it for you or do it yourself. If you're a conscientious worker and you have the time and access to a search facility (see below), I recommend that you do the search yourself in order to make sure that it is done thoroughly and in your desired time frame. In addition, this will save you money and enable you personally to accumulate valuable information, as suggested above.

However, you may have very good reasons for having a professional searcher, e.g., you live far from any search facility or you don't have enough time. Also there's the procrastination factor: half the time the only way some of us will ever get a job done, even though we're capable of doing it, is to turn it over to a pro. If for geographical or other reasons you choose to hire a searcher, you'll find advice on choosing one in Section F below. Even if you do use a searcher, read through the instructions on do-it-yourself searching (Sections 7-10 below) in order to understand what you're paying for and to be able to recognize whether the searcher has done a thorough job.

Some inventors, because of the importance of the reasons for searching listed above, prefer to do the search themselves and also have a professional search done, just to double-check their work. I don't recommend this since I've found that an inventor's diligent search is usually adequate. Still, if you feel insecure about your search, you might want to use a computer search as a rough double-check (Section I below). Don't rely on the computer completely, however, unless your invention is in a very new field, such as biotechnology. This is because computer searches go back only about 15 years. From my experience, most manual searches produce many relevant prior-art references that were published in much earlier periods, some even as far back as the 1800's.

If you do the patentability search yourself, there are two sub-possibilities:

1. You can search in the PTO in Arlington, Virginia (definitely the best place); or

2. You can search in a local Patent Depository Library, or even a regular library which has the Official Gazettes.

Read through Sections I and K below to compare these alternatives.

D. How to Make a Preliminary Look

IF YOU DON'T LIVE NEAR THE PTO in Arlington, Virginia, I recommend that you conduct a brief preliminary search before spending the money or time for a formal search. Sometimes you will quickly "knock out" your invention and save yourself cost and effort. If you haven't made the preliminary search already as part of your commercial evaluation (see Chapter 4), do so now by looking in stores, catalogs, reference books, product directories, etc. for your invention. An hour or two in your local library, and perhaps a visit or two to likely stores or suppliers, should be sufficient. If you don't find anything in your preliminary search, and if your invention doesn't fall into the category discussed under "When Not to Search" (Section B above), you're ready to make the formal search.

E. The Quality of a Patent Search Can Vary

LIKE ANYTHING ELSE, the quality of your patentability search can vary from very bad to near perfect. It can never be perfect since, because of their confidential status, there is no way to search pending patent applications (As stated in the last chapter, a patent application which was filed before your date of

invention is valid prior art against your application, even if a patent issues after you file).

Other reasons why your search may not be perfect are:

- some prior art references can be missing (stolen or borrowed) from the area you're searching ("class and subclass"—see Section G below);

- the area in which you're searching may not contain foreign, non-patent, or exotic references (e.g., theses);

- very recently issued patents may not have been placed in the search files yet; or

- a relevant reference (patent or non-patent) may not have been classified in the proper class or in a way which conforms to your view of reality (i.e., because of human variability, it may be classified where you wouldn't expect it to be).

F. How to Hire a Patent Professional

SUPPOSE YOU DECIDE to "let the pros handle it." Here are some suggestions for how to find a patent professional and what your role in the process should be.

1. Lay Patent Searchers

Many patent searchers can be located in the Yellow Pages of local telephone directories under "Patent Searchers." Others advertise in periodicals such as the Journal of the Patent Office Society, a publication for patent professionals edited and published by a private association of patent examiners. I have had far better results with patent attorneys and agents than with lay searchers. Attorneys and agents understand the concept of unobviousness (see previous chapter) better and thus

dig in more places than might at first appear
necessary. However lay searchers have one big
advantage: they charge about half of what most
attorneys and agents charge. Nevertheless before
hiring a lay searcher, I would find out about the
searcher's charges, technical background, on-the-job
experience, and usual amount of time spent on a
search. Most importantly, I would also ask for the
names of some clients, preferably in your city, so that
you can check with them.

2. Patent Agents

A "patent agent" is an individual with some
technical training (generally an undergraduate
degree in engineering) who is licensed by the PTO to
prepare and prosecute patent applications. A patent
agent can conduct a patent search and is authorized
to express an opinion on patentability, but cannot
appear in court; cannot handle trademarks; and
cannot handle licensing or infringement suits. All
other things being equal, I recommend using an
attorney rather than an agent for searching (and
patent application work) since attorneys' experience
in licensing and litigation will usually lead them to
make wider and stronger searches for possible use in
adversarial situations.

3. Patent Attorneys

A "patent attorney" or "patent lawyer" is licensed to
practice both by the PTO and the attorney licensing
authority (e.g., state bar, state supreme court, etc.) of
at least one state. Thus patent attorneys must be
licensed by two authorities. A lawyer licensed to
practice in one or more states but not the PTO is not
authorized to prepare patent applications or use the
title "patent attorney."

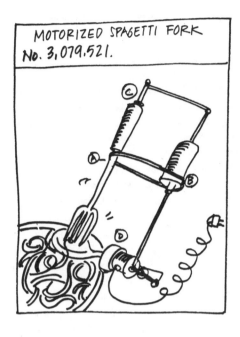

MOTORIZED SPAGETTI FORK
No. 3,079,521.

4. Finding Patent Agents and Attorneys

All patent agents and attorneys are listed in the
PTO's publication Attorneys and Agents Registered
to Practice Before the U.S. Patent and Trademark
Office (A&ARTP). This is available in all medium-
to large-sized public libraries as well as the Patent
Depository Libraries (see Section K below) and
government bookstores.

For patent search purposes, you will want to find
an attorney or agent in the Washington, DC area.
Most patent attorneys and agents who do searching
in the PTO can be found in the District of Columbia
section, or the Virginia section of A&ARTP under
zip code 22202. Pick one or more these and then call
or write to say you want a search made in a particular
field. (Generally, hiring an attorney in your locality
to do the search is a very inefficient and costly way to
do the job since the attorney or agent will have to
hire an associate in or travel to Arlington to make
the search for you. This means you'll have to pay two
patent professionals or travel expenses for the
search.)

Using "Discount" Patent Professionals

Active patent professionals are either in private practice (a law firm), or employed by a corporation or the government. Most patent professionals in private practice charge about $90 to $180 an hour. But many corporate-employed or semi-retired patent professionals also have private clients and charge considerably less. If you want or ever need to consult a local patent professional, you'll save money by using one of these "discount" patent professionals; their services are usually just as good or better than those of the full-priced law firm attorneys. Also since they have much less overhead (rent, books, secretaries) they'll be more generous with their time (except that patent professionals employed by the federal government are not allowed to represent private clients). Look in the geographical section of A&ARTP for corporate-employed or retired (but still active) patent professionals in your area; the latter can usually be identified by having a corporate address or an address in a residential rather than downtown neighborhood

Of course, finding a good patent professional often involves more than checking a list. The best way is by personal referral. Ask another inventor, your employer, your local inventors' organization, a general attorney whom you like, a friend, etc. If you do find someone who seems good, make a short appointment to discuss the broad outlines of your problem. This will give you a feel for the attorney, whether the chemistry's good between the two of you, whether the fees are acceptable, etc. Ask what undergraduate degree the attorney has (almost all have undergraduate degrees in engineering or a science); you don't want to use an mechanical engineer to handle a complex computer circuit.

Your next question should be, will the professional help you help yourself or demand a traditional attorney-client relationship (attorney does it and you pay for it)? Many corporate-employed and retired patent professionals will be delighted to help you with your search, preparation, and/or prosecution of your patent application. Using this approach, you can do much of the work yourself and have the professional provide help where needed at a reasonable cost.

When it comes to fees, you should always work these out in advance. Some patent professionals charge a flat fee for searches (and also for patent applications and amendment; others charge by the hour. If you plan to do much of the work yourself, you'll want hourly billing. Also, be sure it's clear who will pay for other costs associated with prosecuting a patent, such as copies, postage, drafting, filing fees, etc.

G. How to Prepare Your Searcher

YOU'LL WANT TO USE your patent searcher to maximum efficiency. Do this by sending your searcher a clear and complete description of your invention, together with easily understandable drawings. You won't compromise any trade-secret status of your invention by such a letter since by law it's considered a confidential communication. If you wish any type of particular emphasis applied to any aspect of your search, be sure to inform the searcher of this fact. If your notebook record of your invention or your invention disclosure is clear enough, you can merely send the searcher a copy. Whether you send a copy of your notebook entries or a separate disclosure (Form 3-2), I recommend that you blank out all dates on any document you send to anyone: this will make it more difficult for any potential invention thief (extremely rare) who might gain access to your disclosure to antedate you. Fig. 6-A is an example of a proper search request letter from an inventor and Figs. 6-B (a, b, c) are copies of the attachments to the search request letter of Fig. 6-A.

Millie Inventress
1901 JFK Blvd.
Philadelphia, PA 19103

1988 Jan 22, Tue

Samuel Searcher, Esq.
2001 Jefferson Davis Highway
Arlington, VA 22202

Patentability Search: Inventress: Napkin-Shaping Ring

Dear Mr. Searcher:

As we discussed on the phone yesterday, you were highly recommended to me as an excellent searcher by Jacob Potofsky, Esq., who is a general attorney here and a cousin of my friend, Shirley Jaschik. You said that you would be able to make a full patentability search on my above invention, including an examiner consultation and a search in the examiner's files to cover foreign and non-patent references, for $300, including patent copies and postage. I have enclosed this amount as full payment in advance, per your request. You said that you would mail the search report (without an opinion on patentability) and references me within three weeks from the date you received this letter.

Enclosed are three sheets of drawings from my notebook (I have properly signed, witnessed, and dated records elsewhere); these sheets clearly illustrate my napkin-shaping ring invention. As you can see from the prior-art Figs 1 (A and B), previous napkin rings were simple affairs, designed merely to hold a previously rolled or folded napkin in a simple shape.

In contrast, the napkin ring of my invention, shown in Fig 2, and made of metal or plastic, has a heart-shaped outer member 12, an inner leg 14, and two curved-back arms 16. As shown in Fig 3, it is used by introducing a corner 8 of a cloth napkin 10 between an end 4 of leg 14 and the adjacent portion of outer member 12. When napkin 10 is pulled partially through the ring, as indicated in Fig 4, it will be forced to assume the shape of the space between arms 16 and outer portion 12, as indicated.

Thus my napkin-shaping and holding ring can be used to make a napkin have an attractive, graceful shape when it is laid flat and placed adjacent a place setting, as indicated in Fig 5. The extending portion of the napkin can also be folded up and around, as indicated in Fig 6A, so that the napkin and its ring can be stood upright.

In addition to the specific shape shown, you should of course search the broader concept of my invention, namely a ring-shaped outer member with an inwardly-extending tongue or leg which can be used to shape napkins pulled partially through the structure. I believe that I have provided you with sufficient information to fully understand the structure and workings of my invention so that you can make a search, but if any further information is needed, please don't hesitate to call me.

Most sincerely,

Millie Inventress

Millie Inventress (215 776-3960)   (My file: 60:Search.ltr)

Encs.: $300 check; 3 sheets of drawings

Fig. 6-A—Inventor's Search Request Letter to Patent Searcher

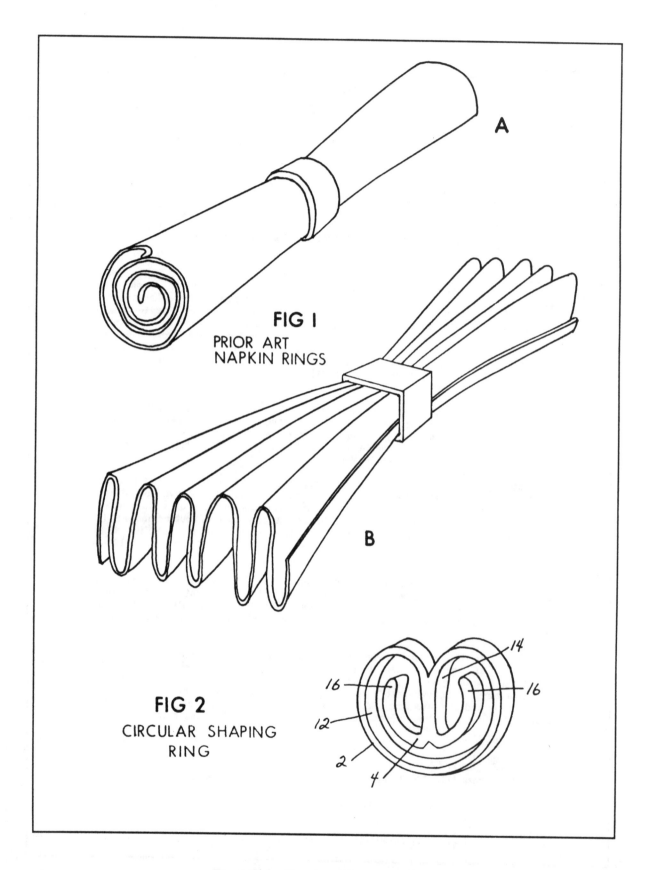

FIG 1
PRIOR ART
NAPKIN RINGS

A

B

FIG 2
CIRCULAR SHAPING
RING

Fig. 6-B(a)—Drawing of Invention, Part a

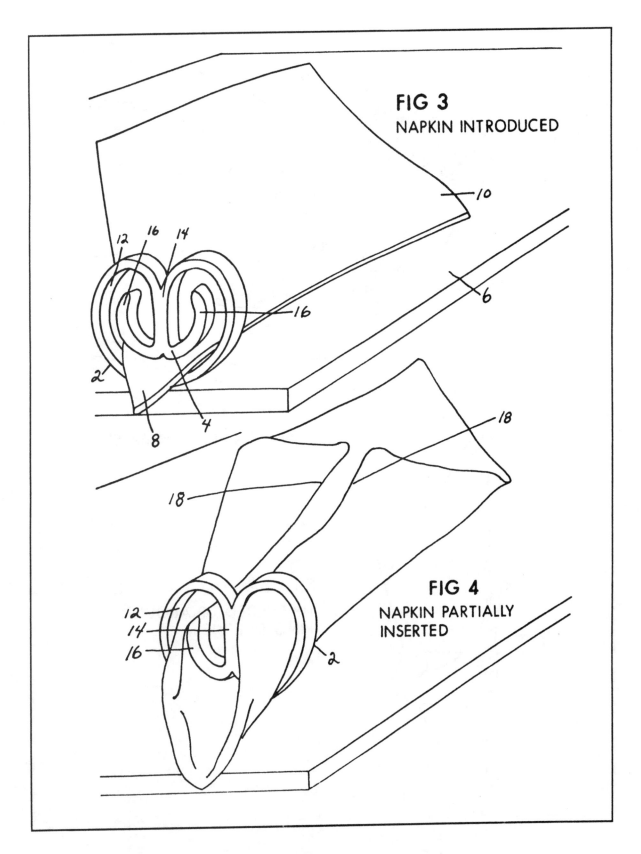

FIG 3
NAPKIN INTRODUCED

FIG 4
NAPKIN PARTIALLY
INSERTED

Fig. 6-B(b)—Drawing of Invention, Part b

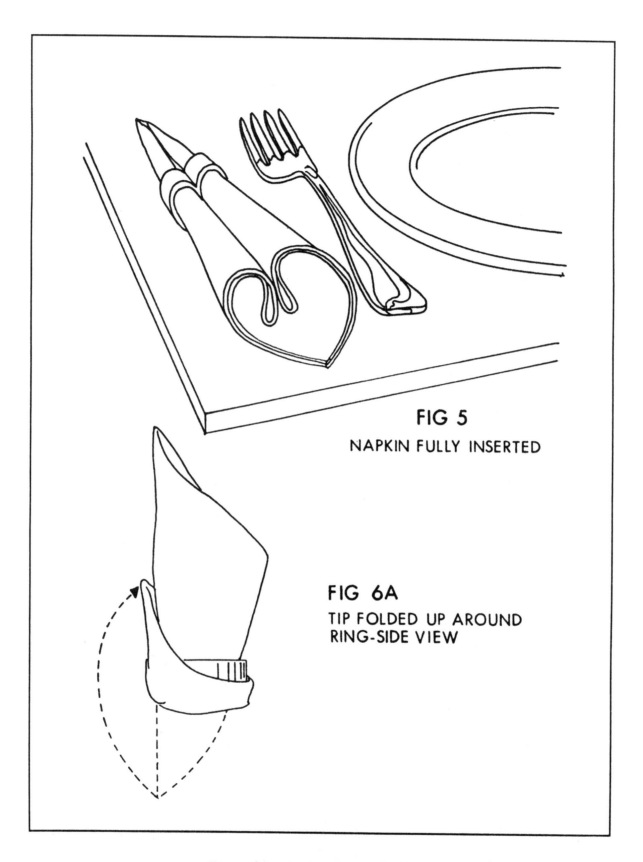

FIG 5

NAPKIN FULLY INSERTED

FIG 6A

TIP FOLDED UP AROUND
RING-SIDE VIEW

Fig. 6-B(c)—Drawing of Invention, Part c

H. Analyzing the Search Report

AFTER YOU SEND OUT your search request, the searcher will generally take several weeks to perform the patentability search, obtain copies of the patents and other references which the searcher feels are relevant, and report back. Most search reports have four parts:

1. A description of your invention provided by the searcher to assure you that your invention has been understood correctly and to indicate exactly what has been searched.

2. A list of the patents and other references discovered during the search.

3. A brief discussion of the cited patents and other references, pointing out the relevant parts of each.

4. A list of the classes and subclasses searched and the examiners consulted, if any.

The searcher will enclose copies of the references (usually U.S. patents, but possibly also foreign patents, magazine articles, etc.) cited in the search report and enclose a bill. Most searchers charge separately for the search, the reference copies, and the postage. If you've paid the searcher a retainer, you should be sent a refund unless your retainer was insufficient. In this case, you'll receive a bill for the balance you owe.

Examples:

• Fig. 6-C is an example of a typical, competently-done search report sent by Sam Searcher, Esq. in response to Millie Inventress' letter of Fig. 6-A;

• Fig. 6-D(a) is a copy of page 1 (the drawing) of the Gabel patent cited in the search report;

• Fig. 6-D(b) is a copy of page 2 of Gabel (the first page of Gabel's specification), and;

• Fig. 6-D(c) is a copy of page 1 of the Le Sueur patent cited in the search report.

I haven't shown the other cited patents and the rest of the Gabel and Le Sueur patents as these aren't necessary for our patentability determination.

You should now read the searcher's report and the references carefully. Then, determine whether your invention is patentable over the references cited in the search report. Let's use Millie's search report as an example of how to do this.

First, note from Fig. 6-B that the napkin-shaping ring of the invention has an annular (ring-shaped) outer member with an inwardly-projecting leg. The leg has flared-back arms at its free end. When a napkin is drawn through the ring, tip first, the arms and annular member will shape the napkin between them in an attractive manner, as indicated in Fig. 6-Bc.

Of the four previous patents cited, let's assume that only Gabel and Le Sueur are of real relevance. Gabel, a patent from 1930, shows a curtain folder comprising a bent sheet metal member. A curtain is folded slightly and is drawn through the folder which completes the folding so that the curtain can be ironed when it is drawn out of the folder. Le Sueur, a patent from 1976, shows a napkin ring with a magnetized area for holding the letters of the name of a user.

SAMUEL SEARCHER
Patent Attorney
2001 Jefferson Davis Highway
Arlington, VA 22202

703 521-3210

1988 Feb 21, Thu

Ms. Millie Inventress
1901 JFK Blvd.
Philadelphia, PA 19103

Search Report: Inventress: Napkin-Shaping Ring

Dear Ms. Inventress:

In response to your letter of Jan 22, I have made a patentability search of your above invention, a napkin-shaping ring comprising an outer portion with an inwardly-extending leg and flared back arms at the end of the leg. I have also searched the broader concept of an annular member with an inward cantilevered leg for shaping a napkin which is drawn therethrough. My bill for $300, the total cost of this search, including the references and postage, is enclosed and is marked "Paid"; I thank you for your check.

I searched your invention in the following classes and subclasses in the actual examining divisions: 40/21; 40/142, D44/20, and 24/8. In addition I consulted Examiner John Hayness in Group Art Unit 353 regarding this invention. In my search I thought the following references (all U.S. Patents) were most relevant and I enclose a copy of each: **Bergmann,** 705,196 (1902); **Gabel,** 1,771,328 (1930); **Hypps,** 3,235,880, (1966); and **Le Sueur** 3,965,591 (1976)

Bergmann shows a handkerchief holder which comprises a simple coiled ring with wavy portions.

Gabel is most relevant; she shows a curtain folder comprising a folded metal device through which a curtain (already partially folded) is inserted and then pulled through and ironed at the exit end.

Hypps shows a necktie tieing and holding device.

Le Sueur shows a napkin ring with magnetically-attachable names.

I could not find any napkin-shaping devices as such and Examiner Hayness was unaware of any either. However be sure to consider the **Gabel** patent carefully as it appears to perform a somewhat similar function, albeit for curtains.

It was my pleasure to serve you. I wish you the best of success with your invention. Please don't hesitate to call if you have any questions.

Most sincerely,

Sam Searcher

Samuel Searcher   Encs: Bill (Paid); 4 patents

Fig. 6-C—Patent Searcher's Search Report

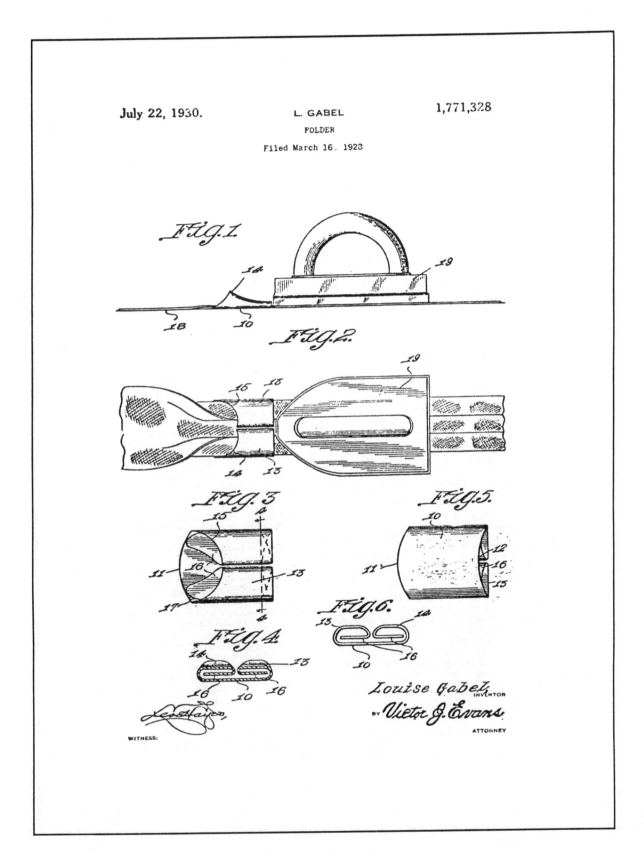

Fig. 6-D(a)—Drawing of Prior-Art Patent

1,771,328

UNITED STATES PATENT OFFICE

LOUISE GABEL, OF COLUMBUS, NEBRASKA

FOLDER

Application filed March 16, 1928.   Serial No. 262,243.

This invention relates to cloth holding devices and more particularly to a device adapted for holding cloth in the form of plaits while ironing and sewing.

Another object of the invention comprehends an enlarged entrance opening in one end of the device within which the cloth may be introduced.

A further object of the invention contemplates tongue members adapted to form creases in the cloth.

An additional object of the invention consists of a portion removed from the discharge end of the device whereby binding action of a sad iron therewith is obviated while pressing the cloth.

With the above and other objects in view, the invention further consists of the following novel features and details of construction, to be hereinafter more fully described, illustrated in the accompanying drawing and pointed out in the appended claim.

In the drawing:—

Figure 1 is a side elevation of the invention while in use and followed by a sad iron.

Figure 2 is a top plan view of Figure 1.

Figure 3 is a top plan view of the invention per se.

Figure 4 is a sectional view taken on line 4—4 of Figure 3.

Figure 5 is a bottom plan view of the invention.

Figure 6 is a front elevation of the invention per se.

Referring to the drawing in detail, wherein like characters of reference denote corresponding parts, the reference character 10 indicates a plate member having a curved outwardly projecting forward end 11 and a concaved inner end 12.

The sides of the plate are bent upon themselves upwardly and inwardly upon the plate to provide horizontally disposed guide members 13.

As illustrated in Figures 1, 3, 4 and 6, the outermost end, namely 11, is flared to provide an enlarged entrance and to accomplish such construction the outermost ends of the guide members 13 are upwardly flared, as indicated at 14 and concaved, as indicated at 15 upon the foremost edges thereof.

Tongues 16, carried by the guide members 13, are extended reversely and disposed in spaced relation to the plate member 10. The tongues being also spaced from the guide members. The foremost portions of the tongues 16 are rounded, as indicated at 17 and projected forwardly for guiding purposes to the adjacent ends of the guide members.

In the use and operation of the invention, lengths of cloth, such as indicated at 18, of a desired width, are partially folded along the side edges thereof and the strip per se laid upon the upper side of the plate member 10. The folded portions of the strip being adapted to repose upon the upper sides of the tongues 16 and to be projected within the spaces as defined between the tongues and the guide members. Due to the fact that the outermost end of the device is flared, an enlarged entrance is provided by means of which the cloth may be readily introduced and fed. The rounded portions 17 for the tongues also permit ease in the drawing of the cloth through the device or the sliding of the device upon the cloth. As illustrated in Figures 1 and 2 of the drawing, a sad iron, such as indicated at 19, may travel upon the cloth 18 immediately behind the device to press the folded side edges or plaits of the cloth. By the same token, the invention could be used in the formation of different kinds of braids and etc., and to effectively feed the cloth or strip to a sewing machine, in the event the plaits are to be held against displacement from the strip per se.

The concaved portion 12, upon the innermost end of the strip 10, is adapted to prevent binding action of the sad iron 19 therewith when the latter closely pursues the plate member. Such construction will also prevent injury to the strip and plaits.

Although I have shown, described and illustrated my invention as being primarily adapted for use in the manufacture of plaits, it is to be obviously understood that the invention could be effectively employed for

[Callout box overlaying text:] In the use and operation of the invention, lengths of cloth, such as indicated at 18, of a desired width, are partially folded along the side edges thereof and the strip per se laid upon the upper side of the plate member 10. The folded portions of the strip be -

Fig. 6-D(b)—Specification of Prior-Art Patent

United States Patent [19]

Le Sueur

[11] **3,965,591**

[45] June 29, 1976

[54] NAPKIN RING

[75] Inventor: **Alice E. J. Le Sueur**, Cobble Hill, Canada

[73] Assignee: **The Raymond Lee Organization,** New York, N.Y. ; a part interest

[22] Filed: **Nov. 26, 1974**

[21] Appl. No.: **527,216**

[52] U.S. Cl. .... 40/21 R
[51] Int. Cl.² G09F 3/14
[58] Field of Search .... 40/142 A, 63, 21 A, 40/21 B, 10; 63/2; 24/8

[56] **References Cited**
UNITED STATES PATENTS

198,065 12/1877 Annin .... 63/1 X

2,600,505 6/1952 Jones .... 40/142 A
2,655,402 9/1953 Bonagura........................ 40/21 A

FOREIGN PATENTS OR APPLICATIONS

1,308,888 10/1962 France 40/142

Primary Examiner—Louis G. Mancene
Assistant Examiner—Wenceslao J. Contreras
Attorney, Agent, or Firm—Howard I. Podell .

[57] **ABSTRACT**

An open cylindrical napkin ring fitted with magnetic means for attaching an identifying name or set of initials in a recess on the outside of the ring.

3 Claims, 4 Drawing Figures

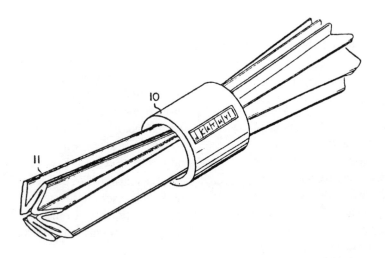

Fig. 6-D(c)—Abstract Page of Prior-Art Patent

Now, as part of analyzing this sample search report, we'll use the master flowchart of Fig. 5-B. To save you from having to turn the pages repeatedly, I've reproduced it here. If any part of this chart confuses you, reread the part of Chapter 5 that explains each box in detail.

Okay, now let's work our way through the chart:

Box A: Millie's napkin-shaping ring can be classified within a statutory class as an article (or even a machine since it shapes napkins).

Box B: It clearly has usefulness since it provides a way for unskilled hostesses or hosts to give their napkins an attractive, uniform shape.

Box C: We must now ask whether the invention is novel, i.e., physically different from any single reference. Clearly it's different from Le Sueur because of its inwardly-extending leg 14. Also it's different from Gabel because, comparing it with Gabel's Fig. 6, it's rounder and it has a complete outer ring with an inwardly-extending leg, rather than a folded piece of sheet metal. It's important to compile a list of the differences which the invention has over the prior-art references, not the differences of the references over your invention.

Box D: We bypass Box D since we don't have any new-use invention.

Box E: The question we must now ask is, do the novel features (the roundness of the ring, the inwardly extending leg, and the flared-back arms) provide any new and unexpected results? After carefully comparing Gabel with Millie's invention, we can answer with a resounding "Yes!" Note that Gabel states, in her column 2, lines 62 to 66, that the strip of cloth is first partially folded along its side edge and then it is placed in the folder. In contrast, Millie's shaping ring, because of its roundness and leg, can shape a totally unfolded napkin—see Millie's Figs. 3 and 4. This is a distinct advantage since Millie's shaper does all of the work automatically—the user does not have to prefold the napkin. While not an earthshaking development or advance, clearly Millie's ring does provide a new result and one which

is unexpected since neither Gabel, Le Sueur, or any other reference teaches that a napkin ring can be used to shape an unfolded napkin. Thus we take the "Yes" output of Box E.

Box F: We next check the secondary factors (1 to 21) listed in Boxes I and M.

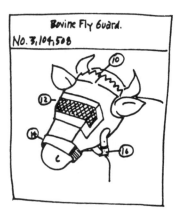

Boxes I-M: Reading through these factors, we find first that #2 in Box I applies, i.e., the invention solves a problem (the inability of most persons to quickly and neatly fold napkins so that they have an attractive shape) which was never before even recognized. Also, we can provide affirmative answers to factors #8 and #11 since the invention provides an advantage which was never before appreciated and it solves a long-felt, but unsolved need—the need of unskilled persons to shape napkins quickly and gracefully (long felt by the more fastidious of those who hate paper napkins, at least).

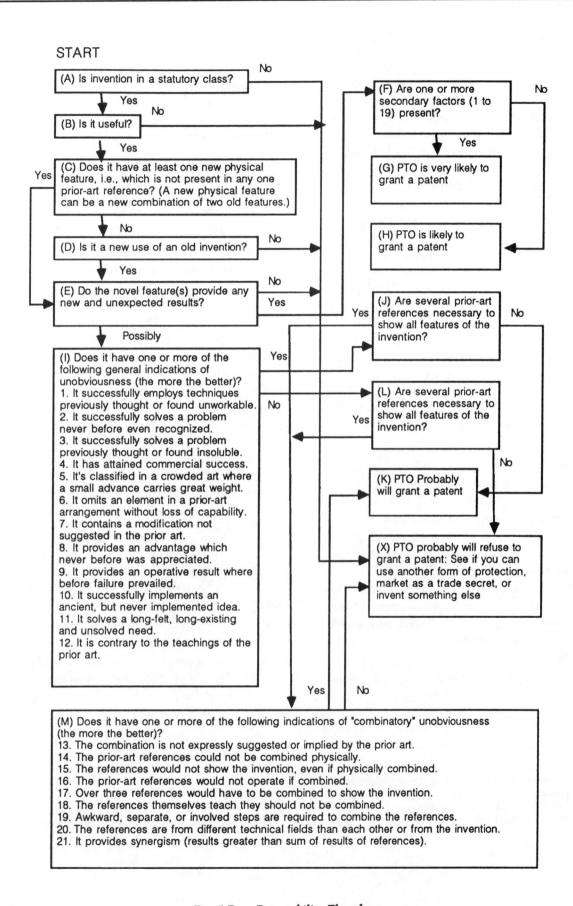

Fig. 5-B — Patentability Flowchart

Box M: Since two references are present, and each shows some part of Millie's invention, we have to consider the possible effect that a combination of these would have on the question of obviousness ("combinatory unobviousness"). In Box M we see that factors 13, 15, 18, 19, and 21 can reasonably be argued as relevant to Millie's invention. The combination of the two references is not suggested (#13) by the references themselves; even if the two references were combined, Millie's inward leg would not be shown (#15). The references are complete and fully functional in themselves and hence impliedly teach that they should not be combined (#18). And it would be awkward, requiring redesign and tooling to combine the references (#19). That is, synergism is present (#21) since the results (automatic folding) are greater than the sum of the references. Thus we can with conviction state that several of secondary factors are present so we take the "Yes" output of Box F to Box G. Here we see that the PTO is very likely to grant a patent and our determination on patentability is accordingly positive.[1]

Note: Although I've analyzed the search report to determine whether Millie's invention was patentable, it's important to remember that a weak patent isn't much better than no patent. Put differently, a very weak patent and $2.00 will get you a cable car ride in San Francisco. So in addition to reaching a decision on patentability, you should also walk the extra mile to determine whether your patent is likely to be of broad enough scope to make it economically worthwhile. I tell you how to do this in Section J of this chapter.

Note that we have done our own patentability evaluation—the four-part list above—and that the search report of Fig. 6-C didn't include an opinion on patentability. There are several reasons for this.

First, if your searcher is a layperson (not a patent attorney or agent), the searcher is not licensed to give opinions on patentability since this constitutes the practice of law.

Second, even if your searcher is an attorney or agent, the searcher usually won't provide an opinion on patentability because most searchers are used to working for other patent attorneys who like to form their own opinions on patentability for their clients.

Third, if the searcher's opinion on patentability is negative, a negative written opinion might be damaging to your case if you do get a patent, sue to enforce it, and the opinion is used as evidence that your patent is invalid. This would occur, for example, if your court adversary (the defendant-infringer) obtains a copy of the opinion by pre-trial discovery (depositions and interrogatories), shows it to the judge[2], and argues that since your own search came up with a negative result, this militates against the validity of your patent. However, a negative written opinion can be "worked" in court, i.e., distinguished, explained, rebutted, etc., so if you want the searcher's opinion on patentability in addition to the search, most patent attorney/agent searchers will be glad to give it to you without extra charge, or for a slight additional cost of probably not more than $50 to $100.

Fourth, armed with the knowledge you've gained from Chapter 5, you should be able to form your own opinion on patentability by now; the exercise will be fun, educational, and insightful to your invention.

In any case, don't hesitate to ask any questions about the searcher's practices in advance and be sure to specify exactly what you want in your search. It's your money and you're entitled to buy or contract for whatever services you desire.

---

[1]This exercise is a real case: An examiner initially rejected an application to the napkin-shaping ring as unpatentable over Gabel and Le Sueur. However, he agreed to grant a patent (#4,420,102) after I filed an argument (see Chapter 13) forcefully stating the above considerations.

---

[2]You may be able to keep the opinion from the judge under a rule of evidence known as the "attorney-client privilege."

I. Do-It-Yourself Searching in the PTO

ALL PRELIMINARY SEARCHES should be made primarily in patent files. This is because patents are classified according to a detailed scheme (discussed later in this chapter) and since there are about 10 times as many devices and processes in the patent files than in textbooks, magazines, etc.[3] All PTO examiners make their searches in the patent files for these reasons, so you should also. However, if you have access to a good non-patent data bank, e.g., a good technical library in the field of your invention, you can use this as a supplement to your search of the patent files.

1. Getting Situated at the PTO

As we have said, the best place to make a search of the patent files is in the PTO in Arlington.[4] This is because the PTO's search facilities have all of the U.S. patents arranged by subject matter (e.g., all patents which show bicycle derailleurs are physically grouped together, all patents which show transistor flip-flop circuits are together, all patents to diuretic drug compositions are together, etc.) Also the PTO has foreign patents and literature classified along with U.S. patents according to subject matter (but only in the examiner search areas).

The PTO is located in a complex of modern medium-rise buildings in Arlington, Virginia, informally called "Crystal City." Although all mail must be addressed to the Commissioner of Patents and Trademarks, Washington, DC 20231, the PTO is physically located at South 26th Street and U.S. 1 (Jefferson Davis Highway) in Arlington, about half a mile due west of the Washington National Airport. It receives over three million pieces of mail a year, more than any other governmental agency.

The PTO is technically part of the Department of Commerce (which is headquartered in Washington) but operates in an almost autonomous fashion. The first floor of the main PTO building contains the main search room. The various examining divisions and administration departments are upstairs.

The PTO employs about 1200 examiners, all of whom have technical undergraduate degrees in such fields as electrical engineering, chemistry, or physics. Many examiners are also attorneys. The PTO also has about an equal number of clerical, supervisory, and support personnel. The Commissioner of Patents and Trademarks is appointed by the President, and most of the higher officials of the PTO have to be approved by Congress. Most patent examiners are well paid; a journeyman examiner (10 years experience) usually makes $40,000 to $70,000 a year.

Assuming you do go to the PTO in Arlington, here's what you'll find. There are two places you can make the search:

- The public search room; and

- The examiners' search files in the actual examining division.

Most searchers make their search in the public search room because it's more convenient—it's on the first floor, there are search tables, and it's large and well-lighted. However, I recommend putting up with a little inconvenience and going upstairs to the examiner's search files. There are several reasons to do this: the examiners are there to assist you; literature and foreign patents are available; it's much

---

[3]To put it bluntly, many more hare-brained schemes are patented than are published elsewhere. This is due to the fact that the requirements for getting a patent, while stricter in the area of required degree of difference (the invention must be unobvious) are practically non-existent in the area of commercial practicality.

[4]Some large companies maintain their own search files in certain fields. If you can get access to one of those in your field (e.g., you or a close friend or relative works for the company), you're probably even better off than if you go to the PTO.

quieter; the patent files are likely to be more intact; and finally, of at least minor importance, the chairs are more comfortable. To get into the search room or upstairs, you must apply for a user pass, which will take only a few minutes. You must also ask permission from a primary examiner or clerk before commencing your particular search.

If you need help with your search, you can ask any of the search assistants in the search room or (even better) an examiner in the actual examining division. You won't be endangering the security of your invention if you ask any of these people about your search and give them all the details of your invention. They see dozens of new inventions every week and they are quite used to helping searchers and others. I've never heard of an examiner or search assistant stealing an invention.[5]

2. How to Do the Search

Okay, now that you know something about the PTO and where to search, what do you do next? There are four basic steps to take when conducting a patent search:

Step 1: Articulate the nature and essence of your invention, using as many different terms as you can think of to describe it.

Step 2: Find the relevant classification(s) for your invention.

Step 3: Note relevant prior art (patents and other publications) under your classification.

Step 4: Carefully review the prior art to see whether it anticipates your invention or renders it obvious.

---

[5]They could not do so personally, anyway, because employees of the PTO are not allowed to file patent applications. True, in theory a PTO employee could reveal an invention to a friend or relative who could file, but it's absurdly unlikely to occur.

Let's take these one at a time.

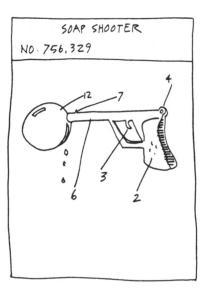

Step 1: Write Out the Nature and Essence of Your Invention

As with any other classification or indexing system, your success will depend on the degree to which the words and phrases you use to define your invention coincide with the terms used by the classifier or indexer. For this reason, you should first figure out several ways to describe your invention. Start by writing down all the physical features of your invention in brief, concise format so that you'll know exactly what to look for when searching.

For example, if you're searching a bicycle with a new type of sprocket wheel, write down "bicycle, sprocket wheel," and add briefly the details. If you're searching an electronic circuit, write down in a series of phrases like the foregoing or, in a very brief sentence, the quintessence of your invention, such as "flip-flop circuit with unijunction transistors" or some other very brief and concise description. Do the same for chemical inventions.

Form 6-1 is a Searcher's Worksheet which you can use to facilitate your searching, and Fig. 6-E is a completed version of Form 6-1 that you might produce if you had searched Millie Inventress' invention. Note that the invention description part of the worksheet contains a concise description of the invention for easy reference.

Once you've written a concise description of your invention, think of some alternative words or phrases to add to your description. Don't hesitate to define your invention in still additional ways that may come to you during your search. Then, take your worksheet with this brief description and the drawing(s) of your invention to the public search room. Even if you're not going to do your search there, use that room to find out how your invention is classified.

Step 2: Find the Proper Classification for Your Invention

To find the place to search your invention, you'll need its most relevant search classification (called class and subclass). To obtain this, first look at the searcher's "tools" or reference publications. These consist of:

- the Index to the U.S. Patent Classification,
- the Manual of Classification, and
- the Classification Definitions.

Again, let's slow down and look at each of these in detail.

The Index to the U.S. Patent Classification

While bearing an awkward title, this book will be your main reference tool. It's a paperbound 8.5" x 11" book which alphabetically lists all possible subject

areas of invention, from "abacus" to "zwieback," together with the appropriate class and subclass for each. The *Index* also lists the classes alphabetically. Let's assume that you've invented a gymnastic exercising apparatus. The first thing to do is to look in the *Index* under "Gymnastic Devices." We come to page 96 (Fig. 6-F), a typical page from the *Index*. It shows, among other things, that "Gymnastic Devices" are classified in class 272, subclass 109.

Manual of Classification

Now that we've found the class and subclass numbers, it's time to turn to the *Manual of Classification*. This loose-leaf book lists the classes of invention numerically (as stated, there are about 300 classes). Each class is on its own page(s), together with about 200 to 300 subclasses under each class heading, for a total of about 70,000 subclasses. The *Manual* lists design as well as utility classes; the classes are not in any logical order. To see where class/subclass 272-109 fits, let's look at the first page that covers class 272. Fig. 6-G is a copy of this page. It shows the first part of "Class 272—Amusement And Exercising Devices." Note that subclass 93 in this class covers "Exercising Equipment," and that subclass 109, which is indented under subclass 93, states "Gymnastic"; thus, class-subclass 272-109 covers gymnastic exercising equipment. Note that under 272-109 are further subclasses which may be on interest; these cover trapezes and rings, horizontal bars, etc.

As I'll explain below, this manual is used as an adjunct to the Index, to check your selected classes and to find other, closely-related ones.

Searcher's Worksheet

Sheet _1_ of _1_

Inventor(s): _Millie Inventor_

Invention Description (use key words and variations): _Napkin folder – Annular member with inner leg with flared-back arms._

Selected Search Classifications

Class/Sub	Description	Checked	Comments
40-21	Napkin holders	✓	Very relevant – the right place
40-142	Misc. tableware	✓	mostly utensils – not too good
D44-20	Napkin holders	✓	somewhat relevant
24-8	Shaping devices	✓	N. G.

Patents (and Other References) Thought Relevant

Patent #	Name or Country	Date	Class/Sub	Comment
705,196	Bergmann	1902	40-142	Plain ring
1,771,328	Gabel	1930	40-21	Curtain shaper ☆
3,235,880	Hypps	1966	40-21	Fancy ring
3,965,591	LeSueur	1976	40-21	Ring w/ magnetic ltrs
				Consulted Exr.
				Hayness, GAU
				353

Searched At _PTO / Exam. GAU 337_   On _84-1-12_   By _S. Searcher_

Fig. 6-E—Completed Searcher's Worksheet

	Class	Subclass
Water pistol	222	79
Well tubing perforator	175	2
Y gun	89	1R
Gussets		
Garment	2	275
Pocketbook body construction	150	30
Gut or Gut Treatment	8	94.11
Splitter	83	926A*
Guttapercha	260	709
Gutter	405	119
Eaves trough	52	11
Electric conductor underground structure	174	39
Road and pavement	404	2
Support design	D 8	363
Guy	52	146
Bed spring and frame	52	272
Gymnastic Devices	272	109
Coin controlled apparatus	194	
Gypsum	423	554
Calcining	106	100
Coating or plastic compositions containing	106	109
Alkali metal silicate	106	77
Gyrating		
Reciprocating sifter		
Actuating means	209	366
Horizontal and vertical shake	209	326
Horizontal shake	209	332
Gyratory Crusher		
Jaw crushers rotary component	241	207
Parallel flow through plural zones	241	140
Series flow through plural zones	241	156
Gyro Stabilized		
Article support	248	183
Furniture for ships	114	119
Gyroplane (See Aircraft)	244	17.11
Gyroscope	74	5R
Aerial camera combined	354	70
Aircraft control	244	79
Ammunition digest	102	DIG.3
Direction indicator	33	318
Gimbals	248	182
Gun sight combined	33	236
Gyroscopic compass	33	324
Telemetric system combined	340	870 7
Monorail rolling stock	105	141
Suspended	105	150
Rotors	74	572
Rotors and flywheels	74	
Ship antiroll	114	112
Ship stabilizer	114	122
Ship steering	114	144R
Speed responsive devices	73	504
Torpedo	114	24
Torpedo steering	114	24
Toy	46	50
Transmission	74	64

H

	Class	Subclass
H Acid	260	509
Haberdashery Item	D 2	378
Habitat, Submarine	114	314
Hack Saw	145	31R
Combined	7	149
Design	D 8	96
Hanging	83	783
Hackling		
Combing	19	115R
Decorticating	19	5R
Hacks Tree	30	121
Haemocytometer	356	39
Testing lenses	356	124
Hair		
Artificial furs	428	85
Artificial structure	132	5
Beauty parlor equipment	D28	10
Brush	15	160
Carried hat fasteners	132	60
Clippers	D28	52
Coating compositions	106	155
Curlers		
Curling iron	D28	38
Electrically heated	219	222
Fluid fuel heated	126	408
Cutters	30	
Design	D 8	57
Design clippers	D28	52

	Class	Subclass
For inside ear or nose	30	29.5
Hair planers	30	30
Drying on head		
Apparatus	34	96
Processes	34	3
Supports for	34	101
Dye applicator	D28	7
Dyeing and dyes	8	405
Fasteners	132	46R
Design	D28	39
Fertilizer from	71	18
Hairpiece	D28	92
Inserters	128	330
Jeweled fastener	63	2
Net	132	49
Pins (See hairpins)	132	50R
Planers	30	30
Removing (See notes)	30	32
Burial preparation	27	21
Butchering	17	1D
Coarse or water hair from fur	69	24
Cutters for inside ear or nose	30	29 5
Depilating untanned skins	8	94.16
Depilatories	8	94.16
Electric needle	128	303.18
Electric needle supports	128	303.19
Fiber liberating	19	2
Fur treatment	69	24
Process	17	47
Razors	30	32
Surgical instruments	128	355
Tweezers	128	354
Shampooing apparatus	4	515
Shearing, fur finishing	26	15R
Thinning shears	30	195
Springs	368	175
Strand making		
Covering by spinning etc	57	4
Spinning etc	57	28
Textile spinning etc	57	29
Thinners	30	195
Design	D28	52
Toilet preparations	424	70
Treating process	132	7
Vulcanizable natural hydrocarbon gums with	260	748
Waving	132	7
Hairpin	132	50R
Design	D28	39
Dispenser	132	1A
Hat fastener cord or loop and	132	61
Making	140	87
Packaging	227	25
Half Belts	2	309
Half Wave		
Gas rectifier	313	216
With hot cathode	313	211
High voltage rectifier	313	317
With emissive cathode	313	310
With thermionic cathode	313	310
Rectifier system	363	13
Circuit interrupter for	200	
Dynamoelectric machine	310	10
Electronic tube for	313	317
Gas tube type	363	114
Power packs	307	150
Unidirectional impedance for	357	
Vacuum tube type	363	114
With filter	363	39
With voltage regulator	363	84
Halftone		
Blanks and processes printing	101	401.1
Etching	156	654
X-art	156	905*
Photographic process	430	396
Photographic screens	350	322
Chemically defined	430	6
Printing plates	101	395
Halides (See Material Halogenated)		
Hydrocarbon	570	101
As azeotropes	203	67
Electromagnetic wave synthesis	204	163R
Electrostatic field or electrical discharge synthesis	204	169
Metal	423	462
Electrolytic synthesis	204	94
Nitroaromatic	568	927
Nonmetal inorganic	423	462
Organic acid	260	543
Rubber	260	772

	Class	Subclass
Hydrohalide	260	771
Mixtures containing	260	735
Hall Effect Means in an Amplifier	330	6
Haloamines	564	114
Acyclic	564	118
Hydroxy or ether containing	564	119
Plural difluoramine groups	564	121
Unsaturated	564	120
Alicyclic	564	117
Amidines	564	116
Halogen Compounds (See Material Halogenated)		
Halogenated Carboxylic Acid Esters	560	1
Acyclic acid esters	560	226
Of phenols	560	145
Acyclic amino acid esters	560	172
Acyclic carbamic acid esters	560	161
Acyclic oxy acid esters	560	184
Acyclic polycarboxylic acid esters	560	192
Acyclic unsaturated acid esters	560	219
Alicyclic	560	125
Aromatic amino acid esters	560	47
Aromatic carbamic acid esters	560	30
Aromatic polycarboxylic acid esters	560	23
Oxybenzoic acid esters	560	65
Phenoxyacetic acid esters	560	62
Halogenation (See Halides)	260	694
Halohydrin	568	841
Halowax	570	181
Halter		
Brassiere type garment	128	425
Feed bags supported on	119	66
Harness	54	24
Design	D30	19
Poke with bar and	119	141
Snap releasers	119	114
Hamburger		
Cookers	99	422
Grinders	241	
Molding and shaping (See briquetting meat)		
Hames	54	25
Collar combined	54	18R
Design	D30	23
Traces and connectors	54	30
Tugs	54	32
Hammer	145	29R
Automobile fender straightening	72	705
Burglar alarm	116	88
Claw	254	26
Combined with additional tools	7	143
Design	D 8	75
Drop forging	72	435
Earth boring tool combined	175	135
Firearm	42	
Forging	72	476
Heads for piano actions	84	254
Impact clutch type	173	93.5
Implement combined	81	463
Awl or prick punch	30	358
Internal combustion charge igniter rocking electrode	123	157
Leather compacting	69	1
Magazine	227	133
Making		
Forging dies for	72	470
Processes of	76	103
Metal bending	72	462
Mills	241	185R
Parallel material flow	241	138
Perforated discharge	241	86
Process	241	27
Series material flow	241	154
Musical instruments		
Piano	84	236
Stringed instrument	84	323
Tuning	84	459
Nut cracker	30	120.1
Pile driver	173	90
Punching machine	83	
Riveting	72	476
Road rammer	404	133
Rock drilling	175	135
Rod encircling type	145	30.5
Saw stretching machine	76	26
Scale removing	29	81D
Shoe lasting stretcher and	12	109
Stoneworking combined	125	6
Impact tools	125	40
Tool driving	173	90

Fig. 6-F—Sample Page of Index to Classification

Classification Definitions

To check our selected class and subclass still further, we next consult a third publication, known as the *Classification Definitions*. This is a series of loose-leaf books which contain a definition for every class and subclass in the *Manual*. At the end of each subclass definition is a cross reference of additional places to look which correspond to such subclass. Fig. 6-H shows the classification definition for 272-109. This definition is actually a composite which I've assembled from several pages of the *Definitions*; i.e., it includes definitions for class 272 per se and the superior class/subclass 272-93. Note that the class definition (272 per se), as well as many of the subclass definitions (see, e.g., 272-110) contain cross references to other classes and subclasses. You should consider these when selecting your search areas.

Be sure to spend enough time to become confidently familiar with the classification system for your invention. Check all of your subclasses in the *Manual of Classification* and the *Class Definitions* manual to be sure that you've obtained all of the right ones. Usually, two or more subclasses will be appropriate.[6] For example, suppose your gymnastic device uses a gear with an irregular shape. Naturally you should search in the gear classes as well as in the exercising device classes. Note that the cross references in the exercising device classes won't refer you to "gears," since this is too specific—the cross references in the PTO's manuals are necessarily general in nature. It's up to you to consider all aspects of your particular invention when selecting search categories.

Another excellent example of using your imagination in class and subclass selection for searching is given in the paper, "The Patent System—A Source of Information for the Engineer," by Joseph K. Campbell, Assistant Professor, Agricultural Engineering Department, Cornell University, Ithaca, New York, which was presented at the 1969 Annual Meeting of the American Society of Agricultural Engineers[7], North Atlantic Region.

Professor Campbell postulates a hypothetical search of a machine that encapsulates or pelletizes small seeds (such as petunia or lettuce seeds) so they may be accurately planted without by a mechanical planter. To find the appropriate subclasses, he first looks in the *Index of Classification*, under the "seed" categories. He finds a good prospect, "Seed-Containing Compositions" and sees that the classification is Class 47 (Plant Husbandry), Sub 1.

---

[6]You can get a free, informal mail-order classification of your invention for search purposes by sending a copy of your invention disclosure, with a request for suggestions of one or more search subclasses, to Search Room, Patent and Trademark Office, Washington DC 20231. However, unless you're really stuck in obtaining subclasses, I don't recommend using this method: since you have the interest and familiarity with your invention to do a far better job if only you put a little effort into it.

[7]The ASAE's address is P.O. Box 229, St. Joseph, MI 49085.

272-1

CLASS 272 AMUSEMENT AND EXERCISING DEVICES

DECEMBER 1983

1 R	AMUSEMENT	53.1	..With actuating means
1 A	..Sand boxes	53.2	..With movable elements
1 B	..Water sports	54	.Seesaws
1 C	..Aircraft simulation	55	..One person
1 D	..Stick horses and body supported devices	56	..Rocking support
		56.5 R	.Slides
1 E	..Birling devices	56.5 SS	...Ski slopes
2	.Houses	85	.Swings
3	.Arenas	86	..Motor operated
4	..Racing	87	..Hand and foot operator
5	...Horse	88	..Hand operator
6	.Elevating devices	89	...Horizontally reciprocating
7	..Combined roundabouts	90	....Cable grasp
8 R	.Illusions	91	....Pulley mounted
8 M	...Mirrors	92	..Foot operator with separate suspender
8 N	..Novelties	93	EXERCISING EQUIPMENT
8 F	...Fire	94	.For head (e.g., jaws, neck, etc.) or foot
8 D	...Display		
8 P	...Projected light	95	..Face (e.g., jaws, lips, etc.)
8.5	..By transparent reflector	96	..Foot
9	..Stage	97	.For simulating skiing
10	...Special projected picture or light effects	98	.For thrusting a simulated weapon (e.g., fencing foil, etc.)
11	...Settings	99	.For improving user's respiratory function
12	....Rapid movement		
13	...Mirror	100	.For track or field sports
14	...Sound imitation	101	..For jumping, vaulting or hurdling
15	...Rain, snow and fire	102	...Cross-bar or support therefor (e.g., hurdle, etc.)
16	...Trip simulation		
17	...In passenger-carrying devices	103	....Means to facilitate adjustment of cross-bar height
18	....With projected picture scenery		
19	..Maze or labyrinth	104	...Vaulting pole or stop therefor
20	..Pyrotechnic display	105	..Starting block for track runner
21	.Stage appliances	106	.For throwing (e.g., javelin, hammer, etc.)
22	..Shifting scenery		
23	...Guides, braces and clips	107	...Discus
24	...Aerial suspension devices	108	...Shot put
25	..Properties	→ 109	.Gymnastic
26	...Stage tanks	61	..Trapezes and rings
27 R	.Initiating devices	62	..Horizontal bars
27 B	...Blowing and cards	63	...Parallel bars
27 W	...Pins and water jets	64	..Vaulting horses
27 N	...Novelties	65	..Projectors
28 R	.Roundabouts	66	...Spring boards
28 S	...Vertical shaft mounting	110	..Swinging tower or pole
29	..Combined with transporting vehicle	111	..Balancing bar or rope
30	..Combined seesaw	112	..For hanging or climbing by the arms
31 R	..Toy	113	..Playground climber (e.g., for use by children)
31 A	....Aircraft		
31 B	....Helicopter	67	.Hand and wrist
31 P	....Phonograph driven	68	..Grips
32	..Marine	69	.Tread mill
33 R	..Occupant propelled	70	.Walking or skating
33 A	....Bowl shape	70.1	..Stilts
33 B	...Bicycle type	70.2	...Steps
34	..Auto-propelled carriages	70.3	..Occupant propelled frame
35	..Free carriage	70.4	...Armpit engaging
36	..Vertical and horizontal axes	71	.Swimming
37	..Plural vertical axes	72	.Rowing
38	..Plural horizontal axes	73	.Bicycle
39	..Vertical axis only	74	.Skipping
40	...Suspended vehicle or rider support	75	..Ropes
41	....Circular swings	76	.Striking
42	....With rotating platform	77	..Striking bags
43	...Rotating vehicle or rider support and stationary track or platform	78	..Supports
		114	.For user locomotion (e.g., pogo stick, etc.)
44	....Vertically undulating track or platform		
		115	..Translatable with user inside
45	....Horizontally undulating track or platform	116	.User-manipulated force-resisting mechanism or element
		117	..User-manipulated weight
46	...Rotating disk, ring or bowl	118	...Including guide for vertical array of weights
47	....Concentric rings or disks		
48	....With vehicle or rider supports	119	...Worn on user's body
49	..Horizontal axis only	120	...Utilizing user's body weight
50	..Gyrating axis	121	...Part of user's body
51	..Inclined axis	122	...Dumbbell or barbell
52	.Hobby horses	123	....Barbell
52.5	..Combined or convertible		

Fig. 6-G—Sample Page of Manual of Classification

Date: March 1973   CLASSIFICATION DEFINITIONS
                                                        Page 272-1

Class 272, AMUSEMENT AND EXERCISING DEVICES

CLASS DEFINITIONS.

This class is generic for amusement and exercising, and includes devices whose purpose is amusement, recreation, exercising, gymnastics, or athletics, unless by analogy of structure or by other functions they are classified in other classes. It includes apparatus used at amusement parks and in theaters, unless otherwise classified, also houses, arenas, and elevators where the sole function is amusement.

SEARCH CLASS:

9, Boats, Buoys and Aquatic Devices, appropriate subclasses for buoyant structure disclosed, but not claimed with features simulating birds, fish, fowl, etc. See (3) Note in Class Definitions of Class 9 for statement of the line.

46, Amusement Devices, Toys, for species of amusement devices commonly called toys which are principally for the amusement of children.

104, Railways, subclass 53+, for amusement railways.

182, Fire Escapes, Ladders, Scaffolds, subclass 137+, for a body catcher or life net.

187, Elevators, appropriate subclasses, for elevators of general utility.

273, Amusement Devices, Games, for species of amusement devices commonly called games which involve skill or competition.

280, Land Vehicles, appropriate subclasses for various types of land vehicles, particularly subclass 1.1+ for simulations, especially progressive hobby horses; subclass 11.1 for skates; subclass 12+ for sleds; and subclass 47.1 for person supporting bodies connected to wheels so as to effect body rocking as the wheels rotate.

404, Road Structure, Process and Apparatus, subclasses 17+ and 71 for pavement and road structure.

Subclasses.

■ ■ ■

93. Apparatus under the class definition intended to be operated by a user of such apparatus for the purpose of developing the muscles of the user's body by repetitive or continuous activity of the user, such activity being facilitated by such apparatus.

(1) Note. Patents placed into this and indented subclasses clearly show that the disclosed purpose is to condition or develop the user's own body. Apparatus that is used by one person to move another person's body will be found in Class 128, Surgery (see the Search Class Note below). Apparatus that is used for moving a user's body for a purpose other than exercising (e.g., transport) will be found in classes that are exemplified in the Search Class notes found under the definition of this class.

(2) Note. The following terms, used in subsequent definitions of the subclasses hereunder, are defined and explained herein: CONTRACTION (i. e., of a muscle) is the physiological effort of the muscle which produces a force that tends to result in shortening of the muscle tissue. It does not necessarily result in an actual me-

chanical change in the length of the muscle, but rather in a tendency to change. The effort may result in shortening of the muscle ("concentric" contraction), or may occur during lengthening of the muscle ("eccentric" contraction), or may occur while the muscle is constrained to remain at substantially the same length ("isometric" contraction), but the tendency to shorten the muscle tissue is generally termed "contraction". Force is the result or effect of the effort exerted by a generator of such effort upon an object. As used in this schedule and in the definition of these subclasses, the term "force" replaces the previously-used terms "push" and "pull", because push and pull are easily confused. As commonly used in this context, "push" refers to a force exerted by a person away from the person's body and "pull" refers to a force exerted by a person towards the person's body; however, in a physiological context, all muscles pull when exercised in the sense that they tend to shorten the muscle tissues when contracted. The term "force" also includes a twisting or turning effort, i.e., a torque as well as a push or pull effort.

■ ■ ■

109. Apparatus under subclass 93, wherein significance is attributed to the use of said apparatus for acrobatic purposes.

(1) Note. The terms "gymnastic" and "acrobatic" have come to denote and describe various pieces of equipment such as trapezes, bars, vaulting horses, etc., that are used in the physical activities known by such names. These activities are characterized by extreme movements of the user, who used the equipment as a fulcrum or starting area to launch his/her body through space, or swing therefrom, or perform other such physical activities thereon. As in previously-described athletic activities, the significance of the apparatus is more in the activity for which the apparatus is used than in the structural differences between the apparatus.

110. Apparatus under subclass 109, including an elongated slender rod, of which one end is secured to the ground and the other end serves as a support for the user as his/her body is exercised thereon.

SEARCH THIS CLASS, SUBCLASS:
104, for a structurally similar flexible pole that is used to help launch a pole vaulter over a high bar.

111. Apparatus under subclass 109, including a relatively slender, horizontally-positioned member, on which member a user supports his/her body with the center of gravity of said body above the member while attempting to maintain the body in a state of equilibrium.

112. Apparatus under subclass 109, including equipment that is grasped by a hand or the hands of the user, and from which equipment the user suspends his/her body or ascends the equipment using only his/her arm(s).

Fig. 6-H—Sample Page of Classification Definitions

After checking this class/subclass in the Manual of Classification to see where it fits in the scheme of things and in the Class Definitions to make sure that it looks OK (it does), he would start his first search with Class 27, Sub 1. Then, using his imagination, Professor Campbell also realizes that some candies, such as chocolate-covered peanuts, are actually encapsulated seeds. Thus, he also looks under the candy classifications, and finds several likely prospects in Class 107: "Bread, Pastry and Confection-Making." Specifically, sub 1.25, "Composite Pills (with core)"; sub 1.7, "Feeding Solid Centers into Confectionery"; and sub 11, "Pills" look quite promising. Thus he adds class 107, subs 125, 1.7, and 11 to his search field.

The moral is this. When you search, look not only for the obvious places, but also in analogous areas, as Prof. Campbell does. Fortunately, the cross references in the Class Definitions manual will be of great help here—note (Fig. 6-H) the copious cross references at the top of Class 272. Also, the PTO and all Patent Depository Libraries now have a computerized class-locating system called CASSIS (see below) which will be of great assistance.

Note how Sam Searcher, Esq. has completed the "Selected Search Classifications" section of the search worksheet with appropriate classes to search for prior art relevant to Millie Inventress' invention.

Step 3: Note Relevant Prior Art (Patents and Other Publications) Under Your Classification

After obtaining a list of classes and subclasses to search, find the actual examining division (or location in the search room if you choose to stay downstairs) where these classes and subclasses are actually located.[8] Then go to your search area and look through all the patents in your selected subclasses.

In the public search room, you'll have to remove bundles of patents from slotlike shelves in the stack area. Bring them to a table in the main search area, and search them by placing the patents in a packet holder and flipping through them. In the examiner's search room, the patents are found in small drawers, called "shoes" by the examiners. You should remove the drawer of patents, hold it in your lap, and flip through the patents while you're seated in a chair; generally no table will be available.

As you flip through the patents, you may at first find it very difficult to understand them and to make your search. I did when as an examiner I made my first search in the PTO. Don't be discouraged! After just a few minutes the technique will become clear and you may even get to like it! You'll find it easier to understand newer patents (see Le Sueur—Fig. 6-Dc) since they have an abstract page up front that contains a brief summary of the patent and the most relevant figure or drawing.

You'll find that the older patents (see Gabel—Fig. 6-Da) have several sheets of unlabeled drawings and a closely-printed description, termed a "specification," after the drawings. However, even with older patents, you can get a brief summary of the patent by referring to the summary of the invention, which is usually found in the first or second column of the specification. Near the end of each patent, you'll find the claims (Chapter 9). See any utility patent, or Fig. 6-K below for some examples of claims. These are formally-worded, legalistic sentence fragments which come after and are the object of the heading words "I [or "We"] claim." As mentioned in the last chapter, and as

---

[8]The PTO recently implemented several automated search systems; ask the search assistants in the main search room if they have one in your field. If so, you're lucky.

you'll learn in detail in Chapter 9, the claims define the legal scope of protection afforded by the patent.

You should generally *not* read the claims of searched patents; this is because claims are written and are used solely to define the scope of an invention to determine whether an infringement exists—not for use in anticipation or obviousness situations. Also because they're usually difficult to understand, they'll probably confuse you. In any event, they merely repeat what is in the narrative description. Nevertheless, some searchers do like to read claims of older patents when searching to get a quick "handle" on the patent's technical content.[9]

If you do read the claims, keep in mind three important considerations:

1. If a prior-art patent shows (i.e. describes) but doesn't formally claim your invention, this doesn't mean you're free to claim it.

2. A patent contains much more technical information than what's in its claims; all of this technical information can be used as prior art, just as if the patent were an article in a technical magazine. Thus you should use the claims only to get a "handle" on the patent; you should not regard them as a summary or synopsis of the patent's disclosure.

3. The scope of coverage you will likely be able to obtain for your invention (see Sec. J below) will usually be narrower than the scope of the claims of the closely-relevant prior-art patents you uncover. (See Ch. 9 to see how to determine the breadth of claims.)

▲

Common Misconception: *If your invention is covered by the claims of a prior patent, you will be liable as an infringer if you file a patent application on the invention.*

Fact: *Neither a patent application nor its claims can infringe a prior patent. Only the manufacture, use, or sale of an invention in physical form can infringe. In fact, most examiners in the PTO rarely read the claims of patents that they cite against pending patent applications; this is because the claims merely repeat, in stilted legalese, what is already clearly described in better detail in the specification. If you do have an improvement invention which, if manufactured, might infringe another person's patent, file a patent application anyway if you think the invention is valuable and patentable. You can almost always get a license from the patent owner for a reasonable fee. This will permit you to make, use, and sell your invention provided you pay royalties to the owner of the patent.*

Alternatively, you may be able to work out a cross-license with the patent owner. In a cross license, each patent owner gives the other patent owner permission to make, use, and sell his or her invention without payment, or for a reduced payment.

Don't think about obviousness as you search since this may overwhelm you and detract from the quality of your investigation. Rather, at this stage, try to fish with a large net by merely looking for the physical features of your invention.

As you search, keep a careful record of all patent classes and subclasses you've searched, as indicated in Fig. 6-E above. If you find relevant patents or other art, write their numbers, dates, names, or other identification, and order copies later; again see Fig. 6-E. Although you need only the number to order a patent, I recommend that you write the issue date, first inventor's name, and classification as well to double-check later in case you write down a wrong number.

If you do find an important relevant reference, don't stop; simply asterisk it (to remind you of its

_____

[9]Also, if you make an Official Gazette search in a Patent Depository or regular library (see Section I below), you'll have to rely on claims for the most part, since most of the OGs contain only a single claim of each patent.

importance) and continue your search to the end. When you note a relevant reference, also write down its most relevant features to refresh your memory and save time later.

If you still don't find any relevant patents, double check your search classes using Class Definitions, the Manual of Classification, and some help from a patent examiner or assistant in the search room. If you're reasonably sure you're in the right class and still can't find any relevant references, write down the identifications of the most relevant ones you can find, even if they're not close. This will establish that you made the search, what the closest art is, how novel your invention is, and you'll have references to cite on your Information Disclosure Statement (see Ch. 13) to make the PTO's file of your patent look good; you should never finish a search without coming up with at least several references. If you do consult any examiners, write their names in the comments section of the worksheet.

In each subclass, you'll find patents which are directly classified there, and "cross-references" (XRs), patents primarily classified in another subclass, but also classified in your subclass because they have a feature which makes the cross-reference appropriate. Be sure to review the cross-references as well as the regular patents in each subclass.

The public search room has copiers for making instant copies of patents for a per page fee, but if you don't need instant copies, you can buy a complete copy of any patent for one patent copy coupon, or use two coupons per patent for rush service. To do this purchase an adequate supply of coupons from the PTO's cashier (see Fee Schedule in Appendix); then write down the number of each patent you select on a coupon, add your name and address, and deposit them in the appropriate box in the search room. The patents you request will be mailed to you, generally in several weeks if you use one coupon per patent, or in several days if you staple two coupons together per patent.

Step 4: Carefully Review the Prior Art to See Whether It Anticipates Your Invention or Renders It Obvious

After you've made your search and obtained copies of all the references you thought were pertinent, study them carefully at your leisure. You may want to write a brief summary of each relevant patent if it doesn't have an abstract, or if the abstract is inadequate. Then, determine if your invention is patentable over the patents you've found. Follow the steps described earlier in this chapter (Section H) for analyzing the search report when your search was done by someone else).

J. The Scope of Patent Coverage

ALTHOUGH YOU'D PROBABLY LIKE things to be simpler, the determination of whether your invention is patentable will rarely be a "yes" or "no" one, unless your invention is a very simple device, process, or composition. Many inventions are complex enough to have some features, or some combination or features, which will be different enough to be patentable. However, your object is not merely to get a patent, but to get meaningful patent coverage, i.e., offensive rights which are broad enough so that competitors can't "design around" your patent easily. As I've said elsewhere, designing around a patent is the act of making a competitive device or process which is equivalent in function to the patented device, but which doesn't infringe the patent.

Many "modern" inventions are actually old hat, i.e., the basic idea was known many years before and the real invention was actually just an improvement on an old one. For example, the first computer was a mechanical device invented in the 1800's by Charles Babbage. The ancient Chinese used a soybean mold to treat infections. One J. H. Loud received a patent on a ballpoint pen in 1888 and the first 3-D film was shown in 1922!

Simply put, you'll often find that your invention, while valuable, may be less of an innovation than you thought it was. You'll thus have to determine whether or not your invention is sufficiently innovative to get meaningful patent protection. In other words, your scope of coverage will depend upon how close the references which your search uncovers are to your invention, i.e., how many features of your invention are shown by the references, and how they are shown. In the end, your scope of coverage will actually depend upon the breadth of the claims which you can get the PTO to allow, but this is jumping the gun at this stage; I cover claims in Chapter 9.

For an example, let's take a simple invention. As stated, in a simple invention patentability will usually be a black or white determination, and you won't have much of a problem about your scope of coverage. Suppose you've just invented the magnetically-operated cat door of Chapter 2, i.e., you provide a cat with a neck-worn magnet that can operate a release on a cat door. Your search references fail to show any magnetically-operated pet release door. Thus, the neck magnet and the

magnetic door release are the novel features of your invention. To get a patent, your invention would have to be limited to these specific features since neither could be changed or eliminated while producing the same result. However, there is no harm in limiting the invention to these features since it would be difficult for anyone to "design around" them, i.e., it would be difficult or impossible for anyone to provide the same result (a cat-operated door release) without using a neck magnet and a magnetic release.

With other inventions, however, your scope of coverage won't be so broad, i.e., it won't be as difficult for someone to design around it. For example, suppose you invented the centrifugal vegetable juicer mentioned previously in Chapter 5, i.e., a juicer with a sloping side basket permitting the solid pulp to ride up and out so that juicing could continue without having to empty the pulp from the basket.

If the prior art were not "kind" to you, i.e., your search uncovered a patent or other publication which showed a juicer with a basket with sloping sides and with a well at the top to catch and hold the pulp, your patent would not be granted if you claimed just the sloping sides (even though it would be superior to the prior art due to the complete elimination of the pulp). To get the patent, you would have to also claim another feature (say, the trough shape). Thus, by having access to the prior art you would know enough to claim your invention less broadly.

Also, suppose you've invented the napkin-shaping ring of Fig. 6-B. Suppose further that Gabel did not exist and that your search uncovers only the Le Sueur patent, which shows a plain, circular napkin ring. You'd be entitled to relatively broad coverage, since your novel features are themselves broad: namely, a ring with inner parts which can shape a napkin when it is pulled through the ring.

However, assuming the Gabel patent does exist and your search uncovers it as well as Le Sueur, what are your novel features now? First, your device has a circular ring with a leg extending inwardly from the ring; neither Gabel nor Le Sueur, nor any possible combination of these references, has this combination. Second, your invention has the flaring arms which shape the napkin; these are attached to the end of the inner leg; the references also lack this feature. Thus to distinguish over Le Sueur and Gabel, you'll have to rely on far more specific features than you'd have to do if only Le Sueur existed. Hence your actual invention would be far narrower, since you'll have to limit it to the novel features which distinguish it from Gabel as well as Le Sueur. Unfortunately, this will narrow your scope of coverage because competitors can design around you more easily than they could do if only Le Sueur existed.

As you've probably gathered by now, your scope of coverage will be determined by what novel features you need to use to distinguish your invention over the prior art and still provide new results which are different or unexpected enough to be considered unobvious. The fewer the novel features you need, the broader your invention or scope of coverage will be. Stated differently, if you need many new features, or very specific features, to define over the prior art and provide new results, it is usually relatively easy for a competitor to use fewer or alternative features to provide the same result without infringing your patent.

You should make your scope of coverage determination by determining the fewest number or the broadest feature(s) you'll need to distinguish patentability over the prior art. Do this by a repetitive narrowing trial and error process: First see what minimum feature(s) you'll need to have some novelty over the prior art, i.e., enough to distinguish under Section 102 (Box C of Fig. 5-B), and then see if these would satisfy Section 103 (Boxes E, I, and M), i.e., would they provide any unexpected new results?

If you feel that your minimum number of features are enough to ascend the novelty box

(pictured in Fig. 5-A in Chapter 5), but would not be sufficient to climb the big unobviousness box, i.e., you don't have enough features to provide new and unexpected results, then try narrowing your features or adding more until you feel that you'll have enough to make it to the patentability summit.

Note of Sanity: This is another one of those aspects of patent law that may have your head spinning. Fortunately, the material covered here under determining the scope of your protection is also discussed in the different context of drafting your claims (see Chapter 9). By the time you read this book thoroughly, you will understand all of this a lot better.

After you evaluate your search results, you'll have a pretty good idea of the minimum number of novel features which are necessary to sufficiently distinguish your invention over the prior art. If you're in doubt that you have enough such features, or if you feel that you'd have to limit your invention to specific features to define structure which would be considered unobvious over the prior art, it probably isn't patentable, or even if patentable, it isn't worth filing on, since it would be easy to design around. One possibility, if you can't make a decision, is to pay for a professional's opinion.

On the other hand, if you've found nothing like your invention in your search, congratulations. You probably have a very broad invention, since, of the 4.5 million patents which have issued thus far, one or more features of almost all inventions are likely to be shown in the prior art.

K. Searching It Yourself In a Patent Depository Library

IF YOU CAN'T SEARCH IN THE PTO, the next possibility, although somewhat inferior, is to search your invention in one of the Patent Depository Libraries listed below as Fig. 6-I, all of which currently receive all patents issued by the PTO. Before going to any PDL, call to find out their hours of operation and what search facilities they have.

Why is searching at a PDL less useful than searching at the PTO? Simply because not all PDL's have all patents issued from No. 1 to the present and none have them physically separated by subject matter into searchable classifications, as does the PTO in Arlington. Using a PDL is therefore more difficult and time-consuming than if you use the PTO. You should carefully balance the large expenditure of your time and the inferiority of the search materials against the $150 to $300 or so you would spend for a professional searcher to do the job at the PTO. Of course, as I suggested, the optimum solution is to visit Arlington yourself.

I like to assign percentage values to the various types of searches: I roughly estimate a good examiner's search at 90% (i.e., it has about a 90% chance of standing up in court), a good search by a non-examiner in the PTO at 80%, and a good search in a PDL at 70%. If your invention is in an active, contemporary field, such as a computer mouse, you should reduce the value of the two non-examiner types of searches somewhat, due to the fact that patent applications in this field are more likely to be pending.

If you do make a search at a PDL, you should go through the same four steps given above. First, articulate your invention (in the same manner as before) and second, use the reference tools to find the relevant classes and subclasses. The third step is a review of the patents in the selected classes and subclasses. And finally, you should analyze all relevant prior art references for their effect on your invention's patentability.

In a PDL these steps are more difficult than in the PTO. First, you must get a list of the patents in your selected classes and subclasses. Most of the PDLs have lists of patents in each class and subclass on microfilm. But if your PDL doesn't have such a list, you'll have to order one from the PTO in Arlington;

this may take several weeks to arrive. The staff will show you how to do this.

Fig. 6-J is a typical sample of a microfiche printout—a list of patents which are classified in our old friend, class 272-109, gymnastic devices.

Once you've obtained a list such as this for the first of your selected classes and subclasses, you'll have to locate each patent (or an abstract of it) individually and examine it. There are two ways to access each patent:

1. Look at an abstract of the patent in its OG volume, or

2. Look at the entire patent in a numerically-arranged stack.

If you make an OG search (this will be much easier), each patent entry you find will contain only a single claim (or abstract) and a single figure or drawing of the patent, as indicated in Fig. 6-K (a typical page from an OG).

Note that for each patent, the OG entry gives the patent number, inventor's name(s) and address(es), assignee (usually a company which the inventor has transferred ownership of the patent to), filing date, application serial number, international classification, U.S. classification, number of claims, and a sample claim or abstract. If the drawing and claim look relevant, go to the actual patent, or order a copy of it, and study it at your leisure.

Remember that the claim found in the *Official Gazette* is not a descriptive summary of the technical information in the patent. Rather, it is the essence of the claimed invention. The full text of the patent will contain far more technical information than the claim. So don't assume that if a patent's *Official Gazette* claim doesn't precisely describe your invention the rest ofpatent isn't relevant.

Reference Collections of U.S. Patents Available for Public Use in Patent Depository Libraries

The following libraries, designated as Patent Depository Libraries, receive current issues of U.S. Patents and maintain collections of earlier issued patents. The scope of these collections varies from library to library, ranging from patents of only recent years to all or most of the patents issued since 1790.

These patent collections are open to public use and each of the Patent Depository Libraries, in addition, offers the publications of the U.S. Patent Classification System (e.g. The Manual of Classification, Index to the U.S. Patent Classification, Classification Definitions, etc.) and provides technical staff assistance in their use to aid the public in gaining effective access to information contained in patents. With one exception, as noted in the table following, the collections are organized in patent number sequence.

Facilities for making paper copies from either microfilm in reader-printers or from the bound volumes in paper-to-paper copies are generally provided for a fee.

Owing to variations in the scope of patent collections among the Patent Depository Libraries and in their hours of service to the public, anyone contemplating use of the patents at a particular library is advised to contact that library, in advance, about its collection and hours, so as to avert possible inconvenience.

State	Name of Library	Telephone Contact
Alabama	Auburn University Libraries	(205) 826-4500 Ext.21
	Birmingham Public Library	(205) 226-3680
Alaska	Anchorage Municipal Libraries	(907) 264-4481
Arizona	Tempe: Science Library, Arizona State University	(602) 965-7607
Arkansas	Little Rock: Arkansas State Library	(501) 371-2090
California	Los Angeles Public Library	(213) 612-3273
	Sacramento: California State Library	(916) 322-4572
	San Diego Public Library	(619) 236-5813
	Sunnyvale: Patent Information Clearinghouse*	(408) 730-7290
Colorado	Denver Public Library	(303) 571-2122
Delaware	Newark: University of Delaware Library	(302) 451-2238
Florida	Fort Lauderdale: Broward County Main Library	(305) 357-7444
	Miami–Dade Public Library	(305) 375-2665
Georgia	Atlanta: Price Gilbert Memorial Library, Georgia Institute of Technology	(404) 894-4508
Idaho	Moscow: University of Idaho Library	(208) 885-6235
Illinois	Chicago Public Library	(312) 269-2865
	Springfield: Illinois State Library	(217) 782-5430
Indiana	Indianapolis–Marion County Public Library	(317) 269-1706
Louisiana	Baton Rouge: Troy H. Middleton Library, Louisiana State University	(504) 388-2570
Maryland	College Park: Engineering and Physical Sciences Library, University of Maryland	(301) 454-3037
Massachusetts	Amherst: Physical Sciences Library, University of Massachusetts	(413) 545-1370
	Boston Public Library	(617) 536-5400 Ext. 265
Michigan	Ann Arbor: Engineering Transportation Library, University of Michigan	(313) 764-7494
	Detroit Public Library	(313) 833-1450
Minnesota	Minneapolis Public Library & Information Center	(612) 372-6570
Missouri	Kansas City: Linda Hall Library	(816) 363-4600
	St. Louis Public Library	(314) 241-2288 Ext. 390
Montana	Butte: Montana College of Mineral Science and Technology Library	(406) 496-4283
Nebraska	Lincoln: University of Nebraska-Lincoln, Engineering Library	(402) 472-3411
Nevada	Reno: University of Nevada Library	(702) 784-6579
New Hampshire	Durham: University of New Hampshire Library	(603) 862-1777
New Jersey	Newark Public Library	(201) 733-7815
New Mexico	Albuquerque: University of New Mexico Library	(505) 277-5441
New York	Albany: New York State Library	(518) 474-5125
	Buffalo and Erie County Public Library	(716) 856-7525 Ext. 267
	New York Public Library (The Research Libraries)	(212) 714-8529
North Carolina	Raleigh: D. H. Hill Library, N.C. State University	(919) 737-3280
Ohio	Cincinnati & Hamilton County, Public Library of	(513) 369-6936
	Cleveland Public Library	(216) 623-2870
	Columbus: Ohio State University Libraries	(614) 422-6286
	Toledo/Lucas County Public Library	(419) 255-7055 Ext. 212
Oklahoma	Stillwater: Oklahoma State University Library	(405) 624-6546
Oregon	Salem: Oregon State Library	(503) 378-4239
Pennsylvania	Cambridge Springs: Alliance College Library	(814) 398-2098
	Philadelphia: Franklin Institute Library	(215) 448-1227
	Pittsburgh: Carnegie Library of Pittsburgh	(412) 622-3138
	University Park: Pattee Library, Pennsylvania State University	(814) 865-4861
Rhode Island	Providence Public Library	(401) 521-8726
South Carolina	Charleston: Medical University of South Carolina Library	(803) 792-2372
Tennessee	Memphis & Shelby County Public Library and Information Center	(901) 725-8876
	Nashville: Vanderbilt University Library	(615) 322-2775
Texas	Austin: McKinney Engineering Library, University of Texas	(512) 471-1610
	College Station: Sterling C. Evans Library, Texas A & M University	(409) 845-2551
	Dallas Public Library	(214) 749-4176
	Houston: The Fondren Library, Rice University	(713) 527-8101 Ext. 2587
Utah	Salt Lake City: Marriott Library, University of Utah	(801) 581-8394
Washington	Seattle: Engineering Library, University of Washington	(206) 543-0740
Wisconsin	Madison: Kurt F. Wendt Engineering Library, University of Wisconsin	(608) 262-6845
	Milwaukee Public Library	(414) 278-3247

All of the above-listed libraries offer CASSIS (Classification And Search Support Information System), which provides direct, on-line access to Patent and Trademark Office data.

*Collection organized by subject matter.

Fig. 6-I—List of Patent Depository Libraries

LISTING	CLASS 272		REEL NO. 7	PAGE 19
105	* 107	* 109	* 109	* 110

	105	107	109	109	109	110
30	1,701,026	1,134,008	1,747,352X	3,659,844X	3,734,758	
53	1,709,832	1,492,976	1,779,905	3,735,979X	3,834,695	
01	1,785,968	1,570,185	1,914,555	3,764,446X	3,835,641X	
98X	1,793,898	1,947,025X	1,918,559X	3,778,054X	3,880,422	
13X	1,990,497	1,958,807X	1,928,089X	3,785,642X	3,896,858X	
28X	2,004,172	2,223,091	2,048,587X	3,835,252X	3,923,302	
86	2,144,962	2,640,699X	2,107,377	3,857,561X	4,084,814	
22X	2,165,749X	2,864,201X	2,167,696X	3,857,563X		
44	2,323,510	3,312,472X	2,169,710X	3,874,657X	111	
31	2,341,473	4,121,826X	2,262,761X	3,891,207X		
97	2,505,784		2,324,970X	3,895,795X	D 175,729X	
92	2,534,159	108	2,496,748	3,912,262	159,301	
62	2,890,048		2,572,149X	3,915,451	971,003	
16X	2,900,187	450,759	2,595,111X	3,937,461	1,001,300X	
18	2,937,871	807,770	2,652,966X	3,947,023X	1,407,642	
10X	2,978,692X	1,036,138	2,671,229X	3,966,200	1,419,191X	
09	3,010,321X	1,805,121X	2,706,632X	3,969,871X	1,537,686X	
22	3,244,421X	1,986,687	2,722,360X	3,971,561X	1,747,721	
06	3,400,928	1,997,958X	2,738,189X	3,981,500X	2,000,250	
82	3,401,931	2,117,938	2,771,615X	4,014,057X	2,197,600X	
91	3,494,615	3,548,420X	2,795,423	4,026,547	2,343,204X	
88X	3,608,897	3,759,513	2,829,892X	4,037,834X	2,646,280X	
64	3,665,452X	3,884,465X	2,858,132X	4,125,257	2,855,201	
64	3,724,843	4,121,826X	2,859,967X	4,137,583X	2,939,704	
00X	3,731,298X		2,885,233X	4,147,129X	3,062,542	
81	3,746,335	(109)	2,897,013X	4,147,828X	3,083,964	
67	3,799,542		2,944,815	4,204,710X	3,173,415X	
32	3,809,392X	D 262,394X	2,953,376X	4,210,322	3,339,920	
17	4,089,519	RE 25,843	3,006,645	4,216,958	3,404,884	
12	4,134,583X	9,695	3,044,773X	4,225,131	3,416,792X	
48		174,499	3,085,357	4,274,626	3,485,493X	
04	106	233,273	3,105,082X	4,275,880X	3,545,747X	
24		233,274	3,106,395X	4,325,546X	3,547,435	
94X	610,131X	233,540	3,204,259X	4,340,215	3,570,847	
16X	649,885	233,541	3,205,888X	4,340,216	3,570,848	
80	1,122,157	451,411	3,207,511X	4,344,617	3,580,568	
88	1,552,442	649,914	3,211,452X	4,350,721X	3,582,068	
68X	1,569,395	664,414	3,242,509X	4,410,175X	3,589,716	
34X	1,731,686	802,338	3,251,076X		3,616,126X	
	1,994,089	811,417	3,262,134X	110	3,658,325	
	2,036,524X	907,075	3,284,819X		3,722,881	
	2,044,092	932,413	3,319,273X	D 198,923X	3,754,757	
66	2,122,023	932,902	3,372,926	D 199,984X	3,781,931X	
01	2,180,384	998,634	3,379,439X	1,085,505X	3,806,118X	
54	2,196,610	1,003,797	3,391,414X	1,100,180X	3,837,644X	
26	2,214,464	1,013,687	3,399,407	1,141,292X	3,850,428X	
22X	3,163,421X	1,015,208	3,405,939X	1,462,910X	3,944,654	
00	3,181,864X	1,126,082	3,409,294X	1,865,095X	3,990,697	
78	3,746,334	1,128,201	3,419,270X	1,907,451	4,105,201	
65	3,942,793X	1,130,813	3,432,163	1,916,809X	4,133,524	
20X	4,084,813	1,142,137X	3,433,477X	2,198,537	4,183,521X	
99	4,333,643	1,177,473X	3,459,611X	2,723,855X	4,197,839X	
37X	4,337,940	1,204,329X	3,526,911X	2,906,531X	4,204,719X	
14X	4,404,053X	1,256,734	3,580,569	2,949,298X	4,258,915	
07		1,479,830	3,598,406X	3,246,893X	4,272,073	
70	107	1,501,823X	3,628,790X	3,250,532	4,278,250X	

CLASS 272	
112	* 113

112	113
D 155,940X	D 173,173
D 208,924X	D 176,999
D 272,021X	D 187,138
D 214,572X	D 187,380
239,970X	D 187,381
450,187	D 187,656
775,309	D 198,532
786,672	D 218,455
950,100	D 218,460
1,485,135	D 218,765
1,585,748	D 224,029
1,670,390	D 224,796
1,676,061	D 227,381
2,240,407	D 227,792
2,303,223X	D 231,555
2,365,117	D 232,4..
2,429,939	D 238,694
2,706,632X	D 250,723
2,800,105X	D 250,783
2,838,307	D 250,784
2,929,627	170,495
2,977,118	209,511
3,032,344	7..,550
3,090,617X	796,159
3,156,465	821,391
3,342,484	1,126,082
3,445,108	1,185,176
3,483,999X	1,351,053
3,501,140X	1,471,465
3,506,261	1,488,244
3,526,399	1,488,245
3,547,435X	1,488,246
3,563,539X	1,707,854
3,598,406X	1,765,361
3,606,315	1,822,786
3,638,602X	1,877,833
3,642,277	1,901,964
3,771,784X	1,917,018
3,782,718	1,929,822
3,794,316	2,126,636
3,837,642X	2,151,403
3,982,754X	2,206,581
4,018,437X	2,222,119
4,077,403X	2,500,425
4,116,433X	2,584,742
4,149,712	2,620,185
4,159,113	2,648,538
4,161,998X	2,648,539
4,272,073X	2,704,667
4,278,250X	2,720,430
4,335,538X	2,723,853
4,355,633X	2,768,823
4,372,552	2,795,423
	2,843,379
113	2,883,192
	2,886,317

Fig. 6-J—List of Patents in Class 272-109 from Microfilm Printout

AUGUST 24, 1976   GENERAL AND MECHANICAL   1395

a plurality of bristles slidingly extending through said sleeve about said head for locking said electrode into heart tissue,

a slide assembly disposed within said sleeve and about said hose for movement toward and away from said head, said

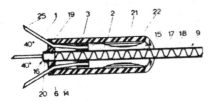

assembly mounting said bristles for movement in and out of said sleeve as said assembly is moved by a catheter inserted in the rear of said sleeve to engage said slide assembly, and

means adjacent the rear of said sleeve for preventing withdrawal of said assembly from said sleeve.

3,976,083
BRASSIERE HAVING SIMULATED NIPPLES
Jakob E. Schmidt, 934 Monroe St., Charlestown, Ind. 47111
Filed Feb. 27, 1975, Ser. No. 553,779
Int. Cl.² A41C 3/00

U.S. Cl. 128—425   8 Claims

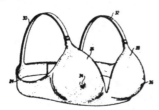

1. An improved brassiere comprising a pair of breast receiving cups, said cups comprising an outer layer of flexible fabric material; and a simulated nipple attached to said outer layer of each cup, said simulated nipple comprising a stud element having distal and proximal ends and an exterior profile simulating the profile of the nipple of a human female breast, said stud element being positioned interiorly of said cup with a portion of said outer layer deformed over said distal end, and means cooperating with said proximal end and said outer layer for attaching said stud element to said breast receiving cup, whereby the exterior profile of said stud element is noticeable exteriorly of said brassiere to enhance the appearance of bralessness when said brassiere is worn beneath outer garments.

3,976,084
RETHRESHER BLOWER FOR A COMBINE
Wilbert D. Weber, Mississauga, Canada, assignor to Massey-Ferguson Industries Limited, Toronto, Canada
Filed Apr. 23, 1975, Ser. No. 570,901
Int. Cl.² A01F 12/18

U.S. Cl. 130—27 F   4 Claims

1. An agricultural combine harvester including a frame; a thresher housing; grain threshing, separating and cleaning apparatus including sieves mounted on the frame in the thresher housing; a conveyor mounted below said sieves for receiving tailings that overflow the sieves and conveying the tailings to one side of the combine; and a tailings rethreshing and blower assembly mounted on the frame for receiving tailings from the conveyor mounted below the sieves, for rethreshing the tailings and for blowing the rethreshed tailings

back into the thresher housing for cleaning, the rethresher and blower assembly including a chamber defined by two end walls and arcuate wall sections joining the two end walls, a blower rotatably mounted between the two end walls, said blower including vanes extending outwardly from the axis of rotation of the blower, rasp bar sections detachably mounted on the outer ends of the blower vanes, stationary threshing elements mounted on at least a portion of the arcuate wall of the chamber to cooperate with the rasp bar sections on the blower vanes to rethresh tailings, a tailings inlet aperture in

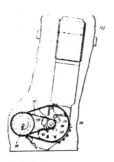

said arcuate wall sections, conveyor means connecting the conveyor mounted below the sieves and the tailings inlet aperture to direct tailings from conveyor mounted below said sieves through the tailings inlet aperture and into said chamber, a threshed tailings outlet in said arcuate wall sections, conveyor means connecting the threshed tailings outlet to the thresher housing to direct rethreshed tailings from the tailings outlet in said arcuate wall sections up and into the thresher housing, at least one air intake in one of the two end walls which define the chamber, and drive means to rotate the blower.

3,976,085
AUTOMATIC CIGARETTE FEED MACHINE
Floyd Vameda Hall, Durham, N.C., assignor to Liggett & Myers, Incorporated, Durham, N.C.
Continuation of Ser. No. 353,372, April 23, 1973, abandoned. This application Sept. 26, 1974, Ser. No. 509,702
Int. Cl.² A24C 5/35

U.S. Cl. 131—25   11 Claims

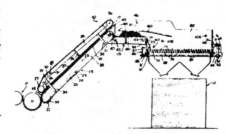

1. In combination in a one-to-one relationship,

a cigarette making machine for producing cigarettes at a high rate of speed, wherein said making machine includes a means for dispensing cigarettes individually in a spaced relationship,

a cigarette packaging machine for packaging cigarettes at approximately said high speed rate, and

a cigarette feed machine connected between said making machine and said packaging machine to convey the cigarettes made in said making machine directly to said ciga-

---

Fig. 6-J—Page of *Official Gazette* showing Various Patent Abstracts

Example: When recently performing a PDL search, a client of mine passed over a patent listed in the OG because the single drawing figure appeared to render the patent irrelevant. In fact, another drawing figure in the passed-over patent (but not found in the OG) anticipated my client's invention and was used by the PTO to reject of his application (after he had spent considerable time, money, and energy preparing and filing it). The moral? Take an OG search with a grain of salt. Note well that a figure of the patent which isn't shown in the OG may be highly relevant; thus it's best to search full patents.

To make an OG search of the patents in Class 272, Subclass 109 (Fig. 6-J), start with the first patent in this list, D-262,394X. The "D" means that the patent is a design patent and the X means that this patent is a cross reference. To locate patent D-262,394, look in the back of any OG for the section with the design patents, and see what range their numbers cover. If the patent numbers are too low, go to a later OG. If those are too high, go to an earlier OG. You'll be able to locate the right OG in a few minutes. You'll find that the present patent, D-262,394, was issued in 1980. Look at the figure from the patent printed in the OG and, if you find it relevant, write its identifying data down on your search worksheet, Fig. 6-1.

The second patent in the list, RE-25,843, is a reissue patent. Reissues are discussed in Chapter 14. For now all you have to know is that reissues are listed in the front of each OG. Find the right OG, and in the same manner as you did with the design patent, locate the patent and list it on your worksheet if you feel it's relevant.

All of the rest of the patents in subclass 109 are regular utility patents in numerical and date order. Start with patent 9,695, which issued in the middle 1800s. You'll be able to locate it easily since the outside of the binder of each OG volume usually lists the numbers of the utility patents which that volume contains. (If your PDL doesn't carry OGs this far back, they'll probably have it on microfilm; ask your librarian for assistance.) Once you locate the OG (or

microfilm reel), look at the patent in the usual manner to see if it's relevant. If so, write its data on your worksheet.

If your PDL has full copies of patents readily accessible (each patent usually consists of several pages), you can look at the full text of each patent, one by one, in a similar manner as you looked at their abstracts in the OGs. If you find that the patent is relevant, usually you'll have access to a photocopy machine where you'll be able to make a copy of the whole patent, or just its relevant parts, on the spot.

Alternatively, if you don't want to interrupt the flow of your searching, you can save your patent numbers and order copies later. The PDL may have PTO patent copy order coupons. If not, you can order patent copies from the PTO by writing a letter listing the numbers of the patents you want (be sure they're accurate!) to "Commissioner of Patents and Trademarks, Washington, DC 20231" with a check for the price per patent (see Fee Schedule in Appendix) times the total number of patents you've ordered.

If you can't get to a PDL, note that most large public or university libraries at least subscribe to the *Official Gazette*. If in addition to this resource a local library also has an *Index of Classification*, you can make a cursory search there.

After you've completed Step Three, the review of patents, then perform Step Four, the analysis and decision, in exactly the same manner as outlined above for the PTO search.

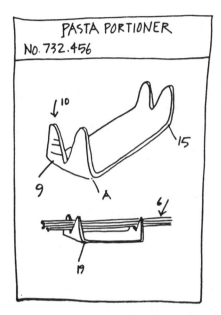

these places. However, to make a computer search, you would select a combination of key words to describe your invention. Here you should use "bicycle" and "carbon fiber alloy." You then send these words to the computer and tell it to look through its data bank for any patent which contains all of these words. When it finds any patents which contain your key words in the combination you specify in your search request, it will identify these patents regardless of their classifications.

If the computer reports too much data for you to conveniently examine—say it's found 200 patents with your words in combination—you should first look at one or two of the patents (the computer will show you the relevant text) to see if your invention is shown in an earlier patent (i.e., your invention has been "knocked out"). If so, your search is over. If not, you'll need to narrow your search. This is easy. Simply add one or more additional key words, say "frame," or some details of the alloy, and re-do the search with these increased key words until you've few enough patents to manually review conveniently. Also, you can narrow the search by using narrower (more specific) key words.

If you get extremely specific, the computer is likely to report no patents, or just one or two. If this occurs, you'll need to broaden your search. This is just as easy. Merely remove one or more key words, or broaden your present key words, and re-do the search until you get back what you want. For example, you could eliminate "bicycle" or substitute "frame" for "bicycle" to broaden the search. If you have difficulty with the concept of broadening and narrowing a search, see Chapters 9 and 13 for more on this subject in the context of drafting your patent claims.

The data which you search by computer (i.e., the texts or claims of patents) is made available to anyone (for a fee) by several computer search service firms. These are private companies which in turn get this data in the form of machine-readable tapes as a by-product of the patent printing process from the Government Printing Office, which prints all

L. Computer Searching

SINCE COMPUTER SEARCHING IS still in its infancy, I don't recommend its use in place of a full manual search. The main reason for this is that the patents in most computer search data banks usually go back to only 1975 (some to 1950). Nevertheless, I cover it here because it does have some advantages that make it useful as a supplement to a manual search. Also, since I believe that computer searching will eventually virtually replace manual searching, you may be able to avoid a bad case of "future shock" by becoming familiar with computer searching now.

Computer search systems don't show the drawings of any prior patents. Nor do they use the PTO classification system. Instead, they search solely for combinations of key words in the texts—claims, abstract or title—of prior patents. For example, suppose you've invented the bike with a frame made of a certain carbon-fiber alloy. To make a manual search, you would look through the patents in the bike and metallurgical (carbon-fiber alloy) classifications, hoping that if a relevant patent exists, someone would have classified it in either or both of patents. However, to produce a truly effective patent search report, these companies will have to find some way, such as using optical character recognition technology, to incorporate the data from earlier patents into their data banks. It doesn't take much imagination to realize what will happen when all patents and possibly other literature are added to the data banks, when the computers can also display and print out the drawings of patents, and when more terminals become available in libraries, service centers, etc. When this occurs, computer searching will be faster and more thorough, as well as being independent of the PTO's classification system, which is subject to human error and troubled by missing patents.

Computer searching is presently used by the PTO's examiners to supplement their searches. As a result, we're getting better examinations and stronger patents. When computer searching is perfected and completed, I believe that patent application pendency time will be reduced from its present level of about 1.5 to 3 years to about six months or less and that, more importantly, hardly any patent will ever be questioned for validity, i.e., almost all patents will be virtually incontestable. (See Chapter 15 for more on patent validity.)

1. Available Computer Search Resources

Now that you get the general idea, how do you go about supplementing your manual search with a computer search? There are two ways to gain access to a computer search service's data bank:

- Via a personal computer (or terminal) with a modem—in this case you'll have to make a suitable agreement with a service; or

- Via an existing terminal which is dedicated to patent searching, such as at a PDL, large company, or law firm.

While all of the six computer search services listed below provide the capability of searching the prior art to determine whether an invention is novel, they also provide other "patent search" capabilities, such as searching for all of the patents issued to a specific inventor, searching for all of the patents assigned to (legally owned by) a specific company, searching for a list of all of the patents in a specific search class, etc. Hence, it's important to be sure, when you sign up for a "patent search," that you make it clear that you need the capability of making a novelty search, the process I describe in this chapter.

Here are the six computer services which provide patent searching, together with their telephone numbers:

Dialog (Claims): 800-334-2564
Bibliographic Research Service (BRS): 800-833-4707
Mead Data Central (Lexpat): 800-543-6862
Permagon: 800-336-7575
Questel: 800-424-9600
Systems Development Corp. (SDC): 800-336-7575

Call any of these and they will be delighted to send you information about their patent search services, or advise you where you can access a terminal near you. If you only need to make a one-time search, clearly it won't pay to make an ongoing agreement with a company for their services. If you want to use your personal computer, or a company computer, as a patent search terminal, you'll have to sign a contract with one of the services. The good news is that all of the services except one have no sign-up or monthly minimum charge, although some services levy a one-time charge of about $50 for their instruction manual. The bad news is that the one with the sign-up and monthly minimum charge, Mead Data Central, is by far and away the best service, since it searches and can send back the full text of any patent in its data base; all of the others search only the abstract or claims of the patents.

All services charge for the amount of time which you use to search their database; the fee runs from about $10 per hour for off-time (night/weekend) searching to $100 per hour for weekday use. These fees include telephone toll costs. A typical search will cost about $50 to $100.

While I'm reluctant to recommend any vendors in this book because of changing conditions, at present there is no question in my mind that Mead is the unqualified "best." If you can gain access to a terminal with a Lexpat search capability near you, go no further. If not, all of the other databases are in my opinion about even in their capabilities.

2. Vocabulary Associated With Computer Searches

How do you use a database? Assuming you're going to do the search yourself, first thoroughly study the service's instruction manual so that you'll be able to conduct your search in as little time as possible, thereby minimizing user-time charges. While every system is different, and while space constraints preclude coverage of them all, the following usage terms are common to all systems. If you're going to do any patent searching, you should learn these terms now.

- A *File* is the actual name of the patent search database provided by the service; for example LEXPAT is the name and trademark for Mead Data General's patent search database; CLAIMS is Dialog's patent search file.
- A *Record* is a portion of a file; the term is used to designate a single reference, usually a patent within a database.
- A *Field* is a portion of a record, such as a patent's title, the names of the inventors, its filing date, its patent number, its claims, etc.
- A *Term* is a group or, in computerese, a "string," of characters within a field, e.g., the inventor's surname, one word of the title of a patent, etc., are terms.

- A *Command* is an instruction or directive to the search system which tells it to perform a function. For example, "Search" might be a command to tell a system to look for some search key words in its database.
- A *Key Word* or a *Search Term* are the words which are actually searched. "Bicycle" and "carbon fiber alloy" are the key words for our example above.
- A *Qualifier* is a symbol which is used to limit a search or the information which the search displays for your use. Normally no qualifier would be used in novelty searches, but if you're looking for a patent to a certain inventor, you could add a qualifier which limits the search to the field of the patentee's name.
- A *Wild Card Symbol* is an ending (familiar to users of sophisticated word processing programs) which is used in lieu of a word's normal ending in order to to broaden a *key word*. The wild card cuts off immaterial endings so that only word roots are searched. For example, if we were searching Millie's annular napkin shaping ring, we would want our search to include the words "annular" and "annulus." Thus instead of using both key words and the *Connector Symbol* "or" (see below), we might search for "annul*" where "*" was a wild card symbol which tells the computer to look for any word with the root "annul" and any ending.
- *Connector Words* are those (such as "or," "and," and "not") which tell the computer to look for certain defined logical combinations of *key words*. For instance, if you issued a *command* telling the computer to search for "annulus or ring and napkin," the computer would recognize that "or" and "and" were connector words and would search for patents with the words "annulus" and "napkin," or "ring" and "napkin," in combination.

Obviously, the use of more *key words* joined by the "and" connector will narrow your search since it will add more *key words* to the search; this will cause the computer to pull out fewer patents, because only patents with all of the *key words* connected by "ands" will satisfy your search request. However the use of more *key words* joined by the "or" connector will broaden your search, since any patent with any one of the *key words* joined by an "or" will be selected. The "not" connector is seldom employed, but it can be used to narrow a search when you want to eliminate a certain class of patents which contain an unwanted *key word*.

- *Proximity Symbols* are those which tell the computer to look for specified *key words*, provided they were not more than a certain number of terms apart. Thus, if you told the computer to search for "napkin w/5 shaping" it would look for any patent which contained the words "napkin" and "shaping" within five words of each other, the symbol "w/5" meaning "within five words of." If no proximity symbol is used and the words are placed adjacent each other, e.g., "napkin shaping," the computer will pull out only those patents which contain these two words adjacent each other in the order given. However if a *connector word* is used, e.g., "napkin and shaping," the computer will pull out any patent with both of these words, no matter where they are in the patent and no matter in what order they appear.

3. Think of Alternative Search Terms

Before you even approach the computer, no matter what search system you use, be well prepared with a well-thought-out group of key words and all possible synonyms or equivalents. Thus, to search for Millie's napkin-shaping ring, in addition to the obvious key words, "ring," "annular," "napkin," and "shaping," think of other terms from the same and analogous fields. In addition to napkin, you could use "cloth." Or, in addition to shaping, you could use "folding" or "bending." In addition to "annulus" or "ring," you could try "device," etc.

4. Using the Computer

From here on, simply follow the instructions in the service manual for operating the computer and gaining access to the database. Once you find relevant patents, analyze them as instructed earlier in this chapter. Good luck and smooth searching!

Chapter 7

Deciding How to Protect Your Invention

NOW THAT YOU have a pretty good idea of the patentability and commercial status of your invention, it is time to make a plan for protecting it to the maximum extent possible under the law. While you might think that your next step would be to prepare and file a patent application, you would be wrong in doing so without first considering the information in this chapter.

I've provided a Decision Chart (Fig. 7-A) to simplify and organize your alternatives. It consists of twenty-three boxes with interconnecting lead lines. The numbered, light-lined boxes (even numbers from 10 to 40) represent various tasks and decisions on your route to making decisions on available options. The lettered, heavy-lined boxes (A to F and X) represent your actual options.

The numbers in parentheses in the following discussion refer to the boxes on the chart. While there are seven options, several of these can be reached by several routes. Accordingly, the following discussion is divided into more than seven sections.

A. Drop It If You Don't See Commercial Potential (Chart Route 10-12-14-X)

THIS ROUTE HAS already been covered in Chapter 4, but in order to acquaint you with the use of the chart, I'll review it again.

Referring to the chart, assuming that you've invented something (Box 10—Chapter 2) and recorded it properly (Box 12—Chapter 3), you should then proceed to build and test your invention as soon as practicable (Box 12). If building and testing would present appreciable difficulty, you should wait until after you evaluate your invention's commercial potential (Box 14—Chapter 4), or patentability (Box 16—Chapter 5). But always keep the building and testing as a goal; it will help you to evaluate commercial potential and may be vital in

the event an "interference" occurs (different persons seek patents for the same invention). What's more, as you'll see in Chapter 11, you'll find a working model extremely valuable when you show the invention to a manufacturer.

Your next step is stated in Box 14—that is, investigate your invention's commercial potential using the criteria of Chapter 4. Assuming you decide that your invention has no commercial potential, your answer to the commercial question is "No," and you would thus follow the "No" line from Box 14 to the ultimate decision Box X which says "Invent something else," as already covered in Chapter 4. See how easy it is?

B. File Your Patent and Sell It to or License a Manufacturer If You See Commercial Potential and Patentability (Chart Route 14-16-18-20-22-A)

FILING A PATENT APPLICATION and selling rights to the invention is the usual route for most inventors. This is because inventors seldom have the capability to establish their own manufacturing and distribution facilities. If a) your invention has good commercial potential (Box 14), b) your decision on patentability is favorable (Box 16), c) you're able to prepare a patent application (Box 18) (or have one prepared for you), and d) you don't wish to manufacture and distribute your product or process yourself (Box 20), your next step is to prepare a patent application (Box 22). After you prepare the patent application, you should then try to sell your invention (and accompanying patent application) to the manufacturer, as stated in Box A.

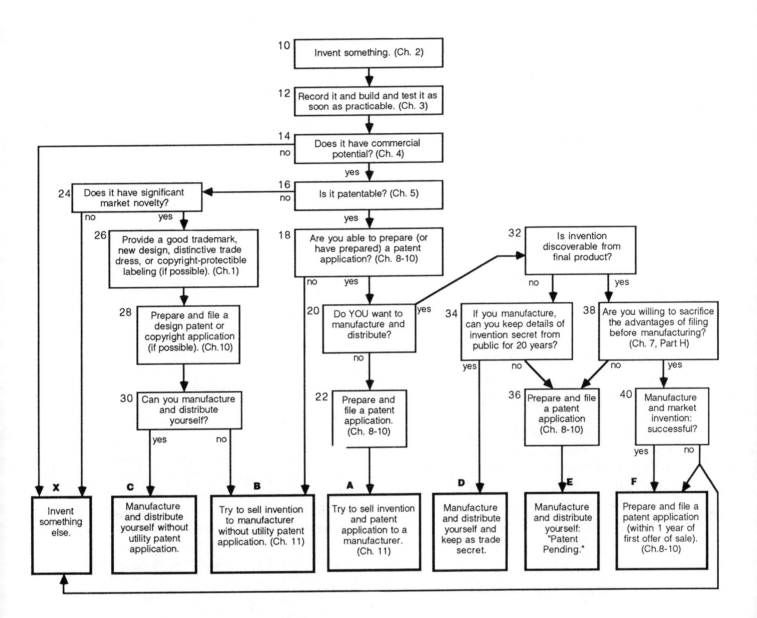

Fig. 7-A—Invention Decision Chart

Why file a patent application before offering the invention to a manufacturer? A good question, which has four good answers. Let's look at each one individually.

1. Create Protection for Your Invention

By preparing and filing a patent application, you've defined your invention and its ramifications in very precise terms, made formal drawings of it, and established your claim to it in the PTO. Thus anyone who later sees the invention wants to steal or adopt it would have to engage in elaborate and (usually) illegal preparations. And, the would-be thief will have filed after you, a serious disadvantage. Thus once you file the application, you may publish details of your invention freely and show it to anyone you think may have an interest in it (unless you've chosen to maintain your invention as a trade secret during the pendency of the patent application process—see Part E below).

2. Respect for Your Invention

A manufacturer to whom you show the invention, seeing that you have thought enough of your invention to take the trouble to prepare and file a patent application on it, will treat it, and you, with far more respect and give it much more serious consideration than if you offer an unfiled invention.

3. You Can Sign a Manufacturer's Waiver

As you'll see in Chapter 11, most manufacturers to whom you offer an invention will not deal with you unless you first waive (give up) certain potential claims that might arise from the transaction (such as

your being able to charge the manufacturer with stealing your idea in the event this occurs). Simply put, signing a waiver if you haven't already filed a patent application will put you at the mercy of the company to whom you show your invention. Fortunately, however, such waivers do not involve your giving up your rights under the patent laws. Thus, having a patent application on file, in this context, affords you powerful rights against underhanded dealing by the manufacturer (assuming the patent subsequently issues).

4. You'll Be Offering More So You'll Get More

Most manufacturers want a proprietary or privileged position, i.e., a position which entitles them to a commercial advantage in the marketplace which competitors can't readily copy and obtain. A patent provides a very highly privileged position: a 17-year monopoly. Thus if you have a patent application which already covers your invention, manufacturers may be far more likely to buy your invention (with its covering patent application) than if you offered them a "naked" invention on which they have to take the time and trouble to file a patent application for you themselves.

An Exception: Although, as stated it's usually best to file your patent application as soon as possible, it may be to your advantage to delay and keep the invention secret or take your chances approaching manufacturers "naked" if your invention is so innovative that it's not likely to be commercialized for many years. Gordon Gould, the inventor of the laser did this unintentionally when he filed his patent application years late because he mistakenly believed he needed a working model to file. His mistake worked to his great advantage, however, since his delay postponed his monopoly period so that it coincided with the laser's commercial period, thereby turning an otherwise worthless patent into pure gold.

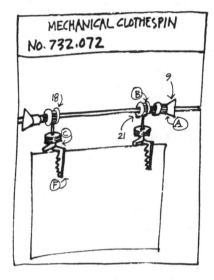

C. Sell or License a Manufacturer Without Filing, If You Have Commercial Potential Without Patentability (Chart Route 16-24-26-28-30-B)

IF YOUR INVENTION ISN'T PATENTABLE (i.e., the decision in Box 16 is negative), don't give up; there's still hope. Many fortunes have been made on products which weren't patentable. For instance, the Apple computer made its designer-promoters, Jobs and Wozniak, multimillionaires, yet lacked any significant inventive concepts and never was awarded a major patent. Ditto for Henry Ford's automobile and George Eastman's camera.

Thus you should now decide, on the basis of your commercial potential and patentability evaluations, whether your invention nevertheless possesses "significant market novelty" (Box 24). If so, it may in fact be quite profitable if introduced to the market.

Put differently, if your patentability search produces close prior art, but not a dead ringer, this indicates that probably no one has tried your specific, particular idea before, although someone has come close enough to preclude you from getting a patent. However, if you feel, looking back on your commercial-potential and patentability evaluations, that it doesn't have significant market novelty, i.e., there's little chance of commercial success, there isn't much hope and you'll have to try again (Box X).

Assuming that your invention does have significant market novelty (Box 24) but does not qualify for protection under a utility patent, there are several ways that you can use to obtain proprietary rights on your invention and make it more attractive to potential purchasers. Let's take a closer look at some of these.

1. Provide a Clever Trademark

One good way to make your invention more attractive is to provide a clever trademark for it (Box 26). As stated in Chapter 1, Section N, a trademark is a brand name for a product. An excellent type of brand name is one which suggests the function of the product in a very clever way. A clever trademark can be a very powerful marketing tool, i.e., a tool which will greatly enhance the value and salability of your invention and give you added proprietary rights to sell to a manufacturer. Examples of clever, suggestive trademarks are *Water Pic* for an oral irrigation device and *Hushpuppy* for shoes. Also consider *Sunkist* citrus fruit, *Shasta* soft drinks, *Roach Motel* roach traps, *Heavyhands* exercise weights, *Sun Tea* beverage containers, and *Walkman* portable tape players.

2. Obtain a Design Patent

If the invention that fails to qualify for a utility patent is a tangible product, the second trick to obtaining proprietary rights is to give it a distinctive design (Box 26). Then, perhaps, a design patent can be obtained. By distinctive design, I mean a shape or layout that is unique and different from anything you've seen so far. The design, in this case, doesn't mean the function or internal structure of the product, but only its outward, non-functional, ornamental, aesthetic shape or layout which makes it distinct visually.

For example, the D-shaped *Heavyhands* weights and Dizzy Gillespie's trumpet with its upwardly-bent bell section are excellent examples of valuable design inventions. If you've invented a computer, a new case shape can be a design invention. For a bicycle, a new frame shape design would be a design invention. From abacuses to zithers, from airplanes to zippers, almost every humanly-made object under the sun can be redesigned or reshaped in a new way so that it can be covered by a design patent.

However, remember from Ch 1 that for a design patent to be applicable, the new features must be for aesthetic or ornamental purposes and not have any significant functional purpose—otherwise the PTO will reject it as non-ornamental, i.e., only a utility patent will be appropriate. Also, the design must be inseparable from the object and not merely surface ornamentation. In the latter case, copyright is the proper form of coverage (see Chapter 1. Section O). For example, the label design on a jar of juice can not be protected by a design patent, but a new shape for the jar would qualify for one. If you do come up with a distinctive design, you should, of course, record it in the same manner as you recorded your invention (see Chapter 3). And as with your invention, you should build a prototype or model as soon as practicable. You should also prepare and file a design-patent application (Box 28) on the ornamental appearance (not workings) of your invention.

As stated in Ch. 6, unless you live near the PTO or a Patent Depository Library, it doesn't pay to search a new design beyond the most cursory look in product catalogs. This is because the cost of the search will greatly exceed the cost and effort to prepare and file a design-patent application. As you'll see in Chapter 10, a design-patent application consists simply of a drawing and a few forms that you fill out; it's very easy and economical to prepare.

3. Provide Distinctive "Trade Dress"

IF YOU CAN'T come up with a new design (or even if you can), you can still enhance the proprietary value of your invention by providing it with a distinctive "trade dress," such as a special, uniform color (as Kodak does with its yellow film packages), a special "certificate of authenticity" (if appropriate) as some manufacturers do with their replicas of antique objects, and/or a unique advertising slogan. This type of enhanced uniqueness is not different or special enough to qualify for a utility patent, design patent, copyright, or trademark. However, you can acquire offensive rights, at least before it is made public, under trade secret law. See Chapter 1, Section P. And the law of unfair competition may provide some rights once it is commercially unveiled (Chapter 1, Section Q). Be sure to record the trade dress properly (see Chapter 3) before showing it to anyone and be sure to use it (or have it used) consistently and as much as possible after marketing.

4. Provide Copyrightable Labeling

Look closely at some of the packaged products that you see in your home or on display in a store for a copyright notice, e.g., "© 1980 S.C. Johnson & Son, Inc." This copyright is intended to cover either the wording on the label or container, the artwork thereon, or both. While relatively easy to design around (i.e., come up with a close but non-infringing alternative), unique labeling with a copyright notice nevertheless provides a measure of offensive rights which is well worth the small effort it takes to invoke. Many market researchers have shown that an attractive label can make all the difference in the success of a product. Accordingly, it can pay, if you're marketing a packaged product, to spend some effort, either on your own or in hiring a designer, to come up with an attractive, unique label, affix a copyright notice, and apply for registration (see Chapter 1 Section O).

D. Make and Sell Your Invention Yourself Without a Utility Patent Application (Chart Route 30-C)

HERE WE ASSUME again that you have an un-patentable invention. If you can make and distribute it yourself (Box 30), it's better to do so (Box C) than to try to sell it to a manufacturer outright. Even if you have a trademark (even a good one), a design patent application, distinctive trade dress, and/or a unique label, the absence of a utility patent

application means a manufacturer does not get a really good privileged position, and so will generally not be as inclined to buy your invention. However, if you decide to manufacture the invention yourself, and you reach the market first, you'll have a significant marketing advantage despite the lack of a utility patent. Also, since you're the manufacturer, you'll make a much larger profit per item than if you received royalties from a manufacturer.

E. Manufacture and Distribute Your Invention Yourself, Keeping It as a Trade Secret (Chart Route 20-32-34-D)

EVEN THOUGH YOUR INVENTION may be commercially valuable and patentable, it isn't always in your best interest to patent it. The alternative, when possible, is to keep an invention a trade secret and manufacture and sell the invention yourself, e.g., by direct mail marketing, broadcast or periodical advertising, possibly eventually working your way up to conventional distributors and retailers. As explained in Chapter 1, Section P, a trade secret has numerous advantages and disadvantages. While an invention can be maintained as a trade secret right up until the time a patent actually issues, once it does, the trade secret is lost through the mandatory public disclosure associated with the patent process. Conversely, if you either don't get a patent or choose to not pay the issue fee, your invention will remain a trade secret as long as you continue to treat it as one.

Remember that you can't maintain trade secret rights on an invention unless it's of the type which can't be discovered from the final product, even if sophisticated reverse engineering is used. One good example of an invention which was kept as a trade secret is the formula used in the *Toni* home permanent wave kit. Its inventor, Richard Harris, manufactured and sold the unpatented invention through his own company for many years, making large profits, and thereafter sold his business for $20 million when he decided to retire.

Although not specifically covered on the chart, there is another possibility in the trade secret category. That is, you may sell your invention to a manufacturer who may chose to keep it as a trade secret. This may occur with either unpatentable or patentable inventions (Chart routines 16-24-26-28-30 B or 16-18- 20-22 A), but you don't have to worry about this alternative since it's the manufacturer's choice, not yours.

If you've already filed a patent application and a manufacturer buys the patent application with a view to using your invention as a trade secret, the manufacturer can simply allow the patent application to go abandoned so it won't be published, thereby maintaining the trade secret. While you may lose the ego boost of a possible patent, your bulging wallet should provide adequate alternative compensation.

Note: You shouldn't refer to your abandoned patent application in any other application that will issue as a patent since anyone can gain access to an abandoned application that's referred to in a patent.

with the notice "patent pending" affixed to the invention (Box E).

While the patent application is pending, you should not publish any details of your invention since, if the patent application is finally rejected, you can allow it to go abandoned and still maintain your trade secret, as discussed above. Remember, by law, the PTO must preserve patent applications in secrecy, and, in practice, is very strict in this regard. Outsiders have no access to any pending patent applications and PTO personnel must keep patent applications in strict confidence.

The patent-pending notice on your product does not confer any legal rights, but it is used by most manufacturers who have a patent application on file in order to deter potential competitors from copying the invention. The notice effectively warns them that you may get a patent on the product so that if they do invest the money and effort in tooling to copy the invention, they could be enjoined from further manufacturing, with a consequent waste of their investment. However, make sure you don't use a patent-pending notice with a product which is not actually covered by a pending application: to do so is a criminal offense.

F. File Patent Application and Manufacture and Distribute Your Invention Yourself (Trade-Secretable Invention) (Chart Route 20-32-34-36-E)

SUPPOSE YOUR INVENTION is not discoverable from your final product (Box 32) so that you can keep it secret for a while, but not for the life of a patent (Box 34). Or, suppose, after evaluating the advantages and disadvantages of a trade secret under the criteria above, you don't wish to choose the trade-secret route, preferring instead to patent your invention. You should then prepare and file a patent application (Box 36) (see Chapters 8 to 10) and then manufacture and distribute the invention yourself

G. File Patent Application and Manufacture and Distribute Invention Yourself (Non-Trade Secretable Invention) (Chart Route 20-32-38-36-E)

THIS WILL BE THE ROUTE followed by most inventors who wish to manufacture their own invention. Assume that the essence of your invention, like most, is discoverable from the final product (Box 32), and assume that it's cheaper to file a patent application than to manufacture and sell products embodying the invention yourself (Box 38). Alternatively, assume that you don't want to sacrifice the advantages of filing before manufacturing. In either case, you should prepare and file a patent application (Box 36) and then manufacture and distribute the invention yourself with the patent-pending notice (Box E).

H. Test Market Before Filing (Chart Route 20-32-38-40-F)

ALTHOUGH I KNOW you'd like to manufacture and test market your invention before filing a patent application on it, I generally don't recommend this for patentable inventions because of the following:

1. You have less than one year to do the test marketing because of the "one-year rule" (Chapter 5).

2. You may get discouraged unjustifiably if you try to market your invention and you aren't successful; i.e., you probably will be too discouraged to file a patent application and therefore you'll lose all rights on the invention forever.

3. You'll lose your foreign rights since most foreign countries or jurisdictions, including the European Patent Office (see Chapter 12) have an absolute novelty requirement, which means that if the invention was made public anywhere before its first filing date, such publication will prevent the issuance of a valid patent.

4. There is a possibility of theft since anyone who sees it can (assuming it's not trade secretable) copy it and file a fraudulent patent application on it.

5. There are business disadvantages when:

- the product has a short or seasonal selling period or limited market life,
- test marketing would disclose an easily-copyable product to competitors,
- the cost of test marketing would be so high as to outweigh the risk of regular marketing,

- the product is merely a response to competition, or
- market conditions in the field are changing so fast that the results of a market test would soon be obsolete.(Wall St. Journal, 1984 Aug. 27, p. 12.)

So, assuming your invention is discoverable from the final product (Box 32), ask yourself whether it's easier and cheaper to manufacture and test market it than to file a patent application. If it is, and if you're also willing to sacrifice the above five advantages of filing before manufacturing (Box 38), and the above business disadvantages don't apply, you can manufacture and market your invention (Box 40) before filing. If you discover, within about nine months of the date you first introduce your product, that it is a successful invention and likely to have good commercial success, begin immediately to prepare your patent application (Box F), so that you'll be able to get it on file within one year of the date you first offered it for sale or used it to make a commercial product.

If your manufacturing and market tests (Box 40) are not successful, you should generally drop the invention and concentrate on something else (Box X), although you still have the right to get a patent on your invention. Thus, if the market test is unsuccessful, but you feel that you don't want to give the invention up forever, by all means follow the line, and prepare and file the patent application within one year of the first offer of sale (Box F). If you do manufacture and market your invention, and then later file a patent application on it, be sure to retain all of your records and paperwork regarding the conception, building, testing, and manufacturing of your invention; these can be vital if you ever get into an interference. See Chapter 13, Section K.

Now that we've covered all possible routes on the chart, I hope you've found one which will meet your needs. If your choice is to file a patent application, move on to Chapters 8 to 10; if you want to try to market your invention first, skip over to Chapter 11. Chapter 10 also covers design patents.

Chapter 8

HOW TO WRITE AND DRAW IT FOR UNCLE SAM

THIS AND THE NEXT TWO CHAPTERS are the heart of this book: they cover the writing and transmittal of your patent application to the Patent and Trademark Office (PTO). This chapter provides an overview of the patent application drafting process, particularly with reference to the specification and drawing. Chapter 9 explains how to draft patent claims (sentence fragments that delineate the precise scope of the patent being sought). Chapter 10 explains how to "final" the application as well as the precise steps involved in transmitting it to the PTO. In addition, Chapter 10 covers design patent applications.

Because these subjects can be difficult to understand in the abstract, I use concrete examples throughout. And, at the end of this chapter, you'll find the specification (including the abstract) and formal drawings of a sample patent application. Similarly, at the end of Chapter 9, you'll find the patent claims of this same application. The completed formal papers for this application appear at the end of Chapter 10.

A. Lay Inventors Can Do It

IT'S A COMMON MYTH that a lay inventor won't be able to prepare a patent application, or prepare it properly. Having worked with many lay inventors I dispute this vigorously. I have found that lay inventors can and have done very good jobs, often better than patent attorneys, by following this book. To prepare a proper patent application, you should be mainly concerned with three considerations:

1. The specification (description and operation of your invention and drawings) should be detailed enough so that there will be no doubt that one skilled in the art will be able to make and use the invention after reading it.

2. The main claim should be as broad as the prior art permits.

3. You should "sell" your invention by stressing all of its advantages.

If you satisfy these three criteria, you'll be home free. All the other matters are of lesser import and can be fixed if necessary. I'll show you how to satisfy these three main criteria in this and the next chapter. Now let's get started by looking at what's contained in a patent application.

B. What's Contained In a Patent Application

A PATENT APPLICATION CONSISTS OF:

1. A self-addressed receipt postcard;

2. A transmittal letter;

3. A check for the filing fee (see Fee Schedule in Appendix)

4. A drawing or drawings of the invention—either formal or informal;

5. A "specification" containing the following sections:

   a. Title of the invention;

   b. Background of the invention. This includes the field of invention, cross-references to related applications, if any, and a discussion of the relevant prior art (previous relevant developments in the same technological area);

   c. Objects and advantages of your invention;

   d. A brief description of the drawing figures;

   e. A list of reference numerals (optional but desirable). (Reference numerals are the numbers which you'll use on your drawings to designate the respective parts of your invention, e.g., 10 = motor, 12 = shaft, etc.);

   f. A narrative description of the structure of the invention;

g. An explanation of how the invention works or operates;

h. A conclusion, discussion of ramifications, and one or more broadening paragraphs (i.e., a summation of the invention's advantages, the alternative physical forms which it can take, and an indication that it shouldn't be limited to the particular form(s) shown);

6. The claims (precise sentence fragments that delineate the exact nature of your invention—see Chapter 9);

7. The abstract (a brief summary of what the invention is and how it works, technically considered part of the specification);

8. A completed patent application declaration form (statement under penalty of perjury that you're the true inventor and that you acknowledge a duty to keep the PTO informed of all material information and prior art related to your invention.);

9. A small entity declaration if you're an individual and you haven't agreed, transferred (or agreed to transfer) ownership or license to a large entity (Form 10-3). If any owner of the patent application is an individual other than the applicant-inventor, a non-profit organization, or one with 500 or fewer employees, an additional declaration for such owner is required—see Chapter 16;

10. An Information Disclosure Statement and List of Prior Art Cited By Applicant., and copies of such prior art. Technically these aren't part of the patent application, but since they're supposed to be sent to the PTO with or soon after the application, I've included them here. These inform the PTO of relevant prior art or any circumstances known to you which may potentially affect the novelty or obviousness of your invention.

C. What Happens When Your Application Is Received by the PTO?

WHEN YOUR APPLICATION ARRIVES at the PTO, their clerical personnel will deposit your check, put all of your papers in a folder (termed a "file wrapper"), assign a filing date and serial number to your application, stamp this information on your postcard, and return it, later send you an official filing receipt, and forward your file to an appropriate examining division. When its turn is reached (within a few months to two years), your application will be reviewed by an examiner who will allow the application (rare) or, more commonly, send you an "Office Action." The Office Action will either object to your specification and/or drawing and/or reject some or all of your claims because of imprecise language and/or because of unpatentability over the prior art. To overcome these objections and/or rejections, you'll have to submit an "Amendment" (Ch. 13) in which you:

1. Make changes, additions, or deletions in the drawings, specification, and/or claims; and/or

2. Convince the examiner that the Office Action was in error.

If the examiner eventually decides to allow the application (either as originally presented or as amended), you'll be given three months to pay an issue fee. Your specification and claims, along with certain other information (your name, address, and a list of all prior art cited by the examiner), will then be sent to the U.S. Government Printing Office; there they'll be printed verbatim as your patent. From filing to issuance, the process usually takes somewhere between one and a half to three years, sometimes longer.

Model Of Invention: You never have to furnish or demonstrate a working model of your invention. (However, in rare cases, if the examiner questions the operability of your invention, e.g., if you claim a perpetual motion or energy machine, one way for you to prove operability is by demonstrating a working model.)

D. Do Preliminary Work Before Preparing Your Patent Application

BEFORE YOU BEGIN the actual writing of your patent application or prepare any of the forms that go along with it, it's wise to make thorough preparations. Having worked on many patent applications, I can tell you that if adequate preparations are made beforehand, the actual writing of the application rarely takes more than several partial days. Here are the basic preparatory steps:

1. Review the Prior Art

Assemble all your prior-art references, including any references gleaned from textbooks, magazines, or journals you've searched or discovered that are relevant to your invention or to the field of your invention. Read each of these references carefully, noting the terms used for the parts or steps that are similar to those of your invention. Write down the terms of the more unusual parts and, if necessary, look them up in your prior-art patents, textbooks, magazine articles, the Glossary at the end of this book, or the *What's What* book (see Appendix) so that you'll be familiar with them and their precise meaning. Also, note the way the drawings in these prior-art references are arranged and laid out, paying particular attention to what parts are done in detail and what parts need be shown only very roughly or generally because they are well-known or are not essential to the invention.

2. Review Your Disclosure

In Chapter 3, I strongly advised that you prepare a description (with sketches) of your invention and have this signed and witnessed, either in a laboratory notebook or on a separate piece of paper, called an invention disclosure. Review this now to be sure you have all of the details of your invention drawn or sketched in understandable form and that the description of your invention is complete. If you haven't done this yet, do it now, referring to Chapter 3 when necessary.

3. Ramifications

Write down all of the known ramifications (potential different uses and methods of operation) and embodiments (potential forms in which the invention can occur). That is, record all other materials that will work for each part of your invention, other possible uses your invention can be put to, other possible modifications of your invention, ways in which its size or shape can be altered, parts (or steps in its manufacture) that can be eliminated, and so on.

The more ramifications and embodiments you can think of, the broader your patent can be claimed, and the more you'll be able to block others from obtaining patents either on devices similar to your invention, or on improvements to it. Also, you'll have something to fall back on if your main or basic embodiment is "knocked out" by prior art which your search didn't uncover or which surfaced after your search.

For instance, if your invention is a delaying device which you use to close the lid of a box automatically a few moments after the lid is opened, another embodiment which could make advantageous use of the delaying device might be in a "roly poly man" toy to make the man stand up again automatically a few moments after he's tipped over.

If you have two related inventions, such as a car radio mount and a housing for the same radio, I suggest that you put both in the same application since the examiner may allow both inventions at once and you'll save fees and effort. However, if the examiner requires you to restrict your application to one invention (Ch. 13, Sec. N), you can easily file a divisional application (Ch. 14, Sec. B) before the original application issues and still get the benefit of your original application's filing date.

4. Sources of Supply

If your invention contemplates the use of any exotic or uncommon materials, or components, or involves unusual manufacturing steps, obtain the names and addresses of potential suppliers and/or identify textbooks or other references outlining how one should obtain or make such unusual elements or procedures. Describe these unusual dimensions, materials or components in detail.

For example, with an electrical circuit, you generally don't have to include the technical values or identifications of components. However, if the

operation of the circuit is at all unusual, or if any component values are critical, write down their names or identifications. With a chemical invention, write down the source or full identification of how to make any unusual components or reactions. With a mechanical invention, if any unusual parts, assembly steps, or materials are required, be sure you provide a full reference as to where to obtain or how to perform them.

The reason why you will need the full details of any special aspects of your invention is simple. Section 112 of the patent laws (35 USC 112) mandates that the specification must be a "complete, clear, and concise" description of the invention such that anyone skilled in the art can make and use it without too much effort. More on this later.

5. Advantages/Disadvantages

List all disadvantages of the relevant prior art that your invention overcomes, referring to the checklist in Chapter 4 (Form 4-2) to make sure your listing is complete. Then list all the advantages of your invention over the prior art, and all of your invention's general disadvantages.

Now that we have reviewed these vital preliminary steps, let's turn to writing the specification.

E. Writing Your Patent Specification to Comply with the Full Disclosure Rules

IN WRITING THE SPECIFICATION of a patent application, your goal is to disclose clearly everything you can think of about your invention. In case of doubt as to whether or not to include an item of information, put it in. The statutory provision that mandates the inclusion of all this information in your patent application is, Section 112 of the patent laws, paragraph 1, which reads as follows:

The specification shall contain a written description of the invention, and of the manner and process of making and using it, in such full, clear, concise and exact terms as to enable any person skilled in the art to which it pertains, or with which it is most nearly connected, to make and use the same, and shall set forth the best mode contemplated by the inventor of carrying out his invention.

As part of doing this, it may help if you keep well in mind the "exchange theory" of patents. The government grants you a patent (that is, a monopoly on your invention) for a limited term of years in exchange for your disclosing to the public the full details of your invention (i.e., how to make and use it) so that they'll get the full benefit of your creativity after your patent expires. Complete disclosure involves disclosing at least the "best mode" of the invention as contemplated by you, the inventor. So, if you have several different embodiments of your invention, make sure you identify the one you currently favor. If you can't decide which embodiment is the best, it's OK to list each embodiment and tell its relative advantages and disadvantages. For example, in the delay device referred to above, its use to close a box lid after a few minutes might be your preferred embodiment, and the delayed "roly poly man" might be an alternative embodiment. In this case you need merely state that the box is your preferred practical application of the delay device.

Another reason for disclosing as much as you can about your invention is, as stated, to block others from getting a subsequent improvement patent on your invention. If you invent something and disclose only one embodiment of it, or only one way to do it, and get a patent which shows only that one embodiment, someone may later see your patent and think of another embodiment or another way to do it that may be better than yours. This person will then be able to file a new patent application on this "improvement invention" and thereby, assuming a patent is issued, obtain a monopoly on the improvement. If this occurs you won't be able to make, use or sell the improvement without a license from the person who owns that patent. This is so even though you have a patent on the basic invention.

What happens if you don't put enough information in about your invention to enable "one skilled in the art" to make and use it without undue effort? Your entire application can either be rejected under Section 112 on the grounds of "incomplete disclosure," or it may be later invalidated if it is issued and challenged by an infringer when you try to enforce it. Also, if your patent application is rejected because of incomplete disclosure, usually there is nothing you can do since you aren't allowed to add any "new matter" to a pending application (see Chapter 13).

As mentioned earlier, you must provide enough information in your patent application to enable anyone working in the field of your invention to be able to build, without undue effort, a working model of your invention from the information contained in your patent application. However, to comply with this section, you ordinarily don't have to put in dimensions, materials, and values of components since the skilled artisan is expected to have a working knowledge of these items. However, as described above, dimensions, materials, or components that are critical to the performance of your invention, or that are at all unusual *must* be included. If in doubt, include this specific information.

Finally, having reviewed many patent applications prepared by laypersons, I find that the most common error in preparing the specification of a patent application is a failure to include enough detail about the invention, or enough ramifications. Thus, if you "sweat the details" like a good professional does, you'll seldom go wrong.

INVENTOR'S COMMANDMENT #7:
Your patent application must contain a description of your invention in such full, complete, clear, and exact terms, including details of your preferred embodiment at the time you file, so that anyone having ordinary skill in the field will be readily able to make and use it, and preferably so that even a lay judge will be able to understand it.

Software Note If your invention includes a microprocessor and an application program for it, either in software or in firmware, you should include a source or object code listing of the program with your patent application. If you don't have one, a detailed flowchart will do, so long as a programmer having no more than ordinary skill would be able to refer to your chart and then be able to write the program and debug it without undue effort or significant creativity, even if the task would take several months.

Microorganism Note: If your invention requires a microorganism or a fusion gene which is not widely available, you must make a deposit of your "special" bug or plasmid in an approved depository; see *MPEP* (*Manual of Patent Examining Procedure*) 608.01(p), referred to in the Bibliography.

F. First Prepare Sketches

BEFORE YOU EVEN BEGIN the actual nuts and bolts preparation of your specification, you should make (or have made for you) pencilled sketches of your invention. These will form the basis of the drawings you'll eventually send to the PTO along with your patent application (see Chapter 10, Part A). Your sketches will also be the foundation of your application. In other words, you'll build from these as you write your specification and claims.

The main reason I discuss sketches at this point is that you have to do your sketches prior to drafting the specification, as well as the other parts of the application. You don't have to worry about planning any layout of your figures on the drawing sheets, or the size of the figures—yet. This will be covered in detail in Chapter 10. For now, merely complete a set of sketches showing all of the aspects of your invention without worrying about size or arrangement; these sketch-figures can even be done very large and on separate sheets. Later on they can be reduced and compiled onto the drawing sheets as part of the "finaling" process (Chapter 10).

After you've completed your sketches, write down a name for each part adjacent to such part in each sketch, e.g., "handlebar," "handgrip," "clamp," "bolt," etc. Write the names of the parts lightly in pencil so that you can change them readily if you think of a better term. Use lead lines to connect each name to its part if the parts are crowded enough to cause confusion. If you have any difficulty naming any part, again, refer to the Glossary, your prior-art patents, or the *What's What* book.

Your drawing should be done in separate, unconnected figures, each one labeled ("Fig. 1, Fig. 2," etc.) so that all possible different views and embodiments of your invention are shown. Use as many views as necessary. Look at a relevant prior-art patent to get an idea as to how it's done. The views should generally be perspective or isometric views, rather than front, side, and top, engineering-type views. If you have trouble illustrating a perspective view, take a photo of a model of your invention from the desired angle and draw the photo, e.g., by enlarging and tracing it. Alternatively you can use a "see and draw" copying device of the type employing a half-silvered mirror in a viewing head on a pedestal; these are available in art supply stores and through gadget mail order houses. Hidden lines should be shown in broken lines, as shown in Fig. 8-A. For complicated machines, exploded views are desirable as shown in Fig. 8-B.

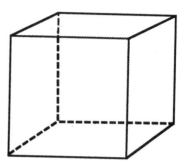

Fig. 8-A—Isometric View with Hidden Lines

You can use any reasonable symbols for mechanical, electronic/electrical, and chemical parts; the PTO has no requirements in this area, except that the symbols not be outrageous. I suggest you use conventional symbols, such as those approved by the ANSI (American National Standards Institute), those used in conventional texts, or those used in your prior-art patents. In lieu of graphical symbols, labeled boxes are also acceptable, so long as the part represented by the box is standard or conventional.

If possible, one figure of your drawing should be comprehensive enough to show the basic idea of the invention and to be suitable for inclusion in the *Official Gazette* where the details of your patent will be published if it is granted. See Chapter 6, Section K for more on the OG. The other figures can be fragmentary views, partial views, etc.; you don't have to show the same details more than once.

1. Machine Sketches

If your invention is a machine or an article, your sketches should contain enough views to show every feature of the invention, but you don't have to show every feature that's old and known in the prior art. For example, if you've invented a new type of pedal arrangement for a bicycle, one view can show your pedal arrangement in gross view without detail. Other views can show your pedal arrangement in detail, but you don't have to include any views showing the bicycle itself in detail, since it isn't part of your invention. If one figure of your drawing shows a sectional or side view of another figure, it is customary to provide cross-section lines in the latter figure; these lines should bear the number of the former figure. Look at prior-art patents to see how this is done. See the example in Fig. 8-C.

If your machine is complicated, you should show an exploded view of it; as in Fig. 8-B.

2. Chemical Composition Sketches

If your invention is a chemical composition, the PTO won't generally require drawings unless your invention is a material which has a non-homogeneous composition (i.e. it's internally differentiated through layering, etc), in which case you should show it in cross-sectional detail. Also if a step-by-step process is involved, the PTO will require a flow chart, even though the process is fully described in your specification; see next section. The reason: so future searchers will be able to understand your patent more rapidly. Benzene rings and other molecular diagrams can usually be presented in the specification.

3. Computer, Chemical or Mechanical Processes Sketches

If your invention includes a process of the electronic-computer, chemical, or mechanical type, you should provide a flowchart showing the separate steps involved, each described succinctly in a different block. If your blocks are connected, they should all be labeled as one figure; if disconnected, they should be labeled as separate figures. As before, each figure should be labeled, for example, Fig. 1, Fig. 2, Fig. 3, etc.

If you desire, you can provide a short title after each figure giving a general description of the part of your invention shown in the figure, just as you would do if you were writing a scientific article for an engineering magazine or textbook.

If you believe it will help in understanding your invention, you may (and should) include a drawing of the prior art as one figure of your drawings. This figure must be labeled "prior art" to indicate that it isn't part of your invention.

G. Drafting the Specification

ONCE YOU'VE REDUCED your invention to sketches, it's time to begin drafting the specification portion of your patent application. Review the specifications of your prior-art patents or the sample "spec." at the end of this Chapter to find out how they're written. Your specification should be written as one continuous document with separate sections, each with a heading, as in the following sections (except that "Title" should not be a heading). No legal or practical requirement exists which says that you must write like a lawyer, or use legalese. Such syntax is actually undesirable since it only makes your writing stilted and less clear; nothing reads as awkwardly as legal cant. The only legal

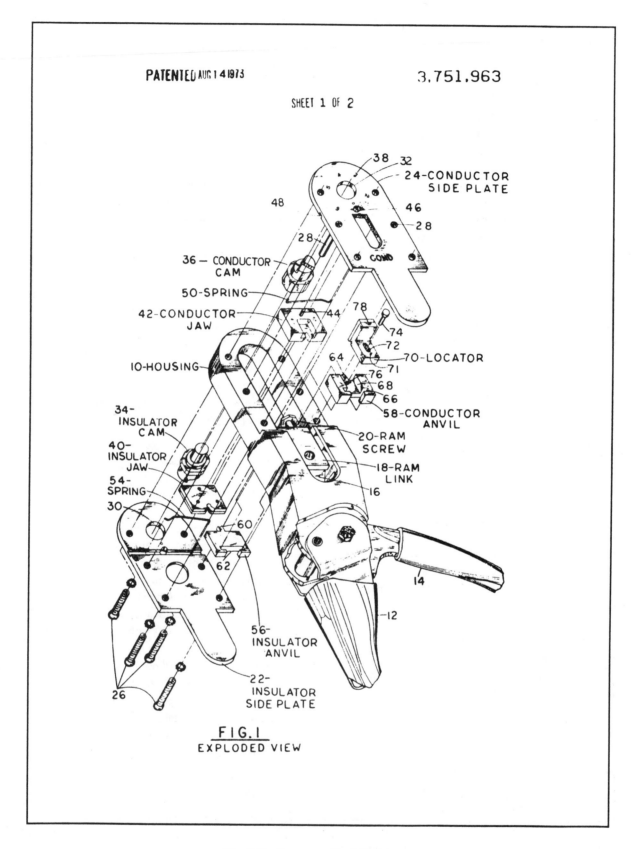

Fig. 8-B—Isometric Exploded View

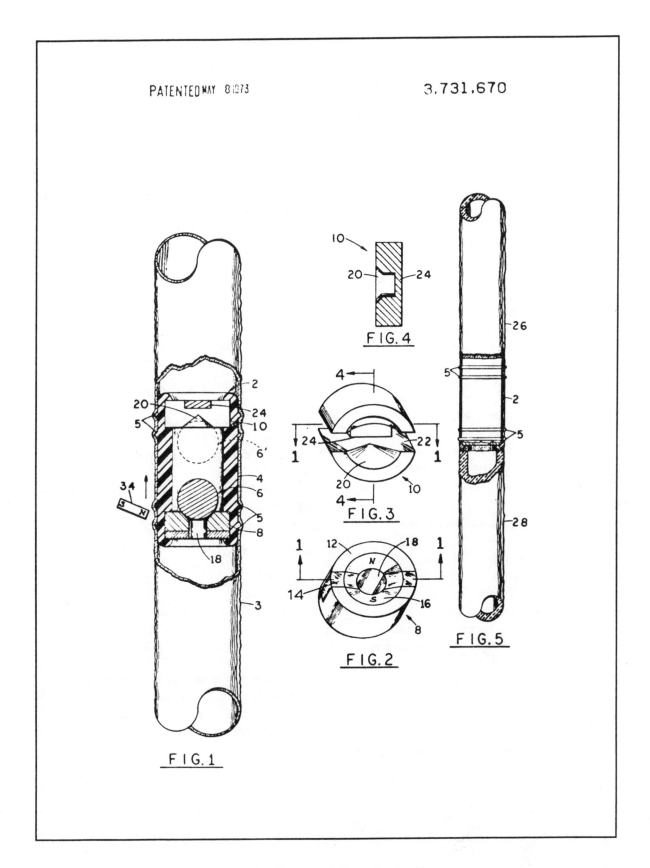

Fig. 8-C—Views with Cross-Section Lines

requirements for a patent specification are that it be a full, clear, concise, and complete description of how to make and use the invention. The claims, however, should be written with extreme clarity and precision, and to do this you may have to use a few "saids" and "wherebys," but I'll explain this fully in Chapter 9.

It's best to write your description in short, simple sentences, with short paragraphs, each paragraph generally being shorter than one page (double spaced) and relating to one part or subpart of your invention. The Cybernetics Institute has found that short sentences communicate best and that 50% of adults can't understand a sentence longer than 13 words anyway. Don't worry about the quality, style of your writing or the beauty of your language, so long as you include all points of substance of your invention and your description is clear and understandable. If you get stuck and don't know how to phrase a description of a part or an operation, simply pretend that you are talking to a close friend, recite it aloud, remember what you said (or make an audio recording), and write it down. Then go back and polish the language. If you attack the job in small chunks or in piecemeal fashion, it usually will go much easier.

Also if you use copious subheadings (e.g., "Fig. 1—Handlebar Attachment"; "Fig. 2—Front Fork Detail"; "Fig. 10—Operation Of Derailleur"; etc.) throughout your specification (as I've done in this book), most people will find it far easier to read. This allows them to take in the information in separate chunks which are easy to digest one at a time. Again refer to the specification at the end of this chapter (Fig. 8-E) to see how it's done.

Getting started is the worst part.

—Roberta Pressman

If you have trouble getting started, don't worry; many writers have blocks from time to time, and lots of inventors initially (and erroneously) lament, "I could never write my own patent application." The words of Lao-Tse will encourage you:

A journey of a thousand miles begins with a single step.

If you still feel daunted, it will help you to know that virtually all inventors who have trouble getting started suffer from *lack of will*, not *ability*. I had a client who came to the U.S. from Hong Kong with little money or English, but with a great invention and tremendous drive. He wrote and filed his own application and got a valuable patent, after I fixed his English. If he could do it, surely you, with probably a much better command of English, can do so also.

If you feel that you can't write that well, I suggest that you give it your best shot and then have a writer, college English major, high school English teacher, etc. edit your draft.

Now let's get to the nitty gritty of preparing the specification portion of a patent application.

Prior to starting, in order to guide your path, you will find it helpful first to make an outline, which should be the same as the headings of parts 2 to 10 below, except that you may want to make headings 8 and 9 (Description and Operation) more specific and/or break them into several more specific headings each, in accordance with your figures and specific situation.

1. Title

Have your title reflect the essence of your invention without being too long or so specific that it's narrower than your invention's full scope. On the other hand, don't pick a title so broad—such as "Electrical Apparatus"—as to be essentially meaningless. A look at some recently-issued patents in your field should give you a good idea of how specific to make your title.

HELPING HANDS HAT
No. 3,169,192.

2. Background—Field of Invention

Your first sentence should be a brief, one-sentence paragraph stating the general and specific field in to which your invention falls. The sentence might read, for example, "This invention relates to bicycles, specifically to an improved pedal mechanism for a bicycle." The field of your invention should be the technical, product, subject, or scientific area with which it's most nearly connected, such as bicycles, kitchenware, lasers, medical instruments, drugs, skiing, etc.

3. Background—Cross-Reference to Related Applications

If you've filed any other patent applications that are related to your present application, include this section and identify these applications here. If, as will likely be the case, you have no such corresponding applications, omit this section and heading entirely.

4. Background—Discussion of Prior Art

Discuss how the problem to which your invention is directed was approached previously (if it was approached at all), and then list all the disadvantages of the old ways of doing it. You can start this section out with the word "Heretofore." For example, "Heretofore, bicycle pedal mechanisms had . . . ," and then list the constructions which were used in the past and their disadvantages. Again, look at prior-art patents to get an idea of what was done.

INVENTOR'S COMMANDMENT #8:
In your patent application, you should "sell" your invention to the examiner or anyone else who may read the application by (a) listing all the disadvantages of the prior art, and (b) all the advantages of your invention, both in the introduction and in a conclusion.

While the PTO doesn't want needlessly derogatory remarks about the inventions of others, you should, as much as possible, try to "knock the prior art" here in order to make your invention look as good as possible. Keep your statements factual (e.g., "The derailleur in patent 3,456,789 to Prewitt, 1982 May 3, had a limited number of discrete gear ratios.") and not opinionated (e.g., don't say, "Prewitt's derailleur was an abject failure"). If applicable, you should tell why prior-art people didn't think of any solution before and why a solution is needed. You shouldn't discuss any detailed structure or operation of any prior-art in this section since detailed mechanical discussions without the benefit of drawings will be incomprehensible to most people. Occasionally you may have such a completely unique invention that there's really no prior art directly germane to your invention. If so, just state the general problem or disadvantage which your invention solves.

5. Objects and Advantages

In the patent field, the term "objects" means "what the invention accomplishes." Thus, you should list here all the things your invention accomplishes and its advantages over the prior art. You can start this section as follows: "Accordingly, several objects and advantages of my invention are . . . ." Then, include the objects or advantages of your invention. You already know what these are from your commercial evaluation (see Chapter 4) and the previous "Background— Description Of Prior Art" section. (Remember, you're still selling your invention—to the examiner, a potential licensee, and possibly to a judge!)

Include all of the positive factors of your invention from Form 4-2 (Chapter 4) and all the disadvantages of the prior art from your "Background" section. Avoid very narrow objects, such as "it is one object to provide a 14.5 mm carbon fiber lever," since these can be used to limit your invention. Similarly avoid objects which can't be accomplished, such as, "it is one object to provide a totally-safe bike," since this can be also used against you. At the end of this section, add a catch-all paragraph reading as follows: "Further objects and advantages of my invention will become apparent from a consideration of the drawings and ensuing description of it."

While it may seem needlessly repetitious to state the disadvantages of the prior art in the "Background" section and then repeat the converse of these in the "Objects" section, you'll soon find that this is but one instance of many where repetition is used in a patent application. For instance, the objects are effectively repeated again (a third time!) in the concluding paragraph of the specification. Moreover the parts of the invention are actually repeated five times, under "Description," under "Operation," in the Claims, in the Abstract, and in your list of reference numerals.

Why is repetition used so much? Because it's one of the keys to effective communication. There's an apropos old saying:

To communicate effectively to someone, you should first tell your listener what you're going to say, then actually say it, and finally tell the listener what you said.

6. Description of Drawings

Here provide a series of separate paragraphs, each *briefly* describing a respective figure of your drawing, for example, "Fig 1 is a perspective [or plan, side, exploded, or rear] view of my invention"; or, "Fig 2 is a view in detail of the portion indicated by the section lines 2-2 in Fig 1."

7. List of Reference Numerals

Although the PTO doesn't require or even recommend a separate list of the reference numerals and the names of their respective parts in an application, I strongly advise that you include such a list in a separately-headed section, as I've done in the sample specification at the end of the chapter. Why? There are three very important reasons:

- To help you to keep your reference numerals straight, i.e., to avoid using the same number for different parts.
- To help you to keep your nomenclature straight, i.e., to avoid using different terms for the same part.
- To provide a very visible and easy-to-find place where examiners, searchers, and others who read your application or patent can go to instantly identify any numbered part on your drawings.

I find it helpful to compile this list on a separate sheet of lined paper as I write the patent application, and then incorporate the list in the text.; I've provided a suitable worksheet as Form 8-1 in the Appendix.. Also, to keep confusion at a minimum, I advise that you don't use single-digit reference numerals, and that you begin your numbers with a number higher than your highest-numbered drawing figure, e.g., if you have Figs 1 to 12 of drawings, begin your reference numerals with number 20.

Lastly, I advise that you use even-numbered reference numerals when you write the application; in this way, if you later have to add another reference number, you can use an odd number and put it between two logically-related even numbers. See the list in the sample specification at the end of the chapter):

8. Description of Invention

Here you should describe in great detail the static physical structure of your invention (not how it operates or what its function is). If your invention is a process, here describe the procedures or machinery involved in your invention. Begin by first stating what the figure under discussion shows generally, e.g., "Fig 1 shows a perspective view of a basic version of my widget." Then get specific by describing the base, frame, bottom, input, or some other logical starting place of your invention. Then work up, out, or forward in a logical manner, numbering and naming the parts in your drawing as you proceed. Use the part names which you previously wrote on your sketches. To number the parts, write a number near each part and extend a lead line from the reference number to the part to which it refers. Don't circle your reference numerals since a PTO rule prohibits this. The lead lines should *not* have arrowheads, e.g., a bicycle grip might be designated "22———." However, to refer to a group of parts as a whole, e.g., a bicycle, use an arrowhead on the lead line, thus, " 10———>." If you have several closely related or similar parts, you can give them the same reference number with different letter suffixes to differentiate, e.g., "arms 12a and 12b," or "arms 12L (left) and 12R (right)."

Although you may think that the patent examiner won't need to have parts that are clearly shown in the drawing separately described in detail, all patent attorneys provide such a description. This is part of the previously-mentioned repetition technique which is used to familiarize the examiner with the invention and set the stage for the claims (Chapter 9) and operational description. When you mention each part twice, once in the description and again in the operation discussion, the first mention will initially program your reader to relate to the part so that the reader will really understand it the second time around when it counts. This is the same technique as is used in the lyrics of blues songs, where the first two lines are always restated to enhance communication.

As stated, before you begin a description of any figure, refer to it by its figure number, e.g., "Fig 1 shows an overall view of the can opener of the invention." Then as you come to each part or element, give it a separate reference number. E.g., "The can opener comprises two handle arms 10 and 12 which are hinged together at a hinge 14."

Where several figures show different views of an embodiment of your invention, you can refer to several figures at once. E.g., "Fig 1. and 2 show plan and elevational (front) views of a scissors according to the invention. The scissors comprises first and second legs 12 and 14, the second leg being best shown in Fig 2."

Be sure to cover every part shown in your drawings and be sure to use consistent terminology and nomenclature for the parts in the drawing. For example, if gear 44 is shown in Fig 8 and also in Fig 11, you should label it with the reference numeral "44" in both figures.

Lastly, be sure to detail the interconnections or mountings between parts, e.g., "Arm 14 is joined to base 12 by a flange 16."

To understand the technique commonly used to describe the parts and their interconnections, think of the song, "*Dem Bones*." The song details virtually every bone-to-bone connection in the body in logical order, e.g., "The knee bone's connected to the thigh bone. The thigh bone's connected to the hip bone." etc. In a similar manner, your description should also detail every part-to-part interconnection, even if you think the reader would find it obvious from your drawing.

If any special or exotic parts are used, or are needed, don't be coy! Include detailed descriptions of these, including dimensions, relationships, materials, and sources of supply, as applicable, in this section. Putting in such specifics will not limit your invention in any way since the claims (next chapter) will determine its scope. However failing to include these specifics can render your patent application fatally flawed if they are necessary for one skilled in the art to make and use the invention.

Including details at crucial places can prove vital later if you have to rely on these in order to distinguish your claims over a close prior-art reference cited by the examiner. Thus, it's almost axiomatic in patent law that you should make your specification as long, specific, and detailed as possible, and your main claim as short, broad, and general as possible. If you're tempted to skip the details, remember that a few strokes on a keyboard here now can save you from losing many thousands of dollars later.

If any material substance or component of your invention is a trademarked product, e.g., *Teflon*, you may use the trademark as part of your description, provided you identify it as such, tell who the trademark owner is, and give the common (generic) name of the product. An example of this is, "To prevent corrosion, arm 13 is preferably coated with *Teflon*-brand PTFE; *Teflon* is a trademark of E.I. duPont de Nemours & Co., Wilmington, DE." If you don't know the composition of a trademarked material or substance, you can try writing to the trademark owner, or you can simply use a general

(fuzzy) name of the trademarked substance, e.g., "*Ajax* brand developer, manufactured by Goldberger Graphics of San Francisco." However, if the trademarked substance is so critical to the practice of your invention that you'll want to recite it in your claims (see Chapter 9), you'll have to determine its exact composition, since trademarks, being proper adjectives, are too imprecise to be used in claims.

Avoid technical language, insofar as possible, but if you use any technical terms, be sure to define them. Try to make your description as simple as possible, without eliminating any crucial details. Avoid absolute terms, such as "always" and relative terms, such as "dense," "hot," or "hard"—use quantitative values which can be used in claims, such as "having a mass greater than 2.5," "between 70° C and 100° C," or "having a hardness less than 10 durometer," or "having a thickness less than that of member 13."

Computer Programs Note: If your invention involves a computer program, include a program listing (or at least a detailed flow chart). This can be submitted as part of the specification, or as part of the drawings, provided it's contained on ten printout pages or less. In either case, the listing should be a very black, camera-ready copy. If the printout is to be submitted on drawing sheets, these should be of the proper size (U.S. or international, see Chapter 10), with each figure or sheet separately numbered (Fig 1, Fig 2, etc.; or Fig 1A, Fig 1B, etc.). If the printout is to be submitted as part of the specification, it can be on sheets up to 11 by 14 inches, with printing limited to an area of 9 by 13 inches. The printout should be sent in a protective cover. If your program is longer than 10 pages, it should be submitted on microfiche, as an appendix. It will not be printed with the patent, but will be referred to in the patent. The standards for the microfiche are contained in the MPEP (*Manual of Patent Examining Procedure*—see Bibliography), Section 608.05, which is available at all Patent Depository Libraries.

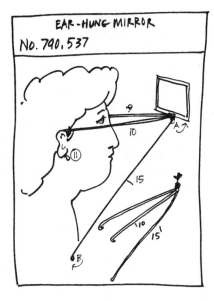

EAR-HUNG MIRROR
No. 790,537

9. Operation of Invention

Here describe in extensive detail the operation or function of the parts covered in your description. Refer to each part by its reference numeral and be sure to include the working or function of every part. Your invention may be of such a nature that it may not be possible to include a physical description and an operational description in separate sections, but you'll find that this mode of description works generally for most inventions, and you should try to adhere to it since it will force you to be complete and comprehensive.

If your invention includes several embodiments and ramifications, you can describe each separately in your description, or you can describe the structure and the operation of each embodiment together. If you choose the latter course, several sets of description/operation sections would be indicated. For example, "Description of Motor/Operation of Motor," "Description of Mounting/Operation of Mounting," etc. Any way that you arrange these sections is satisfactory, as long as you include a very, very detailed description of each and every part of your invention, together with a very, very detailed description of the operation of each part and its relation to the other parts.

I emphasize once again that you should include all reasonably-important embodiments and ramifications so that you'll have more support for broader claims and so that if an infringer is making or selling a ramification, you'll be able to show the judge that you specifically showed that ramification in your application. Although infringement is supposed to be determined mainly by the wording of your claims, when you get to court judges are psychologically influenced in your favor, as a practical matter, if your specification and drawings show and discuss the very embodiment which is being infringed.

If your invention operates by utilizing an interesting or unusual theory, you can include this also, either before or after describing the operation. If you're not sure about the correctness of your theory, you can state this, e.g., "While I believe the reaction occurs because of a catalytic effect of the platinum, I don't wish to be bound by this." You are not required to give any theory of operation in your application, since this isn't necessary to enable one skilled in the art to make or use the invention. However if you can include any theory of operation, you should since it will make your invention more interesting, believable, and likeable to your readers (e.g., your patent examiner or a judge.)

10. Conclusion, Ramifications, and Scope of Invention

After you finish your detailed description of the invention's operation, add a "Conclusion, Ramifications and Scope Of Invention" section to sum things up and to remind the judge who sees your patent that the claims control. Here's an example:

Thus the reader will see that the can opener of the invention provides a highly reliable, lightweight, yet economical device which can be used by persons of almost any age. [Keep selling it!]

While my above description contains many specificities, these should not be construed as limitations on the scope of the invention, but rather as an exemplification of one preferred embodiment thereof. Many other variations are possible. For example [then continue with brief description of possible variations which aren't important enough to show as ramifications in the drawing]. Accordingly, the scope of the invention should be determined not by the embodiment(s) illustrated, but by the appended claims and their legal equivalents.

In the first paragraph quoted above, the objects and advantages of the invention are restated and summarized to hammer home the greatness of your invention. In the "for example" portion of the second quoted paragraph, include a brief description of any alternative embodiments you can think of and which (as stated) you didn't consider important enough to show in the drawing and describe in detail in your description. I usually put exotic, untested embodiments, as well as minor variations in color, size materials, etc. in the broadening paragraph. Look at the sample specification of Fig. 8-E to see how this is done.

That's just about all there is to drafting the specification portion of your application. What's left you ask? The small matter of "Claims," that's what. I'll tell you how to write these in the next chapter.

H. Drafting the Abstract

YOUR ABSTRACT SHOULD COME at the end of your patent application, after the claims. However it will be printed on the first page of your patent and is printed right after the sample specification of Fig. 8E since the claims have been saved for the next chapter. The abstract is relatively easy to do once you've done the specification, and since it's very closely related to the specification, I'll cover it here.

The abstract should be put on a new page with the title and then the heading "Abstract." To do the actual abstract, write one or two brief paragraphs providing a concise summary of your invention in about 250 words. Generally you should spend enough time writing the abstract to make it concise, complete, and clear. This is because the abstract is usually the part of an application that's read first and most frequently consulted. Look at the abstracts of several of your prior-art patents to get an idea of what's involved. To be concise, your abstract should not include throat-clearing phrases like "This invention relates to," but rather, should get right into it and state, for example, "An improved bicycle pedal mechanism having . . . , etc." If you think you may file the application in other countries, you should include reference numbers in the abstract (with each one in parentheses) to comply with the international rules. International filing is covered in (see Chapter 12).

I. Review Your Specification and Abstract Carefully

AFTER YOU'VE COMPLETED YOUR DRAFT, be sure to review it carefully to be sure you've included everything about your invention you can think of, and that there is no possible ground for anyone to say that you haven't included enough to teach one skilled in the art how to make and use your invention. You may have to go through two, three, or more drafts to get it right. Be sure to compare your specification with those of other patents in the field so that yours is at least as complete as theirs. If you allow yourself plenty of time, e.g., a few days to do the drawings, a few days to write the introductory parts of your specification, a few days to do the static description, etc., you won't feel pressured and thus you'll be able to do a better, more readable, more legally adequate job.

Don't do your work in haste. The public won't ask whether it was completed in three days, but whether it's accurate and complete.

—Anon.

Warning: Many prior-art patents are not properly described, especially under today's more demanding standards, so don't absolutely rely on them as a standard; rather, follow the guidelines of this chapter.

Checklist for Your Patent Application Draft

After reviewing many patent applications prepared by laypersons, I've come up with three lists of the most common errors and areas generally needing improvement. The first list (in two parts) follows; it covers the preliminary drawings and draft specification. Before you go on to the claims (Ch.9) or to the finaling process (Ch.10), I suggest that you check this list carefully and make any needed corrections.

Checklist for Preliminary Drawings

[ ] D01. Every significant part in the drawings has its own reference numeral.

[ ] D02. All different parts have different respective reference numerals, i.e., the same reference numeral is never used to indicate different parts.

[ ] D03. The same reference numeral is always used to indicate the same part when such part is shown in different Figs.; i.e., two different reference numerals are never used to indicate the same part.

[ ] D04. Arrowheads are not used on any lead line, unless it refers to an entire assembly of elements.

[ ] D05. Perspective, rather than engineering views, are primarily used.

[ ] D06. The drawings show enough details of your invention to enable it to be fully and readily understood by a lay judge.

[ ] D07. The reference numerals start with a number higher than your highest Fig number.

[ ] D08. Even reference numerals (10, 12, etc.) are used so you can add more numerals in sequence later, if needed.

[ ] D09. The Fig details and reference numerals are large enough to be easily read.

[ ] D10. A descriptive label is placed on or near each component whose function is not apparent.

Checklist for Draft of Specification

[ ] S01. The title indicates the essence of your invention without being too long.

[ ] S02. No sentence is over about 13 words (unless really necessary or unless two independent clauses are used).

[ ] S03. A heading is supplied for approximately every two pages of discussion.

[ ] S04. The Field Of The Invention section is not longer than two sentences.

[ ] S05. Each discussion relates to and explains only its heading.

[ ] S06. Adjacent paragraphs are connected by transitions and no paragraph is longer than about one page, double spaced.

[ ] S07. The introductory sections of the specification do not discuss details of your invention (its general advantages can be stated).

[ ] S08. Every sophisticated term is defined clearly.

[ ] S09. All detailed technical discussions refer to a drawing Fig (humans can't comprehend abstract technical discussions).

[ ] S10. Each prior-art approach you discuss is knocked.

[ ] S11. When any patent is referred to, the inventor's name(s), the patent #, and its issue date are included.

[ ] S12. The Objects And Advantages section states all advantages of your invention. These include the converse of every disadvantage of the prior-art approaches discussed in the Prior-Art section (and the prior-art section includes the converse of every advantage of the O. & A. section).

[ ] S13. The Drawing Description section has just one short sentence for each Fig.

[ ] S14. Every reference numeral in the drawings is used in the specification and every reference in the specification is used in the drawings.

[ ] S15. The description and the operation of the invention are discussed in separate sections.

[ ] S16. Wishy-washy descriptions ("a plastic brace might work here") are eliminated; all descriptions are firm, sure, and positive.

[ ] S17. The dimensions, preferred materials, relationships and/or sources of supply are stated for all exotic or critical parts.

[ ] S18. For ease of reading, a shorter term is used when you refer again to a part with a long name. E.g., First time: "A liquid-overflow check valve 12". Second time: "Valve 12".

[ ] S19. Each trademark is identified as such, typed in caps, used with a generic noun, & its owner is indicated.

[ ] S20. The article "a" (rather than "the") is used to introduce parts in the specification.

[ ] S21. The article "the" isn't used to refer to a part by its name and reference numeral.

[ ] S22. Already-introduced parts are not referred to with the article "a".

[ ] S23. The same parts are referred to by the same respective words throughout.

[ ] S24. No legal words, such as "said" or "means", are used in the specification or abstract.

[ ] S25. The writing doesn't change voices (active to passive, or vice-versa) in a paragraph and you use the active voice as much as possible.

[ ] S26. Metric (or metric followed by British) dimensions are used.

[ ] S27. The discussion discusses one Fig at a time, insofar as possible, and doesn't jump from Fig to Fig too much.

[ ] S28. Your reader is always kept clearly advised which Fig is under discussion.

[ ] S29. The Description and Operation sections contain enough detail to enable your invention readily to be built and used.

[ ] S30. The Operation section does not introduce any part.

[ ] S31. A Summary, Ramifications, and Scope section is provided at the end of the specification.

[ ] S32. The Abstract is technical & terse, without listing too many advantages.

[ ] S33. The Abstract has a reference numeral in parentheses "(12)" after each named part for possible foreign filing.

your disc and file # (optional)

8-9 cm top margin on p. 1

2-3 cm
left
margin

Patent Application of

Lou W. Koppe

for

PAPER-LAMINATED PLIABLE CLOSURE FOR FLEXIBLE BAGS

Background—Field of Invention

one
sentence
for field of
invention

This invention relates to plastic tab closures, specifically to such closures which are used for closing the necks of plastic produce bags.

2-3 cm
right
margin on
8-1/2 x 11
paper

Background—Description of Prior Art

description
of and
knocking
of prior art

Grocery stores and supermarkets commonly supply consumers with polyethylene bags for holding produce. Such bags are also used by suppliers to provide a resealable container for other items, both edible and inedible.

Originally these bags were sealed by the supplier with staples or by heat. However consumers objected since these were of a rather permanent nature: the bags could be opened only by tearing, thereby damaging them and rendering them impossible to reseal.

Thereafter, inventors created several types of closures to seal plastic bags in such a way as to leave them undamaged after they were opened. U.S. patent 4,292,714 to Walker (1981) discloses a complex clamp which can close the necks of bags without causing damage upon opening; however, these clamps are prohibitively expensive to manufacture. U.S. patent 2,981,990 to Balderree (1961) shows a closure which is of expensive construction, being made of PTFE, and which is not effective unless the bag has a relatively long "neck".

type
almost to
bottom of
page so
A4 copies
can be
made with
proper
margin

Thus if the bag has been filled almost completely and consequently has a short neck, this closure is useless. Also, being relatively narrow and clumsy, Balderree's closure cannot be easily bent by hand along its longitudinal axis. Finally his closure does not hold well onto the bag, but has a tendency to snap off.

Although twist closures with a wire core are easy to use and inexpensive to manufacture, do not damage the bag upon being removed, and can be used repeatedly, nevertheless they simply do not possess the neat and uniform appearance of

continue
knocking
the prior
art

a tab closure, they become tattered and unsightly after repeated use, and they do not offer suitable surfaces for the reception of print or labeling. These ties also require much more manipulation to apply and remove.

Several types of thin, flat closures have been proposed—for example, in U.K. patent 883,771 to Britt et al. (1961) and U.S. patents 3,164,250 (1965), 3,417,912 (1968), 3,822,441 (1974), 4,361,935 (1982), and 4,509,231 (1985), all to Paxton. Although inexpensive to manufacture, capable of use with bags having a short neck, and producible in break-off strips, such closures can be used only once if they are made of frangible plastic since they must be bent or twisted when being removed and consequently will fracture upon removal. Thus, to reseal a bag originally sealed with a frangible closure, one must either close its neck with another closure or else close it in makeshift fashion by folding or tying it. My own patent 4,694,542 (1987) describes a closure which is made of flexible plastic and is therefore capable of repeated use without damage to the bag, but nevertheless all the plastic closures heretofore known suffer from a number of disadvantages:

(a) Their manufacture in color requires the use of a compounding facility for the production of the pigmented plastic. Such a facility, which is needed to compound the primary pigments and which generally constitutes a separate production site, requires the presence of very large storage bins for the pigmented raw granules. Also it presents great difficulties with regard to the elimination of the airborne powder which results from the mixing of the primary granules.

(b) If one uses an extruder in the production of a pigmented plastic—especially if one uses only a single extruder—a change from one color to a second requires purging the extruder of the granules having the first color by introducing those of the second color. This a process inevitably produces, in sizeable volume, an intermediate product of an undesired color which must be discarded as scrap, thereby resulting in waste of material and time.

(c) The colors of the closures in present use are rather unsaturated. If greater concentrations of pigment were used in order to make the colors more intense, the plastic would become more brittle and the cost of the final product would increase.

(d) The use of pigmented plastic closures does not lend itself to the production of multicolored designs, and it would be very expensive to produce plastic closures in which the plastic is multicolored—for example, in which the plastic has stripes of several colors, or in which the plastic exhibits multicolored designs.

(e) Closures made solely of plastic generally offer poor surfaces for labeling or printing, and the label or print is often easily smudged.

(f) The printing on a plastic surface is often easily erased, thereby allowing the alteration of prices by dishonest consumers.

(g) The plastic closures in present use are slippery when handled with wet or greasy fingers.

(h) A closure of the type in present use can be very carefully pried off a bag by a dishonest consumer and then attached to another item without giving any evidence of such removal.

Objects and Advantages

reverse of
disadvantages
of prior art =
praise of your
invention

Accordingly, besides the objects and advantages of the flexible closures described in my above patent, several objects and advantages of the present invention are:

(a) to provide a closure which can be produced in a variety of colors without requiring the manufacturer to use a compounding facility for the production of pigments;

(b) to provide a closure whose production allows for a convenient and extremely rapid and economical change of color in the closures that are being produced;

(c) to provide a closure which both is flexible and can be brightly colored;

(d) to provide a closure which can be colored in several colors simultaneously;

(e) to provide a closure which will present a superior surface for the reception of labeling or print;

(f) to provide a closure whose labeling cannot be altered;

(g) to provide a closure which will not be slippery when handled with wet or greasy fingers; and

(h) to provide a closure which will show evidence of having been switched from one item to another by a dishonest consumer—in other words, to provide a closure which makes items tamper-proof.

Further objects and advantages are to provide a closure which can be used easily and conveniently to open and reseal a plastic bag, without damage to the bag, which is simple to use and inexpensive to manufacture, which can be supplied in separate tabs en masse or in break-off links, which can be used with bags having short necks, which can be used repeatedly, and which obviates the need to tie a knot in the neck of the bag or fold the neck under the bag or use a twist closure. Still further objects and advantages will become apparent from a consideration of the ensuing description and drawings.

Drawing Figures

one short
sentence
for each
figure

In the drawings, closely related figures have the same number but different alphabetic suffixes.

Figs 1A to 1D show various aspects of a closure supplied with a longitudinal groove and laminated on one side with paper.

Fig 2 shows a closure with no longitudinal groove and with a paper lamination on one side only.

Fig 3 shows a similar closure with one longitudinal groove.

Fig 4 shows a similar closure with a paper lamination on both sides.

Fig 5 shows a similar closure with a paper lamination on one side only, the groove having been formed into the paper as well as into the body of the closure.

Figs 6A to 6K show end views of closures having various combinations of paper laminations, longitudinal grooves, and through-holes.

Figs 7A to 7C show a laminated closure with groove after being bent and after being straightened again.

Figs 8A to 8C show a laminated closure without a groove after being bent and after being straightened again.

Reference Numerals In Drawings

10 base of closure	12 lead-in notch
14 hole	16 gripping points
18 groove	20 paper lamination
22 tear of paper lamination	24 corner
26 longitudinal through-hole	28 neck-down
30 side of base opposite to bend	32 crease

4

Description—Figs. 1 to 6

A typical embodiment of the closure of the present invention is illustrated in Fig 1A (top view) and Fig 1B (end view). The closure has a thin base **10** of uniform cross section consisting of a flexible sheet of material which can be repeatedly bent and straightened out without fracturing. A layer of paper **20** (Fig 1B) is laminated on one side of base 10. In the preferred embodiment, the base is a flexible plastic, such as poly-ethylene-tere-phthalate (PET—hyphens here supplied to facilitate pronunciation)—available from Eastman Chemical Co. of Kingsport, TN. However the base can consist of any other material that can be repeatedly bent without fracturing, such as polyethylene, polypropylene, vinyl, nylon, rubber, leather, various impregnated or laminated fibrous materials, various plasticized materials, cardboard, paper, etc.

At one end of the closure is a lead-in notch **12** which terminates in gripping points **16** and leads to a hole **14**. Paper layer **20** adheres to base **10** by virtue either of the extrusion of liquid plastic (which will form the body of the closure) directly onto the paper or the application of heat or adhesive upon the entirety of one side of base **10**. The paper-laminated closure is then punched out. Thus the lamination will have the same shape as the side of the base **10** to which it adheres.

The base of the closure is typically .8 mm to 1.2 mm in thickness, and has overall dimensions roughly from 20 m x 20 mm (square shape) to 40 mm x 70 mm (oblong shape). The outer four corners **24** of the closure are typically beveled or rounded to avoid snagging and personal injury. Also, when closure tabs are connected side-to-side in a long roll, these bevels or roundings give the roll a series of notches which act as detents or indices for the positioning and conveying of the tabs in a dispensing machine.

A longitudinal groove **18** is formed on one side of base **10** in Fig 1. In other embodiments, there may be two longitudinal grooves—one one each side of the base—or there may be no longitudinal groove at all. Grove **18** may be formed by machining, scoring, rolling, or extruding. In the absence of a groove, there may be a longitudinal through-hole **26** (Fig 6L). This through-hole may be formed by placing, in the extrusion path of the closure, a hollow pin for the outlet of air.

Additional embodiments are shown in Figs 2, 3, 4, and 5; in each case the paper lamination is shown partially peeled back. In Fig 2 the closure has only one lamination and no groove; in Fig 3 it has only one lamination and only one groove; in Fig 4 it has two laminations and only one groove; in Fig 5 it has two laminations and one groove, the latter having been rolled into one lamination as well as into the body of the closure.

There are various possibilities with regard to the relative disposition of the sides which are grooved and the sides which are laminated, as illustrated in Fig 6, which presents end views along the longitudinal axis. Fig 6A shows a closure with lamination on one side only and with no groove; Fig 6B shows a closure with laminations on both sides and with no groove; Fig 6C shows a closure with only one lamination and only one groove, both being on the same side; Fig 6D shows a closure with only one lamination and only one groove, both being on the same side and the groove having been rolled into the lamination as well as into the body of the closure; Fig 6E shows a closure with only one lamination and only one groove, the two being on opposite sides; Fig 6F shows a closure with two laminations and only one groove; Fig 6G shows a closure with two laminations and only one groove, the groove having been rolled into one lamination as well as into the body of the closure; Fig 6H shows a closure with only one lamination and with two grooves; Fig 6I shows a closure with only one lamination and with two grooves, one of the grooves having been rolled into the lamination as well as into the body of the closure; Fig 6J shows a closure with two laminations and with two grooves; Fig 6K shows a closure with two laminations and with two grooves, the grooves having been rolled into the laminations as well as into the body of the closure; and Fig 6L shows a closure with two laminations and a longitudinal through-hole.

From the description above, a number of advantages of my paper-laminated closures become evident:

(a) A few rolls of colored paper will contain thousands of square yards of a variety of colors, will obviate the need for liquid pigments or a pigment-compounding plant, and will permit the manufacturer to produce colored closures with transparent, off color, or of leftover plastic, all of which are cheaper than first quality pigmented plastic.

(b) With the use of rolls of colored paper to laminate the closures, one can change colors by simply changing rolls, thus avoiding the need to purge the extruder used to produce the closures.

(c) The use of paper laminate upon an unpigmented, flexible plastic base can provide a bright color without requiring the introduction of pigment into the base and the consequent sacrifice of pliability.

(d) The presence of a paper lamination will permit the display of multicolored designs.

(e) The paper lamination will provide a superior surface for labeling or printing, either by hand or by machine.

(f) Any erasure or alteration of prices by dishonest consumers on the paper-laminated closure will leave a highly visible and permanent mark.

(g) Although closures made solely of plastic are slippery when handled with wet or greasy fingers, the paper laminate on my closures will provide a nonslip surface.

Operation—Figs 1, 6, 7, 8

operational description of figures

The manner of using the paper-laminated closure to seal a plastic bag is identical to that for closures in present use. Namely, one first twists the neck of a bag (not shown here but shown in Fig 12 of my above patent) into a narrow, cylindrical configuration. Next, holding the closure so that the plane of its base is generally perpendicular to the axis of the neck and so that lead-in notch **12** is adjacent to the neck, one inserts the twisted neck into the lead-in notch until it is forced past gripping points **16** at the base of the notch and into hole **14**.

To remove the closure, one first bends it along its horizontal axis (Fig 1C—an end view—and Figs 7 and 8) so that the closure is still in contact with the neck of the bag and so that gripping points **16** roughly point in parallel directions. Then one pulls the closure up or down and away from the neck in a direction generally opposite to that in which the gripping points now point, thus freeing the closure from the bag without damaging the latter. The presence of one or two grooves **18** or a longitudinal through-hole **26** (Fig 6L), either of which acts as a hinge, facilitates this process of bending.

The closure can be used to reseal the original bag or to seal another bag many times; one simply bends it flat again prior to reuse.

As shown in Figs 1C, 7B, and 8B (all end views) when the closure is bent along its longitudinal axis, region **30** of the base will stretch somewhat along the direction perpendicular to the longitudinal axis. (Region 30 is the region which is parallel to this axis and is on the side of the base opposite to the bend.) Therefore, when the closure is flattened again, the base will have elongated in the direction perpendicular to the longitudinal axis. This will cause a necking down **28** (Figs 1D, 7C, and 8C) of the base, as well as either a tell-tale tear **22**, or at least a crease **32** (Figs 7A and 8A) along the axis of bending. Therefore if the closure is attached to a sales item and has print upon its paper lamination, the fact that the closure has been transferred by a dishonest consumer from the first item to another will be made evident by the tear or crease.

Figs 7A and 8A show bent closures with and without grooves, respectively. Figs 7C and 8C show the same closures, respectively, after being flattened out, along their longitudinal axes, paper tear **22** being visible.

6

Summary, Ramifications, and Scope

repeat advantages Accordingly, the reader will see that the paper-laminated closure of this invention can be used to seal a plastic bag easily and conveniently, can be removed just as easily and without damage to the bag, and can be used to reseal the bag without requiring a new closure. In addition, when a closure has been used to seal a bag and is later bent and removed from the bag so as not to damage the latter, the paper lamination will tear or crease and thus give visible evidence of tampering, without impairing the ability of the closure to reseal the original bag or any other bag. Furthermore, the paper lamination has the additional advantages in that

• it permits the production of closures in a variety of colors without requiring the manufacturer to use a separate facility for the compounding of the powdered or liquid pigments needed in the production of colored closures;

• it permits an immediate change in the color of the closure being produced without the need for purging the extruder of old resin;

• it allows the closure to be brightly colored without the need to pigment the base itself and consequently sacrifice the flexibility of the closure; it allows the closure to be multicolored since the paper lamination offers a perfect surface upon which can be printed multicolored designs;

• it provides a closure with a superior surface upon which one can label or print;

• it provides a closure whose labeling cannot be altered or erased without resulting in tell-tale damage to the paper lamination; and

• it provides a closure which will not be slippery when handled with wet or greasy fingers, the paper itself providing a nonslip surface.

additional ramifications Although the description above contains many specificities, these should not be construed as limiting the scope of the invention but as merely providing illustrations of some of the presently preferred embodiments of this invention. For example, the closure can have other shapes, such as circular, oval, trapezoidal, triangular, etc.; the lead-in notch can have other shapes; the groove can be replaced by a hinge which connects two otherwise unconnected halves, etc.

broadening paragraph Thus the scope of the invention should be determined by the appended claims and their legal equivalents, rather than by the examples given.

**[CLAIMS FOLLOW, STARTING ON A NEW PAGE
BUT ARE PRINTED IN THE NEXT CHAPTER]**

start abstract on new page, after claims

reproduce title from p. 1

PAPER-LAMINATED PLIABLE CLOSURE FOR FLEXIBLE BAGS

Abstract: A thin, flat closure for plastic bags and of the type having at one edge a V-shaped notch (12): which communicates at its base with a gripping aperture (14). The base (10) of the closure is made of a flexible material so that it can be repeatedly bent, without fracturing, along an axis aligned with said notch and aperture. In addition, a layer of paper (20) is laminated on one or both sides of the closure. The axis of the base may contain one or two grooves (18) or a through-hole (26), either of which acts as a hinge to facilitate bending.

inset reference numerals in paretheses for possible foreign filing

Chapter 9

Now for the Legalese—The Claims

A. What Are Claims?

IF YOU DON'T YET KNOW what patent claims are, or have never read any, you're in for a surprise. The word "claim," in the patent context is definitely a term of art. A "claim" is not what the common dictionary definitions recite—it's not a demand for something due, a title to something in the possession of another, or that which one seeks or asks for. Rather, a "claim," in the arcane world of patents, is a very formally-worded sentence fragment contained in a patent application or patent. Claims recite and define the structure, or acts, of an invention in very precise, logical, and exact terms. They serve as tools to determine whether a patent application is patentable over the prior art, or whether a patent is infringed. Just as a deed recites the boundary of a real estate parcel, patent claims recite the "bounds" or scope an invention for the purposes of dealing with the PTO and possible infringers.

While claims are literally sentence fragments, they are supposed to be the object of the words "I [or We] claim." They are actually interpreted, when in a patent application, as saying to the examiner, "Here is my definition of my invention. Please search to see whether my invention as here defined, is patentable over the prior art." When in a patent, they 're interpreted as little statutes which say to the public, "The following is a precise description of the elements of this invention; if you make, use, or sell anything which has all of these elements, or all of these elements plus additional elements, or which closely fits this description, you can be legally held liable for the consequences of patent infringement."

Since there are only five statutory classes of inventions (see Chapter 5), every claim must be classifiable into one of these classes. Thus there are: (1) process or method claims; (2) machine claims; (3) article or article of manufacture claims; (4) composition of matter claims; and (5) claims showing a new use of any of the other four statutory classes. Again, the line between (2) and (3) is blurred. Fortunately, as mentioned in Chapter 5, you don't have to do the classifying unless the PTO decides that your invention doesn't fit within any class at all.

If all of this sounds a bit formidable, don't let it throw you; it will become quite clear as we progress, after you see some examples. What's more, when it comes to claims, every layperson who "prosecutes" (handles or controls) a patent application has a safety net: So long as you can convince the patent examiner that you have a patentable invention, the examiner is required by law to write at least one claim for you, for free. I discuss this along with several aids to claim drafting in Section G of this chapter. But a word of caution. If knowing this you're tempted to skip this chapter and solely rely on the examiner, you can't. You must provide at least one claim in your application to obtain a filing date. In addition, familiarity with the information I provide here is essential to securing the strongest possible patent on your invention. So I urge you to approach this chapter as if there were no safety net. Take this chapter as I present it, in small, easy-to-digest chunks, and you'll have no trouble. If you don't understand something the first time, go back again so you'll be further down on the learning curve where you'll see things much more clearly.

B. The Law Regarding Claims

THE LAW (statutes and PTO rules) concerning claims are written in only the most general and vague terms. Accordingly, I'll be turning to the real world of everyday practice to actually understand the requirements for drafting claims. Before I do, however, let's at least take a brief look at the law as it is written.

ALL WEATHER SUIT

No. 729,831

1. Patent Laws

The only pertinent statute comprises two paragraphs of our old friend, Section 112 of the patent laws (35 USC 112), which states:

The specification shall conclude with one or more claims particularly pointing out and distinctly claiming the subject matter which the applicant regards as the applicant's invention.

An element in a claim for a combination may be expressed as a means or step for performing a specified function without the recital of structure, material, or acts in support thereof, and such claim shall be construed to cover the corresponding structure, material, or acts described in the specification and equivalents thereof.

The first sentence is the one that mandates the use of claims in patents. It also means that the claims must be specific enough to define the invention over the prior art ("particularly pointing out") and also should be clear, logical, and precise ("distinctly claiming"). This sentence is the most important part of Section 112 and is referred to by patent examiners almost daily because of the frequency with which they reject claims for lack of clarity or for some other similar reason.

The second sentence was enacted to overrule certain Supreme Court decisions (G.E. v Wabash, 304 U.S. 371 (1938) and Halliburton v. Walker, 329 U.S. 1 (1946)). These decisions held certain claims invalid on technical grounds, specifically for "functionality at the point of novelty" because they expressed the essence of an invention in terms of its novel function, rather than reciting the specific structure that performed the novel function. In other words, they contained a broad expression like "means for hardening latex" rather than a specific expression like " a sulfur additive." To enable patent applicants to continue to claim their inventions broadly, in terms of their novel function without getting specific, this paragraph of Sec. 112 was enacted. It authorizes the use of "means" clauses when drafting claims in order to be able to claim the invention broadly without limiting it to any specific material or hardware. More on "means" clauses later in this chapter.

2. Rules of Practice

In addition to the statutes, claims are governed by the PTO's "Rules of Practice". PTO Rule 75, parts (b), (d)(1), and (e) [see Introduction] add these additional requirements:

(b) More than one claim may be presented provided they differ substantially from each other and are not unduly multiplied.

(d)(1) The claim or claims must conform to the invention as set forth in the remainder of the specification and the terms and phrases used in the claims must find clear support or antecedent basis in the description so that the meaning of the terms in the claims may be ascertainable by reference to the description.

(e) Where the nature of the case admits, as in the case of an improvement, any independent claim should contain in the following order, (1) a preamble comprising a general description of all the elements or steps of the claimed combination which are conventional or known, (2) a phrase such as "wherein the improvement comprises," and (3) those elements, steps, and/or

relationship which constitutes that portion of the claimed combination which the applicant considers as the new or improved portion.

(f) If there are several claims, they shall be numbered consecutively in Arabic numerals.

(g) All dependent claims should be grouped together with the claim or claims to which they refer to the extent possible.

Part (b) requires that the claims differ substantially from each other and not be too numerous. In practice, minimal differences will suffice. The rule prohibiting numerous claims is more strictly enforced. If more than about twenty claims are presented, there should be some justification, such as a very complex invention.

Part (d)(1), enforced only sporadically, requires that the terms in the claims should correspond to those used in the specification. It has often been said that the specification should serve as a dictionary for the claims.

Part (e), a newcomer, was introduced to require that claims be drafted, insofar as practicable, in the German or "Jepson" style (from a famous decision of that name). The Jepson-type claim is very easy for examiners to read and understand. It puts the essence of the invention into sharp focus by providing in the first part of the claim an introduction which sets forth the environment of the invention, i.e., what is already known, and in the second part, or body of the claim, the essence of the invention, i.e., the improvement of the current invention. In practice, I've never seen this part of Rule 75 enforced.

Part (f) is self-explanatory and part (g) will be explained in Section J of this chapter.

C. Some Sample Claims

AS MENTIONED, claims boil the invention down to its essence. In their broadest sense, they eliminate everything non-essential to the invention. In fact, many inventors first realize what their invention truly is when they write or see a claim to it, especially after the claim has been rejected in the patent prosecution process. Conversely, you won't be able to draft an adequate claim unless you have a clear understanding of your invention. Although not a patent attorney, the great theatrical producer David Belasco showed that he understood the principle behind claims well when he said, "If you can't write your idea on the back of my calling card, you don't have a clear conception of your idea."

And claims are difficult to write just because they are so short. Blaise Pascal once concluded a letter to a friend as follows: *"I have made this letter a little longer than usual because I lack the time to make it shorter."* Nevertheless, don't get discouraged; if you follow the step-by-step, four-part procedure I give later, you'll find that writing claims is not too much more difficult than writing the specification.

Consider some hypothetical simple claims in the five respective statutory classes of invention. Patent applications containing the first four of these claims would now be rejected since the "inventions" they define are obviously old and in the public domain. The fifth—a "new use" claim—is from a patent.

1. Process or Method Claim

Sewing Fabric

A method for joining two pieces of cloth together at their edges, comprising the steps of:

a. positioning said two pieces of cloth together so that an edge portion of one piece overlaps an adjacent edge portion of the other piece, and

b. passing a thread repeatedly through and along the length of the overlapping portions in sequentially opposite directions and through sequentially-spaced holes in said overlapping adjacent portions.

Note that the first part of this claim contains a title, preamble, or genus, which states the purpose of the method but doesn't use the term "sewing," because sewing is the invention and is assumed to be new at the time the claim is drafted. The claim contains two steps, (a) and (b), which state in sequence the acts one would perform in sewing two pieces of cloth.

2. Machine Claim

Automobile

A self-propelled vehicle, comprising:

a. a body carriage having rotatable wheels mounted thereunder for enabling said body carriage to roll along a surface,

b. an engine mounted in said carriage for producing rotational energy,

c. means for controllably coupling rotational energy from said engine to at least one of said wheels so as to propel said carriage along said surface.

This claim again contains a title in the first part. The second part or body contains three elements, the carriage, the engine, and the transmission. These elements are defined as connected or interrelated by the statement that the engine is mounted in the carriage and the transmission (defined broadly as "means for controllably coupling . . . ") couples the engine to at least one wheel of the carriage.

3. Article of Manufacture Claim

Pencil

A hand-held writing instrument comprising:

a. an elongated core-element means that will leave a marking line if moved across paper or other similar surface, and

b. an elongated holder surrounding and encasing said elongated core-element means, one portion of said holder being removable from an end thereof to expose an end of said core-element means so as to enable said core-element means to be usable for writing, whereby said holder protects said core-element means from breakage and

provides a means of holding said core-element means conveniently.

This claim, like the machine claim, contains a preamble and a body with two elements: (a) the "lead" and (b) the wood. As before, the elements of the body are associated; here the wood ("elongated holder") is said to surround and encase the lead ("elongated core"). Since the purposes of the holder are not readily apparent, a "whereby" clause has been added at the end of the claim to state these. Such clauses are optional, but they are useful since they provide anyone who reads the claim with an explanation of otherwise unexplained elements, and since they drive home the advantages of the invention.

4. Composition of Matter Claim

Concrete

A rigid building and paving material comprising a mixture of sand and stones, and a hardened cement binder filling the volume between and adhering to sand and stones.

This claim, although not in subparagraph form, still contains a preamble and a body containing a recitation of the elements of the composition (sand, stones, and cement binder), plus an association of the elements. (Sand and stones are mixed and binder fills volume between and adheres to sand and stones.)

The height of brevity was reached (and will never be exceeded) in a composition of matter claim some years ago when the PTO issued a patent to Glenn T. Seaborg on a new element, Americum; the claim read simply,

Element 95.

5. New Use Claim

Aspirin to Speed Growth of Swine

*A method of stimulating the growth of swine
comprising feeding such swine aspirin in an amount
effective to increase their rate of growth.*

This claim recites the newly-discovered use of
aspirin and the purpose of the new use in a manner
which defines over and avoids the known, old use of
aspirin (analgesic).

Now that you've read a few claims, I suggest you
try writing a practice claim or two of your own to
become more familiar with the process. Try a simple
article or machine with which you are very familiar.
Write the preamble and then the body. To write the
body, first list the elements or parts of the article or
machine, and then associate or interconnect them.
Don't worry too much about grammar or style, but try
to make the claim clear and understandable.

D. Common Misconceptions
Regarding Claims

IN MY EXPERIENCE, inventors' misconceptions
about claims are more widespread than in any other
area of the patent law, except possibly for the
misconception regarding the "Post Office Patent"
explained in Chapter 3. Consider some of the
following:

▲

Common Misconception: *The more claims which the
PTO (Patent and Trademark Office) allows in your patent
application, the broader your scope of protection.*

Fact: *The scope of your monopoly is determined by the
wording of your claims, not their number. One broad claim
can be far more powerful than fifty narrow claims.*

▲

Common Misconception: *If you want to get broad
protection on a specific feature of your invention, you should
recite that specific feature in your claims.*

Fact: *If you recite a specific feature of your invention in a
claim, that claim will be limited to that feature as recited, and
variations may not be covered, e.g., if you have a two-inch,
nylon gear in your apparatus and you recite it as such in a
claim, the claim may not cover an apparatus which uses a
one-inch gear, or a steel gear. The best way to cover all
possible variations of your gear is to recite it simply as a
"gear," or better yet, "rotary transmission means."*

▲

Common Misconception: *If a claim doesn't recite a
specific feature of your invention, then this feature is
necessarily not covered; thus, if anyone makes your
apparatus with this feature, no infringement will occur. E.g.,
if your invention includes a two-inch, nylon gear and you fail
to recite it specifically in a claim, then anyone who makes
your invention with this gear can't infringe your patent.*

Fact: *The fact that a feature isn't recited doesn't mean that
it isn't covered. An absurd example will make this clear.
Suppose your invention is a bicycle and you show and
describe it with a front wheel having 60 spokes. You don't
mention the spokes at all in a claim; you simply recite a
"front wheel." Any bike which has all of the limitations of the
claim will infringe it. Thus a bike which has any "front wheel"
will infringe, whether it has zero or 600 spokes.*

As I'll explain from time to time, to infringe a
claim, an accused apparatus must have at least all of
the elements of the claim; if it has more elements
than recited in the claim, it still infringes, but if it
has less, then it doesn't infringe.

▲

Common Misconception: *The more about your invention you recite in a claim, the broader that claim will be. (Stated differently, the longer a claim is, the broader it is.)*

Fact: *As will be apparent from the previous common misconception, the less you recite in a claim, i.e., the fewer the elements you recite, the broader the claim will be. This seeming paradox exists because an accused infringing device must have all the elements of a claim to infringe. Thus, the fewer the elements specified in a claim, the fewer the elements an accused infringing device needs to have to infringe. Put differently, infringement is generally easier to prove if a claim is made shorter or has fewer elements. "To claim more, you should recite less" is a concept which is difficult for most inventors to absorb, but which you should learn well if you want to secure the broadest possible protection. See Computer Searching in Chapter 6, Section L for further clarification of this point.*

E. One Claim Should Be as Broad as Possible

Inventor's Commandment #9:
You should write one claim of your patent application as broadly as the prior art permits by using the broadest terms you know and reciting as few elements as you can, thereby gaining coverage for the full scope of your invention.

There are two ways to make a broad claim: (1) *minimize* the number of elements; and (2) *maximize* the scope of these elements. Let's see how this works.

1. Minimize the Number of Elements

Take our machine claim which recites three elements, A, B, and C—that is, the wheeled carriage, the engine, and the transmission. If an accused machine contains just these three elements, it will, of course, infringe. If it has these three plus a fourth, for example a radio, which we'll label D, it will still infringe. But if our accused machine contains only elements A and B, the carriage and engine, it won't infringe since it simply doesn't contain all of the claimed elements, A, B, and C.

If a claim contains many, many elements, say A to M, only devices with all thirteen elements, A to M, will infringe. If just one of the thirteen elements, say G, is eliminated, infringement will be avoided. Thus it's relatively easy to avoid infringing a claim with many elements.

If a claim contains only two elements, A and B, any device with these two elements will infringe, no matter how many other elements it has. The only way to avoid infringement is to eliminate either element A or element B, a relatively difficult task.

Again, the fewer the elements in a claim, the harder the claim will be to avoid, that is, the broader it will be and the more devices it will cover. Therefore, when drafting a claim to your invention, it will behoove you to put in as few elements of your invention as possible.

2. Maximize the Scope of Elements

With regard to the second way of broadening a claim, that is, reciting existing elements more broadly, consider a few examples. Suppose an invention involves a chair, and the chair elements can be drafted broadly as "seating means" or narrowly as a four-legged maple chair with a vinyl-covered padded seat and a curved plywood back. Obviously, a three-legged plastic stool would be a "seating means," and it would infringe the broadly-recited element, but would miss the narrowly-recited maple chair by a country mile. In electronics, a "controllable electron valve" is broader than a "vacuum tube" or "transistor." In machinery, "rotational energy connecting element" is broader than "helically cut gear" or "V-belt."

The best way of reciting elements broadly is to take advantage of the last sentence of Section 112 by reciting an element, wherever possible, as "means" plus a specific function. In this way, any device or means which performs the function will infringe. For example, "means for conveying rotational energy" is broader than and covers gears, belts and pulleys, drive shafts, etc. "Amplifying means" is broader than and covers transistor amplifiers, tube amplifiers, masers, etc.

If you do use the word "means" in a claim, Section 112 requires that the claim recite a "combination," i.e., two of more elements or parts. Claims which recite a single element are not supposed to use the word "means" to describe the single element since this is considered too broad, e.g., "17. Means for providing a continuously-variable speed/power drive for a bicycle" would be an example of a prohibited "single means" claim. However, you can effectively obtain practically the same breadth of protection by adding an immaterial second element to the claim to make it a combination claim. Thus, "17. In combination, a bicycle having a pedal mechanism and means for providing a continuously-variable speed/power drive for coupling rotational energy from said pedal mechanism to a wheel of said bicycle" would satisfy Section 112.

To sum up, while you should write your specification as specifically and with as much detail as possible (Chapter 8), you should make the substance of your claims as general (broad) as possible by (1) eliminating as many elements as is feasible and (2) describing (reciting) the remaining elements as broadly as possible. In other words, make your specification specific and long and your claims general and short.

F. The Effect of Prior Art on Your Claim

NOW THAT YOU'VE LEARNED how to make your claims as broad as possible, it's time for the bad news. What is "possible" is generally much less breadth than you'd like. This is because each claim must define an invention that is patentable over the prior art. Remember the issues of novelty and unobviousness? Well, they (especially unobviousness) are an ever-present factor always to be considered in claim drafting.

Let's go back to Section 102, which deals with novelty (Chapter 5). A claim must define an invention that is novel in view of the prior art. That is, it must recite something which no single reference in the prior art shows, i.e., it must contain something new.

Just as a claim can be made broader by eliminating elements and reciting the existing elements more broadly, it can be made narrower in order to define novel structure, (1) by adding elements, or (2) by reciting the existing elements more narrowly.

For an example of adding elements, suppose a prior-art reference shows a machine having three elements—A, B, and C, and your claim recites these three elements A, B, and C. Your claim would be said to lack novelty over the prior art and would be rejectable or invalid under Section 102. But if you added a fourth element, D, to the claim, it would clear the prior art and would recite a novel invention (but not necessarily a patentable one, because of the unobviousness requirement[1]).

For an example of reciting existing elements more narrowly, suppose the prior art shows a

---

[1]In this example, if the prior art were an in-force patent, the mere addition of the element to actual hardware, while creating a "novel" invention, would not prevent the hardware from infringing the patent's claim.—see Sec. D.

machine having the same three elements—A, B, and C. You could also clear this prior art and claim a novel invention by reciting in your claim elements A, B, and C', where C' would be the prior-art element C with any change that isn't shown in the prior art. For example, if the prior art shows element C as a steam engine, and you recite a gasoline engine (C'), you've obviated any question of lack of novelty, (though probably not obviousness).

In sum, although you'd like to be able to eliminate as many elements as possible and recite all of your elements as broadly as possible, you will usually have to settle for less because there will always be prior art there to make you toe the line of novelty.

Moreover, as I've stressed, novelty isn't enough. The claims must define an invention which would be unobvious to one having ordinary skill in the art. Or to use the paraphrase of the law from Chapter 5, the novel feature(s) of the invention defined by each claim must have one or more new features that are important, significant, and produce valuable, unexpected new results. Thus, when you have to narrow a claim to define over the prior art, you must do so by adding one or more elements or by reciting existing elements more narrowly, and you must be sure that the added or narrowed elements define a structure or step which is sufficiently different from the prior art to be considered unobvious. More on this in Chapter 13.

For the last bit of bad news, note that if the wording of a claim has two possible interpretations, the examiner is entitled to use the one least favorable to you in determining whether the claim clears the prior art.

Now that I've given you the bad news, I suggest you ignore it at this stage. You should try to rewrite the claims as broadly as possible while keeping in mind the prior art which you've uncovered. In case of doubt, you should err on the side of too much breadth since you can always narrow your claims later if your examiner thinks they're too broad.

Conversely, if your examiner allows your narrow claims on your first office action (rare), you'll find it very difficult to broaden them later.

G. Technical Requirements of Claims

AS STATED, in addition to defining adequately over the prior art, each claim must also be worded in a clear, concise, precise, and rational way. If the wording of a claim is poor, the examiner will make a "technical" (non-art) rejection under Section 112. It is this technical aspect of drafting claims that most often serves as a stumbling block to the layperson. Yet claim drafting really won't be that hard if you:

- Study the sample claims listed later in this chapter, plus those of a few patents, to get the basic idea;

- Use the four-step method (preamble-element-interconnections-broaden) set out in Section H below; and

- Are conversant with the appropriate terminology associated with your invention's elements.

Remember also that you needn't write perfect claims when you file the application. Why? Because if you have a patentable invention, you can have the examiner write them for you. A provision of the *Manual of Patent Examining Procedure*, Section 707.07(j), states:

When, during the examination of a pro se [no attorney] case, it becomes apparent to the examiner that there is patentable subject matter disclosed in the application [the examiner] shall draft one or more claims for the applicant and indicate in office action that such claims would be allowed if incorporated in the application by amendment.

This practice will expedite prosecution and offer a service to individual inventors not represented by a registered patent attorney or agent.

Although this practice may be desirable and is permissible in any case where deemed appropriate by the examiner, it will be expected to be applied in all cases where it is apparent that the applicant is unfamiliar with the proper preparation and prosecution of patent applications.

You do have to at least give it a try since you must file at least one claim with your application to get a filing date. But, as indicated, this claim need not be well written or qualified for patent protection. Instead, during the ensuing prosecution stage, you can ask the examiner to write claims for you pursuant to this section if you feel yours aren't adequate. The examiner is bound to do so if your invention is patentable.

If you do choose this option, be sure the claims are broad enough, since it isn't in the examiner's own interest to write broad claims for you. As with any other claim, ask yourself if any elements of the examiner's claim can be eliminated or recited more broadly, and still distinguish adequately over the prior art. If so, amend it as I suggest in Chapter 13, Section E.

Also remember that many patent attorneys and agents will be willing to review or draft your claims at their regular hourly rates. But use this as a last alternative since most patent attorneys in private practice charge $75 to $200 per hour. If possible, you should choose a company-employed patent attorney or a retired patent attorney who works at home, since such attorneys' rates will usually be one-half to one-third of those charged by their downtown counterparts. See Chapter 6, Section F for how to find patent attorneys and agents.

Now that you know there's help out here, let's look at some of the basic rules covering the drafting of claims.

1. Be Precise

Your claims must be precise and logical. One of the most common claim rejections is for improper use of articles, such as "a", "the", and "said". Generally, the

first time you recite an element, use the article "a." If you refer to the same element again using exactly the same words to describe it, use "said." But if you refer to the same element again by using different, but implicitly clear words to describe it, use the article "the," just as you would do in ordinary speech. Here's an example showing how "a," "said" and "the" are properly used in a claim to a table:

Example: An article of furniture for holding objects for a sitting human, comprising:

(a) *a* sheet of rigid material having sufficient size to accommodate use by a human being for writing and working,

(b) *a* plurality of elongated support members of equal length,

(c) *said* support members being joined normally to *the* undersurface of *said* sheet of rigid material at spaced locations so as to be able to support *said* rigid member in a horizontal orientation.

Note that the first time any element is mentioned, the article "a" is used, but when it's referred to again by its original designation, "said" is used. When it's referred to again with a different designation, i.e., the undersurface of the table, "the" is used.

In addition to being precise in the use of articles, you should avoid ambiguous references. For example, if "said elongated lever" is used in a claim and no "elongated lever" has previously been recited, a non sequitur has occurred and a rejection for indefiniteness due to a missing antecedent will be made by the PTO. Or, if the same element is positively recited twice, such as "a lever" . . . "a lever," the claim is unclear. The solution is to change the second "a lever" to "said lever."

Vagueness and indefiniteness can also occur if you use abbreviations, e.g., "d.c." (say "direct current" instead) or relative terms without any reference, e.g., "large" (say "larger than . . ." or "large enough to support three adults."

2. Use Only One Capital, One Period, and No Dashes, Quotes, or Parentheses

Amateurs violate this rule so often that a friend who has a foreign patent translation agency and who wants to show he's professional includes the following blurb in his ad flyer: "We promise never to include more than one period or capital letter in any translated claim, no matter how long it is." While it may be hard for you to accept, and while it may seem silly, the rules are that the only capital letter in a claim should be the first letter of the first word, the claim should contain a period only at its end, and there should be no dashes, quotes, or parentheses.

3. Use Means Clause to Avoid Functionality of Claim

The technical error of "functionality" occurs when elements of the claim are recited in terms of their advantage, function or result rather than in terms of

their structure. The remedy is to recite the elements of the claim as "means" for performing the function or achieving the result.

For example, here are some typical improper functional claims actually written by a layperson:

7. An *additive for paints that makes the paint dry faster.*

8. A *belt buckle which does not tend to snag as much.*

Both of these claims would be rejected under Section 112 because they don't particularly point out and distinctly claim the invention since they recite what the invention *does* rather than what it *is*.

The remedy: use "means" as much as possible. But remember that the claim must be to a combination; a single "means" claim won't pass muster. Thus even if Claim 7 were written as follows, it would violate Section 112:

7. *Additive means for paints for making them dry faster.*

Here's how the above two claims can be properly rewritten to pass muster under Section 112:

7A. A *paint composition comprising:*

(a) a paint compound comprising an oil-based paint vehicle and a suspended pigment in said vehicle, and
(b) additive means admixed with said vehicle for decreasing the drying time of said paint compound.

8A. A *belt buckle comprising:*

(a) a catch comprising two interlocking rigid parts which can be attached to opposite ends of a belt, and
(b) anti-snag means for preventing said interlocking parts from snagging on cloth when placed adjacent said interlocking parts.

A moment's reflection will show you that claiming your invention in terms of its unique structure, rather than its results, effects, or functions, makes logical sense. This is because a monopoly, to be precise and to have reasonable limits, must be defined in terms of its structure, rather than the result such structure produces. In other words, if you recited "a belt buckle which doesn't snag" you would be claiming a result only, so that any belt buckle which fulfilled this result would infringe, regardless of its structure. This "functional" type of claim would accordingly be considered unreasonably broad and therefore would have to be narrowed and made more explicit by the addition of some additional structure in order to be more commensurate with the invention.

Of course, while both of the above claims (as I revised them) would pass Section 112, they would not be novel or patentable under Sections 102 or 103 since they recite nothing new according to our present state of knowledge.

4. Be Complete

Each claim must stand on its own, i.e., it must recite enough elements to make a working, complete invention. The remedy for failing to do this is to add the needed elements, either separately or by rewording other elements to cover any missing function. Examiners and attorneys frequently disagree as to whether a claim is incomplete, the examiner wanting the claim narrowed by the addition of elements and the attorney wanting it to remain broad, that is, not to add any more elements.

5. Keep Language Straightforward and Simple

Properly drafted claims use a minimum number of words to delineate the essence of the invention. Excess wordiness of a claim, termed "prolixity" by the PTO, is a frequent error committed by beginners.

The remedy is to reword the claim in more compact language.

6. All Elements of Invention Must Logically Interrelate

Each of the elements in a claim must be logically related and connected to the other elements. When the elements of an invention don't appear to cooperate and be connected in a logical or functional sense, the PTO will reject the claim. This is a more substantive type of rejection since it's often directed at the underlying invention rather than simply the way the claim is drafted. For example, if you claim the combination of a waffle iron and tape recorder, these elements don't cooperate and hence your claim would be rejected as drawn to an aggregation. But the elements don't have to work at the same time to cooperate; in a typewriter, for example, the parts work at different times but cooperate toward a unitary result.

7. Old Combination

Avoid drafting your claim in terms of an old or well-known combination (e.g., a carburetor and an automobile), when the invention is only in the carburetor. Thus, a claim that recites an automobile, including a chassis, tires, engine, transmission, and a carburetor mounted on the engine which has trapezoidally-shaped Venturi chambers would be rejected as drawn to an old combination since the invention is in the carburetor and since the combination of a carburetor with an automobile is notoriously old. The remedy is to remove the automobile from the claim: carburetors (and fuel injectors) have a recognized status of their own.

8. Use Only Positive Limitations

In the past, all negative limitations (e.g., "non-circular") were verboten, but now only those which make the claim unclear or awkward are proscribed. However, because many examiners still wince when they see negative limitations in claims, it's best to avoid them by reciting what the invention is, rather than what it isn't. For instance, instead of saying, "said engine connected to said wheels without any transmission" say "said engine connected directly to said wheels." You are permitted to recite holes, recesses, etc.; see "Voids" in the Glossary for a list of "hole-y" words.

9. Use Proper Alternative Expressions

As with negative limitations, disjunctive expressions—that is, those using "or" or the like—are permissible so long as the two expressions are just different ways of reciting the same things. If two different things are meant, however, try to find a generic term to cover both, or use two separate claims. Thus, instead of saying, "said amplifying circuit containing a vacuum tube or transistor," say "said amplifying circuit employing an amplifying valve." You can also use two claims, one to recite the tube and the other to recite the transistor.[2]

---

[2]A third, sophisticated way is to recite the disjunctive elements by using a Markush group. A Markush (from a decision with that name) group is a series of related elements joined by "and" which follows these magic words: "selected from the group consisting of". Thus a tube or a transistor could be recited in one claim as follows: "said amplifying circuit containing a device selected from the group consisting of tubes and transistors."

10. Avoid Too Many Claims

If you've put in too many similar claims, even though you've paid for them, you'll have to eliminate some to make the examiner's job easier. If you ever have more than twenty claims, the invention should be complex enough to justify them and the claims should differ substantially.

11. Make Sure Claims Correspond With Disclosure

The literal terms or words of the claim must be present somewhere in the specification. If they aren't, the remedy is to amend the specification by adding the exact terms used in your claims, or to amend the claims by eliminating those terms which aren't literally in the specification.

12. Make Sure Claims Are Supported In Drawing

Nothing is really wrong with the claim here, but remember that the drawings must show every feature recited in the claims. If they don't, amend either the drawing or the claims. A broad recitation in a claim, e.g., "fuel supply means", can be supported by specific hardware, e.g., a carburetor, in the drawings.

13. Claim Computer Program with Some Hardware

If your invention involves (or actually is) a computer program—i.e., a set of instructions for a computer—you must claim it in relation to how it works with some physical structure, even though minimal. You can't simply claim the software as a process. Why not? No one knows for sure why the PTO resisted the first programs. My guess is because of neophobia (fear of the new), lack of understanding, and a wish to avoid work. Anyway, the courts have generally upheld the PTO on this point. See Chapter 5 Section B. for more on this point. Here's an example of some "program" claims drafted to recite enough hardware to pass muster; these claims go about as far as one can go in claiming programs.

9. *A process of operating a general purpose data processor of known type to enable the data processor to execute formulas in an object program comprising a plurality of formulas, such that the same results will be produced when using the same given data, regardless of the sequence in which said formulas are presented in said object program comprising the steps of:*

    (a) examining each of said formulas in a storage area of the data processor to determine which formulas can be designated as defined;

    (b) storing, in the sequence in which each formula is designated as defined, said formulas which are designated as defined;

    (c) repeating steps (a) and (b) for at least undefined formulas as many times as required until all said formulas have been designated as defined and have been stored; thereby producing the same results upon sequential execution of the formulas stored by said process when using the same given date, regardless of the order in which said formulas were presented in the object program prior to said process. (Pardo & Landau, patent 4,398,249; 1983)

10. *A method for evaluating boolean expressions in a computer system comprising:*

    - *forming a first constant from the expression to specify rearrangement of the variables,*

    - *setting said first constant into a work area,*

    - *translating said first constant in said work area using the variables as a translate table,*

    - *forming a second constant from the expressions where the second constant functions to change the values of the variables to position numbers having values one less and two less than the position number of the variable and where the second constant changes the*

zeros between variables into position numbers which point to previous positions in the result string containing values of previously evaluated subexpressions

- *logically combining said translated first constant with said second constant using an exclusive OR operation, and*

- *translating the result of the exclusive OR operation using the result as the translate table as the result is changing during the translation, the result from last translation in the result being the value of the boolean expression being evaluated. (Berstis, patent 4,417, 305; 1983)*

Note that in the Pardo et al. patent, the patent attorney has thrown some hardware, albeit minimal, into the claim: "storage area" is recited in clause (a). In the Berstis patent, "work area" and "exclusive OR" are used to clothe a technically-unpatentable algorithm with some hardware.

The patent court has held, however, that in computer-art cases, the hardware recitation must be fairly specific (as in the Pardo/Landau and Berstis patents) and not solely a "means for" performing specified algorithmic functions. This is because "means" clauses can effectively cover all ways to perform the algorithm: the court doesn't want you to be able to get around their bar on patenting algorithms per se by using only "means" clauses to define the algorithm.

14. Recite Each Element Affirmatively as Subject of Its Clause

For maximum clarity the elements of your invention should be affirmatively and directly recited; don't bring them in by inference or incidentally, e.g., say "A transmission comprising: (a) a gear, (b) a shaft, (c) said gear being mounted on said shaft" [etc.], and not "A transmission whose gear is mounted on its shaft . . ." In other words, each significant element of the claim should be recited for the first time (introduced) with the word "a" so it's the subject of its clause, and not with wording which makes it part of the object or assumes that the reader already knows that it's there. This rule is especially important for do-it-yourselfers to follow in order to write clear and understandable claims.

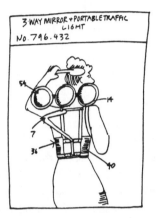

15. Include Structural Support In Recitation of Operation

Assume a claim recites "a lever connected to move said pendulum to and fro at the same rate as said lights flash." The movement of the pendulum at this special rate is too much for the lever to do all by itself. In other words, there's not enough structural support for the operation recited. The remedy? Recite either (a) enough structure to do the job or (b) use a "means" clause. Here's examples of both methods:

(a) "a photoresponsive electromechanical circuit terminating in a lever which is connected to said pendulum and is arranged to move said pendulum at the same rate as said lights flash."

(b) "means, including a lever connected to said pendulum, for moving said pendulum at the same rate as said lights flash."

H. Drafting Your Main (Independent) Claim

THERE ARE TWO basic types of claims: "independent" and "dependent." "Independent claims" are those that don't refer back to any preceding claim; they stand alone. Examples of independent claims are all of those given in the preceding sections of this chapter. Note that these claims don't refer back to any preceding claim and each defines a complete, operative invention by itself.

"Dependent claims," which will be covered in the next section, refer back to a preceding or "parent" claim (this preceding claim can either be independent or dependent) and narrow the subject matter of the preceding claim in either of the two standard ways, i.e., either by adding an additional element(s) or defining one or more elements of the preceding claim more narrowly.

The reasons for providing dependent claims will be covered in the next section also; the main point to remember here is that your independent claims are the important ones since they're the basic and broadest definitions of your invention. If a dependent claim is infringed, its independent or parent claim(s) must also be infringed. If an independent claim is infringed, however, that's enough to win the case. You don't have to worry about your dependent claims.

To draft an independent claim, the easiest and most direct way to do it is to follow these four basic steps:

1. Write a preamble giving the name or title of the invention.

2. List the elements (or steps) of the claim.

3. Interconnect the elements or steps.

4. Broaden the claim.

The claim can be structured so that the elements of the claim appear together, followed by the interconnections. Or, each element can appear in conjunction with its interconnection(s) to adjacent element(s). Most patent attorneys use the latter method—see Claims (2), (3), and (4) in Section C above for examples—but you you may find it easier to recite the interconnections separately. An exception is process claims, where you'll find it easier directly to associate each step with its predecessor.

Start by writing your first claim without regard to breadth, i.e., just get a preamble written, set down the elements of the invention, and interconnect them, paying no attention to how broadly you can recite the invention. In other words, just define your invention as you believe necessary to "get it all down" in a complete manner.

Then, see how many elements (or steps) you can eliminate and how many remaining elements you can broaden so that the result maintains sufficient structure and yet does not tread on the prior art to much. Remember that the broadest way of defining any element is by using "means-plus-a-function" language. Don't forget to refer to your prior-art patents for examples.

To provide a real example which everyone can understand, let's assume you've just invented a table. Since you've already written your specification, you have a name for each part of your invention so this chore is already behind you. (If you believe your part names leave something to be desired, you can get additional part names from your prior-art search patents, the Glossary at the end of this book, or the *What's What* book (see Bibliography). All that remains now is to provide a title or preamble. List the parts, interconnect them, and then broaden your claims.

1. The Preamble

To write the preamble, pick a name or title for the whole unit, remembering that you can't use the word "table" since it hasn't been invented until now. Try to put it in a class to which it belongs. Since a table in an "article of furniture," these words would be fine.

You could also use any other suitable class, such as a "work station device," a "support for holding objects to be handled by a sitting human," etc. I've used "an article of furniture" and I've added the modifier "for holding objects for a sitting human comprising:" to narrow the field a bit and to make my title more meaningful.

2. The Elements

Next, to list the parts of the table, I'll start with the largest, most visible part, the top, and then add the smaller, less apparent parts, the legs. Since the table's just been invented, we'll assume that the words "top" and "legs" are still unknown, but even if they were, we wouldn't want to use "top" anyway, since it's a notoriously vague homonym (it can mean anything from a hat to a bottle cap to a toy). To define the top, then, we need a more meaningful term or phrase. Let's suppose we've made a model of our invention and have used a large sheet of chipboard for the top. All we need to do at this stage is to say so; thus our first and most basic element becomes "(a) a large sheet of chipboard."

Suppose our model table has four legs and we've made them of six-cm diameter circular oak dowels, each 65 cm long. Then our legs would be recited simply as "(b) four oak dowels, each having a circular cross section 6 cm in diameter and each 65 cm long." Our elements are now all recited—wasn't that easy!

3. Interconnections

Lastly, we have to interconnect the legs to the top, an easy task. Suppose our legs are joined at the underside of the top using four metal flanges, attached at the four corners of the top with each having a cylindrical portion with female threads, and with the top sections of the legs being provided with mating male threads which are screwed into the respective flanges so that the legs extend at right angles to the top. Merely recite the flanges positively and add an interconnection clause as follows:

(c) four flanges, each having attachment means for attachment to one side of said sheet of chipboard and each having a cylindrical portion with female threads, and

(d) said four flanges being attached to one side of said sheet of chipboard at four respective corners thereof and said four oak dowels having male threads on a top section thereof and being screwed into the cylindrical portions of said respective flanges so that said dowels extend from said sheet of chipboard at right angles.

Eureka! It's done. You've written a complete independent claim.

Here's how it looks:

11. *An article of furniture for holding objects for a sitting human, comprising:*

(a) a large sheet of chipboard,

(b) four oak dowels, each having a circular cross section 6 cm in diameter and each 65 cm long, and

(c) four flanges, each having means for attachment to one side of said sheet of chipboard and each having a cylindrical portion with female threads; and

(d) said four flanges [etc.]

Note that I always recite the elements and their interconnections in lettered subparagraphs. The PTO and courts prefer (but do not require) this format since it's easier to analyze than a continuous paragraph.

Is there anything wrong with this? Yes! As you probably will have realized by now that this claim is far too narrow, i.e., it has many elements each of which is recited very specifically. In fact it even recites specific dimensions, which you don't generally even need in the specification. Let's broaden it then.

Remember, you broaden a claim by (1) eliminating elements where possible, and (2) reciting the remaining elements as broadly as possible.

Going through the claim to eliminate elements, we see that the top can't be eliminated since it's an essential part. However, we don't need to recite four legs—we can eliminate one of these since three legs will support the top. But better yet, we can even use the word "plurality" since this covers two or more legs. (The term "plurality" means more than one. Used here, it is an example of how you'll sometimes need to search for a word or phrase that most broadly describes a particular element.) Even though two may not be sufficient to support a top, the PTO will usually not object to this word in this context. We could even go further and eliminate the recitation of legs entirely by reciting "support means," but this would include solid supports, such as in a chest or bureau, which would not be suitable for table-type uses.) Lastly, we can eliminate the flanges since these aren't essential to the invention and since there are many other possible ways of attaching legs to a table top.

Next, let's go through the claim to see which elements can be recited more broadly. First, the top. Obviously "a large sheet of chipboard" is a very narrow recitation since plywood, solid wood, metal, and plastic tops would avoid infringement. A broad recitation would be "a large sheet of rigid material" but, as stated above, the word "large" is frowned upon by the PTO as too vague to satisfy Section 112. So let's make the top's size more specific. Since we're interested in providing a working surface for humans, let's merely specify that the top is "a sheet of rigid material of sufficient size to accommodate use by a human being for writing and working."

Next the legs. Obviously the recitation of four circular oak dowels with specific dimensions is very limiting. Let's eliminate the material, shape, and dimensions and recite the legs as merely "a plurality of elongated support members of equal length." This covers square, round, triangular, and oval legs, regardless of their length or material.

Lastly, instead of the flanges, (which we've eliminated as unnecessary,) to join the legs to the top, let's use "means" (to make it as broad as possible) as follows: "means for joining said elongated support members normally or at right angles to the underside of said top at spaced locations so as to be able to support said top horizontally."

The result would look like this:

11. An article of furniture for holding objects for a sitting human, comprising:

(a) a sheet of rigid material of sufficient size to accommodate use by a human being for writing and working;

(b) a plurality of elongated support members of equal length;

(c) means for joining said elongated support members or at right angles to the underside of said top at spaced locations so as to be able to support said top horizontally.

Obviously, Claim 11 is now far broader than our first effort. Your first independent claims should be as broad as possible, but of course, you can't make it so broad that it lacks novelty or unobviousness. Thus, when you eliminate as many elements as possible, and when you broaden the remaining elements in the manner just described, keep in mind that you must leave enough structure or acts to define your invention over the prior art. Put differently, writing claims is like walking on a beam: you can't sway too far on the side of specificity or you'll fall onto the side of worthlessness and you can't sway too far onto the side of breadth or you'll fall onto the prior art. To obtain the broadest possible coverage, you should not draft your main claim primarily to cover your invention; rather draft it as broadly as possible with at least some thought of clearing the prior art, then go back and make sure that it at least covers your invention. Some patent attorneys compare the writing of their first claim to passing through a wall of fire. However, I have found that if I follow the above four steps—(1) write a preamble, (2) recite the elements, (3) interconnect them, and (4) broaden the claims—the going is relatively painless. In case of doubt, err on the side of breadth at this stage since you can always narrow your claims later, but you may not be able to make them broader.

I. Other Techniques In Claim Writing

NOW THAT YOU UNDERSTAND the basics, here are some other tricks you may want to use when writing your claims. Obviously not all apply all of the time, but you will probably find that at least several can be used to improve your claim writing.

- Use "weasel" words like "substantially," "about," or "approximately" whenever possible whenever you specify a dimension or any other specific parameter to avoid limiting your claim to the specific dimension specified.

- At the end of your claim, you may want to add a "whereby" clause to specify the advantage or use of the invention to hammer home to the examiner, or anyone else who reads your claim, the value of your invention. Thus in Claim 11 above, you could add at the end of this claim, "whereby a human can work, eat, and write in a convenient seated position." "Whereby" clauses don't help to define over the prior art, but they do force the examiner to consider the advantages of your invention and thus help to get the claims allowed.

- You may put the drawing's reference numerals in your claims after the appropriate elements, but this is seldom done unless the elements of the claim aren't clear.

- If your invention has an opening, hole, or recess in its structure, you may, as stated, recite the hole directly as such, even though it isn't tangible. For example, the recitation "said member having a hole near its upper end" is permissible.

- Sometimes, instead of using the preamble-elements-interconnections approach, it's desirable to omit the preamble, especially if you feel the preamble will be too restrictive, that is, if the elements of the body of the claims can be used for another function. For example, if we recited "A working surface comprising" as a preamble in the above claim and someone used the actual structure claimed, but turned it upside down and used the legs for a quoit game, it would not infringe since it isn't being used as a working surface. In this case simply start the claim, "In combination:" or "A process comprising:" and then recite the elements or steps and their interconnections.

- With regard to the rarely-enforced Rule 75(e) (quoted above) requiring the use of Jepson (with a preamble containing old elements and body of claims containing improvements of your invention), most patent attorneys recommend that claims not be cast in this style unless the examiner requests it or unless the examiner is having trouble understanding exactly what your inventive contribution is. The reason for this is that a Jepson claim isolates and hence minimizes your improvement, making it easier to invalidate. If you do claim in the Jepson format, draft your preamble so that it includes all the elements or steps and their interconnections that are already known from the prior art, then add a "cleavage" clause such as "the improvement comprising," or "characterized in that," and then recite the elements of your invention and their interconnections.

- A claim which recites a group of elements can be made "open" or "closed." An open claim (the normal case) will cover more elements than it recites, whereas a closed claim is limited to and will cover only the elements it specifically recites. To make a claim open, use "includes" or "comprising," e.g., "said machine comprising A, B, and C." In this case, a machine with four elements A, B, C, and D will infringe. To make a claim closed, use "consist" or "having only," e.g., "Said machine consisting of A, B, and C." In this case, a machine with elements A, B, C, and D will not infringe since, in patent law, the word "consist" is interpreted to mean "having only the following elements."

- After drafting your claim, you or a friend should be able to make enough sense out of it to sketch your invention. If this isn't possible, the claim is unclear and needs to be reworked.

- You can use any technical or descriptive terms which you feel are reasonably necessary to define or describe your invention—the claim does not have to be limited to any special "legalese." One patent attorney I know had a devil of a time defining (to the satisfaction of the examiner) a convex transistor structure with a nubbin on top until he simply called it "mammary-shaped."

- Lastly, many patent attorneys recommend that a claim not appear too short. A claim that is short will be viewed adversely (as possibly overly-broad) by many examiners, regardless of how much substance it contains. Thus, many patent attorneys like to pad short claims by adding whereby clauses, providing long preambles, adding long functional descriptions to their means clauses, etc. The trick here, of course, is to pad the claim while avoiding a charge of undue prolixity under Section 112.

If you do recite any "means," it's desirable to label the means with an adjective in order to provide a mnemonic aid in case you need to refer to the means later. E.g., "distal means," "closing means," etc. Also the "means" must be followed by or be modified by a function. E.g., "plate means" isn't a means plus a function since "plate" isn't a function, but "printing means" is OK since "printing" is a function. Ex parte Klumb, 159 U.S.P.Q. 694 (Pat. Off. Bd. App. 1967)

J. Drafting Dependent Claims

IN SECTION H, I pointed out that there are two basic types of claims—independent claims (that stand on their own) and dependent claims (that incorporate an entire other claim, this other claim can be a previous independent or dependent claim. A dependent claim is simply a shorthand way of writing a narrower claim, i.e., a claim which includes all the elements of the previous claim, and adds one or more additional elements or narrows one or more elements of the previous claim).

Dependent claims are by definition always narrower than the claims on which they depend. You may accordingly be wondering, "If my broad independent claim covers my invention, why do I need any more claims of narrower scope?" True, if all goes well, your broad claim will be all you'll need. However, suppose you sue an infringer who finds an appropriate prior-art reference which neither you nor the PTO examiner found and which adversely affects the validity of ("knocks out") your broad claim. If you've written a narrower claim you can then disclaim the broad claim and fall back on the narrower claim. If the narrower claim is patentable over the prior art, your patent will still prevail. Each claim, whether independent or dependent, is interpreted independently for examination and infringement purposes. If the claim is dependent, it's interpreted as if it included all the wording of its parent (incorporated) claim or claims.

Also, your dependent claims are useful to explain and reify (make real) some of the broad,

abstract terms in your independent claims. For instance, if you recite in a claim "additive means," many judges may not be able to understand what the "additive means" actually covers, but if you add several dependent claims which state, respectively, that the additive means is benzine and toluene, they'll get a very good idea of what types of substances the additive means embraces.

Further, narrower claims can be used to provide a range or spectrum of proposed coverage from very broad to very narrow so that your examiner can, by allowing some narrower claims and rejecting the broader ones, indicate the scope of coverage the examiner's willing to allow.

Finally, providing dependent claims of varying scope and approaches forces the examiner to make a wider search of your invention on the first examination. This will prevent the examiner from citing new prior art against your application on the second office action, which usually must be made "final" action (see Chapter 13).

Thus, when you're satisfied with your first, basic, and broadest independent claim, you should write as many dependent claims as you can think of. Each dependent claim should begin by referring to your basic claim, or a previous dependent claim, using its exact title. If the dependent claim is narrowing one or more elements of the independent claim, it should start, e.g., "The bicycle of Claim 1 wherein" and then continue by narrowing the elements of the independent claim. If the dependent claim is adding additional elements, it should start, e.g., "The bicycle of claim 1, further including..." then continue by reciting the additional feature(s) of your invention. The additional features can be those you eliminated in broadening your basic claims and all other subsidiary features including combinations and permutations of such features of your invention you can think of. The features added or narrowed by the dependent claims can be specific parameters (materials, temperatures, etc.) or other specifics of your invention (specific shapes, additional elements, specific modes of operation, etc.). Refer to your prior-art patents for guidance on how to draft these.

Here are some dependent claims for Claim 11, above. Note that each dependent claim either adds an element to the coverage of Claim 11 or narrows an already recited element.

> 12. *The article of furniture of claim 11 wherein said sheet of rigid material is made of wood.*

> 13. *The article of furniture of claim 12 wherein said wooden sheet of rigid material is made of chipboard.*

> 14. *The article of furniture of claim 13 wherein said sheet of chipboard has a rectangular shape.*

> 15. *The invention of claim 11, further including a set of flanges, each of which joins a respective one of said support members to the underside of said sheet of rigid material.*

> 16. *The invention of claim 15 wherein each of said flanges is made of iron and includes a cylinder with female threads and wherein one end of each of said elongated members has male threads and is threadedly mated with the female threads of a respective one of said flanges.*

Note that a dependent claim may be dependent upon the parent claim or another dependent claim. A dependent claim should be numbered as closely as possible to the number of its parent claim.[3] Note also how I've made a physical indication of claim dependency by indenting each dependent claim under its parent claim(s). This is optional, but makes things clearer for you and the examiner

A dependent claim will be read and interpreted by examiners and judges as if it incorporated all the limitations of its parent claim(s). Thus suppose your

independent and dependent claims read, respectively, as follows:

> 17. *A rifle having an upwardly-curved barrel.*

> 18. *The rifle of Claim 17 wherein said barrel is made of austenitic steel.*

The dependent claim (18) will be treated independently, but with claim 17 incorporated as if it read as follows:

> 18'. *A rifle having an upwardly-curved barrel, said barrel being made of austenitic steel.*

You can make your dependent claims as specific as you want, even to reciting the dimensions of the table top, its color, etc.. However, extremely specific limitations like this, while possibly defining an invention which is novel over the prior art (Section 102), do not recite unobvious subject matter (Section 103) so they'll be of little use to fall back on if you lose your independent claim. Thus you should mainly try to use significant limitations in your dependent claims.

You should try to draft at least one dependent claim with as many parts as possible so as to provide as broad a base as possible for maximizing infringement damages. Also try, insofar as possible, to draft at least one claim to cover parts of the invention whose infringement would be publicly verifiable, rather than a non-verifiable factory process or machine.

As with independent claims, you should not make your dependent claims purely "functional," i.e., each dependent claim should contain enough physical structure to support its operational or functional language. Here are some examples:

Wrong:
> 17. *The bicycle of claim 16 wherein said derailleur operates with continuously variable speed-to-power ratios.* [This claim has no structure to support its operational limitation.]

---

[3]A dependent claim may actually be made dependent upon several previous claims (called "multiple dependent claiming" and common in Europe) but you should not write multiple dependent claims since the PTO's examiners dislike the practice and since there's a surcharge for the privilege. See Fee Schedule in Appendix.

Right:

17. The bicycle of claim 16 wherein said derailleur contains means for causing it to operate with continuously-variable speed-to-power ratios. [The "means" limitation is a recitation of structure which supports the operational limitation.]

Right:

17. The bicycle of claim 16 wherein said derailleur contains a cone-shaped pulley and a belt pusher for causing it to operate with continuously-variable speed-to-power ratios. [The pulley and pusher constitute structure which supports the operational limitation.]

If your independent claim recites a means plus a function, your dependent claim should modify the means and not the function. E.g., if an independent claim recites, "variable means for causing said transmission to have a continuously variable gear ratio." Here are the right and wrong ways to further limit this "means" in a dependent claim:

Wrong:

19. The transmission of claim 18 wherein said continuously variable gear ratio ranges from 5 to 10.

Right:

19. The transmission of claim 18 wherein said variable means is arranged to provide ratios from 5 to 10.

▲

Common Misconception: *If a dependent claim recites a specific feature of your invention, say a two-inch nylon gear, your invention will be limited to this gear so that if any copy of the invention uses a one-inch gear, or a steel gear, it won't infringe on your patent.*

Fact: *Although the copy won't infringe the dependent claim, it will infringe the independent claim so long as it isn't limited to this specific feature. And as long as even one claim of a patent is infringed, the patent is infringed and you can recover as much damages (money) as if 50 claims were infringed.*

If you still don't get the principle of broad and narrow claims, here's three simple claims which everyone can understand:

1. All eye care professionals

2. The persons of Claim 1 who are medical doctors.

3. The persons of Claim 2 who are strasbismologists living in the City of Belvedere

Claim 1 is very broad: it will cover opthamologists, optometrists, and opticians all over the earth. Claim 2 is of intermediate scope: it's longer than #1 but is narrower in scope since it eliminates everyone but MD's. Claim 3 is very narrow: it's still longer than Claim 2, but is far narrower since it eliminates everyone but strasbismologists in Belvedere.

K. Drafting Additional Sets of Claims

AFTER YOU'VE WRITTEN your first independent claim and all the dependent claims you can think of (all numbered sequentially), consider writing another set of claims (an independent claim and a set of dependent claims) if you can think of a substantially different way to claim your invention. See the prior-art patents and the sample set of claims at the end of this chapter (Fig. 9-A) for examples of different independent claims on the same invention. Your second set of dependent claims can be similar to your first set; a word processor with a block copy function will be of great aid here. Writing more sets of claims will not give your invention broader coverage, but will provide alternative weapons to use against an infringer and will give your examiner additional perspectives on your invention. Also your chances of getting your examiner to bite will be increased if you present many flavors to choose from.

In the example above (Claim 11), I might start my second independent claim with the legs instead of the top and I might try to define the top and legs differently, e.g., instead of "elongated members," I might call the legs "independent support means." Instead of calling the top a "sheet of rigid material" I might call it a "planar member having paralleled, opposed major faces."

Here are still other ways to write a different independent claim: ① Rewrite one of the dependent claim from your first set in independent form; ② wait a few days and write an independent claim again, with independent thought, and ③ write the second independent claim by reciting the elements of the first independent claim in reverse or inverse order.

Still another way to write a different independent claim is to provide a method claim if your first independent claim is an apparatus claim, or vice-versa; you're allowed to have both method and apparatus claims in the same case.

Your filing fee entitles you to up to three independent claims and twenty total claims. I generally try to use up my allotment by writing three independent claims and three sets of five to seven dependent claims each. However, if I feel that I can write a fourth, substantially different independent claim and the cost can be borne by my client, I will add it, plus more dependent claims. The PTO charges for each independent claim over three, and for each claim (independent or dependent) over twenty.

On the other hand, for relatively simple inventions, I may not be able to think of any substantially different ways to write an independent claim, so I may submit only one, plus a few dependent claims. I advise you generally not to submit more than the number of claims permitted for your basic filing fee, i.e., three independent claims 20 total claims, unless the complexity of your invention justifies it, or you have some other good reason.

As with the specification, be sure to review your claims very carefully after you've written them.

L. Checklist for Draft of Claims

HERE'S THE SECOND PART of the application checklist which I started in Chapter 8. As before, I suggest you go through this list carefully and make any needed corrections in your claims before going on to Chapter 10.

[ ] CO1. Grammatical articles are used properly in the claims:

"a" to introduce any part,
"the" to refer to a part a second time when using a different (but clearly implied) term as before, and,
"said" only to refer to a part using the IDENTICAL term as before.

[ ] CO2. Two articles together, such as "the said," aren't used.

[ ] CO3. Every part in every claim is shown in the drawings and discussed in the specification.

[ ] CO4. No claim uses any disjunctive ("or") expression (except to recite two equivalent parts or a disjunctive function of a machine).

[ ] CO5. No claim uses any naked functional clause; all claims contain a structural recitation or "means" to support every functional recitation.

[ ] CO6. A function is recited adjacent each "means"; also each "means" has a label (e.g., "lever means") if the function itself doesn't provide a label.

[ ] CO7. "Consisting" isn't used in any claim (except to say "having only").

[ ] CO8. No claim uses any abbreviation, dash, parentheses, or quote.

[ ] CO9 No term is used for the first time in any claim.

[ ] C10 The subparagraph form is used in long claims for ease of reading.

[ ] C11. All claims have just one capital letter and one period.

[ ] C12. All significant parts are affirmatively recited in the claims as the subject and not the object of a clause.

[ ] C13.   All parts recited in claims are connected together.

[ ] C14.   All claims recite enough parts to provide complete assemblage.

[ ] C15.   You haven't submitted over 20 total or over three independent claims unless the case is very complex or extra claims are justified.

[ ] C16.   No independent claim refers to any other claim and all dependent claims refer to a previous claim in line 1 or line 2.

[ ] C17.   You've filed enough dependent claims to cover all features and permutations and you've filed second and third sets of claims (with differently-phrased independent claims) if possible.

[ ] C18.   Every dependent claim starts with either:
"The _____ of claim x wherein..." to narrow existing element(s), or
"The _____ of claim x, further including..." to add new element(s).

[ ] C19.   No dependent claim is used to substitute a different part for any part in its parent claim; each dependent claim either narrows or adds to the existing parts of its parent claim.

[ ] C20.   No dependent claim recites a method limitation if its parent claim is an apparatus claim, and vice versa.

start claims on new page

Claims: I claim:

first
independent
claim

1. In a bag closure of the type comprising a flat body of material having a lead-in notch on one edge thereof and a gripping aperture adjacent to and communicating with said notch, the improvement wherein said closure has a layer of paper laminated on one of its sides.

2. The closure of claim 1 wherein said body of material is composed of polyethyleneterephthalate.

optional
indent for
dependent
claims

3. The closure of claim 1 wherein said body is elongated and has a longitudinal groove which is on said one side of said body and extends the full length of said one side, from said gripping aperture to the opposite edge.

4. The closure of claim 3 wherein said groove is formed into and along the full length of said lamination.

5. The closure of claim 1 wherein said body is elongated and has a longitudinal groove which is on the side of said body opposite to said one side thereof and extends the full length of said one side, from said gripping aperture to the opposite edge.

6. The closure of claim 1 wherein said body is elongated and has two longitudinal grooves which are on opposite sides of said body and extend the full lengths of said sides, from said gripping aperture to the opposite edge.

7. The closure of claim 6 wherein the groove on said one side of said body is formed into and along the full length of said lamination.

8. The closure of claim 1 wherein said body has a paper lamination on both of said sides.

9. The closure of claim 8 wherein a groove is on one side of said body and extends the full length of said one side, from said gripping aperture to the opposite edge.

10. The closure of claim 8 wherein two grooves, on opposite sides of said body, extend the full lengths of said sides, from said gripping aperture to the opposite edge.

11. The closure of claim 10 wherein said grooves are rolled into and along the full lengths of said laminations, respectively.

12. The closure of claim 1 wherein said paper lamination is colored.

13. The closure of claim 1 wherein said body is elongated and has a longitudinal through-hole.

second
independent
claim,
phrased
differently
than first

14. A bag closure of the type comprising a flat body of material having a lead-in notch on one edge thereof, a gripping aperture adjacent to and communicating with said notch, characterized in that one of its sides has a layer of paper laminated thereon.

15. The closure of claim 14 wherein said body of material is composed of polyethyleneterephthalate.

16. The closure of claim 14 wherein said body is elongated and has a longitudinal groove on said one side of said body and which extends the full length of said one side, from said gripping aperture to the opposite edge.

17. The closure of claim 14 wherein said body is elongated and has a longitudinal groove which is on the side of said body opposite to said one side thereof and extends the full length of said one side, from said gripping aperture to the opposite edge.

18. The closure of claim 14 wherein said body is elongated and has two longitudinal grooves which are on opposite sides of said body and extend the full lengths of said sides, from said gripping aperture to the opposite edge.

19. The closure of claim 14 wherein said body has a paper lamination on both of said sides.

   20. The closure of claim 19 wherein a groove is on one side of said body and extends the full length of said one side, from said gripping aperture to the opposite edge.

   21. The closure of claim 19 wherein two grooves, on opposite sides of said body, extend the full lengths of said sides, from said griping aperture to the opposite edge.

22. The closure of claim 14 wherein said paper lamination is colored.

23. The closure of claim 14 wherein said body is elongated and has a longitudinal through-hole.

abstract follows on new page—see Ch. 8

Chapter 10

Finalizing and Mailing Your Application

NOW THAT YOU'VE DRAFTED your patent application, it's time to put it in final form. Since I place great emphasis on thoroughness, this chapter is, accordingly, filled with many picky details. In the event you want to rebel and simply pass over those requirements that are inconvenient, remember that your patent examiner has enormous discretion over whether your application will be approved or rejected. An application that meets the exacting standards of the PTO will have a better chance than one that doesn't.

Fortunately, while you must pay attention to detail, meeting the PTO's standards is relatively easy if you've followed my suggestions in the previous chapters. Because you've reviewed a number of patents in the same field as your own, you'll be familiar with the standards for writing the specification and claims (Chapters 8 and 9). Because you've prepared preliminary drawings (Chapter 8) in basic conformance with the rules for final drawings, putting them in final form will not involve great difficulty. Because you analyzed all relevant prior art known to you and can distinguish it from your invention, you are in a good position to follow through with your application to a successful completion (Chapter 13).

Enough said. Let's get started.

A. The Drawing Choices

YOU HAVE TWO BASIC CHOICES for your drawings. You can file the application with:

1. Formal drawings (generally xerographic copies of ink drawings done with instruments on bristol board or Mylar film and in accordance with all the rules); or,

2. Informal drawings (generally xerographic copies of good pencil or ink sketches).

Further, in each case the drawings can be filed in either:

A. The larger U.S. size (8.5" x 14");

B. The smaller U.S. size (8.5" x 13"); or

C. The A4 international size (210 mm x 297 mm).

Which type of drawing should you submit— formal or informal? I strongly recommend that, if at all possible, you do formal drawings. Formal drawings look much nicer and neater and thus will make a far better impression on the examiner, showing that you're serious about the invention. Remember, "Quality is remembered long after price is forgotten."—Stanley Marcus. However if cost and time are important considerations, you should file informal drawings. If you do, the PTO will examine your application in the same way, but will require you to file formal drawings when and if any claims are allowed. Also you'll have to prepare formal drawings about 11 months after filing if you want to file abroad; see Ch. 12. Neither photographs nor color drawings are ever acceptable (unless they show an aggregate or special structure).

As far as the choice of the U.S. or international sizes is concerned, the U.S. sizes are larger and easier to work with since you won't have to squeeze things in or cramp your drawings as much. However, if you have any serious thoughts about filing abroad, it's better to use the international (A4) size, since you can make good photocopies, file these for your U.S. application, and later use the originals (or another good set of copies) for the international application. (I discuss foreign filing in Chapter 12). If you do use the U.S. size and later decide to foreign file, you can still go to A4 by using a reducing photocopier or a patent drawing service in the Arlington, VA area (about $10 a sheet).

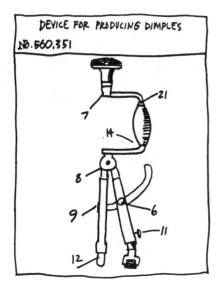

B. Doing Your Own Drawings

YOU MAY NOT WANT to get involved with doing your own drawings at all, but instead hire a draftsperson. If so, skip to Section D below. If you do your own drawings, it's best to use pencil or ink on Mylar film or vellum for most of the work, since this allows you to change your drawings readily and repeatedly. When you're satisfied with the drawings, make the lines very black by going over them with a soft, thin pencil or with India ink.

If you're like me, and can't draw well, neither India ink on bristol board nor Mylar film will be suitable. However, there still is a way for you to do it yourself. If you have, or can get access to, a CAD (computer-aided drafting) machine, you're in luck. Most of the popular home computers with bit-mapped, high-resolution screens and a good CAD or graphics program can be used to make excellent drawings in a very short time. You can use the hard copy output directly, provided you print it out on good paper (at least 20# bond).A printout from a dot matrix printer will be OK for informal drawings, but for formal drawings the lines must be smooth and sharp, so a laser printer will be necessary.

Most commonly, drawings have been done in India ink on bristol board or Mylar film. You should keep the originals and send three good photocopies (on 20# or 24# bond) of each sheet to the PTO. You should not send your originals because you're no longer allowed to borrow them back from the PTO if changes are required. By keeping the originals, you can make required changes and then send in new photocopies.

If you're submitting "informal" drawings, the copies need not be perfectly clean and neat, but if you choose the formal route, the copies must be very clean, neat, and all lines must be sharp and black. Full details about both U. S. and A4 sizes and the margin requirements are shown in the diagrams of Fig. 10-A, below.

If you decide to use international-size drawings, you'll find that some copiers now have A4 size paper and settings. But if not, make copies on legal size sheets and trim them down to 210 mm x 297 mm (8+1/4"x11+11/16"). To get the margins right, you'll probably have to experiment a bit with the position of your original on the copier platen.

Even if you file informal drawings, you must include everything you can in your drawings, since you won't be able to add any "new matter" (any new technical information that is not present in your original sketches) after you file. Be sure to study the drawings of the patents uncovered in your patentability search (Chapter 6) to get an idea of what's customarily done for your type of invention, and to better understand the PTO rules.

I always make my drawings as comprehensive and meaningful as possible, almost to the point that most people can fully understand the invention by looking at the drawings alone. This is because most people are picture, rather than word, oriented and thus can understand an invention far more readily from the drawings, which are at a lower level of abstraction than the text.

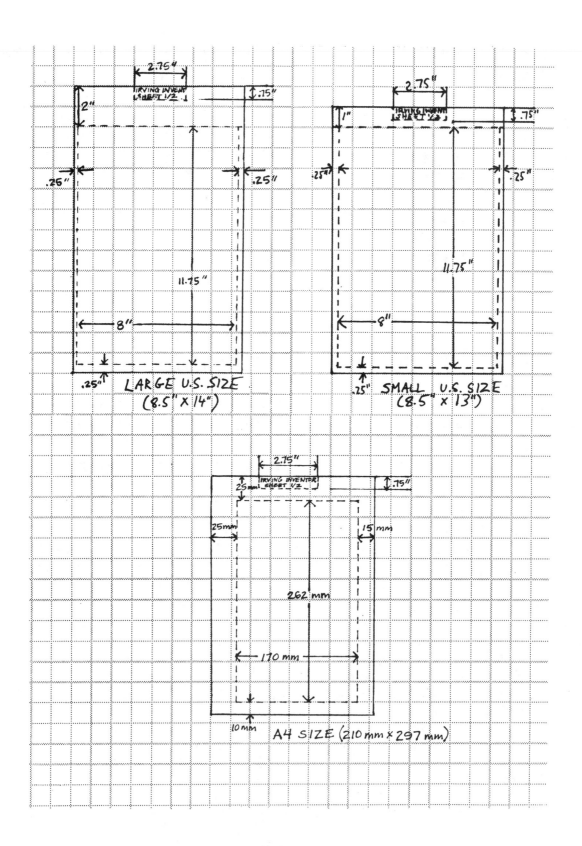

Fig. 10-A—The Three Permitted Drawing Sizes

For example, in electronic schematics, I try to arrange the parts so that the signal progresses from left to right, so that the input sources and output loads are clearly indicated, so that transistor states are indicated (i.e., NNC = normally non-conductive; NC = normally conductive), so that signal waveforms are shown, so that circuits are labeled by function (e.g., "Schmitt Trigger"), etc. In chemical and computer cases, I always try to use a flow chart, if possible. In mechanical cases, I use exploded views, perspective views from several directions, and simplified perspective "action" views, showing the apparatus in operation and clearly illustrating its function.

C. PTO Rules for Drawings

THE PTO HAS A NUMBER OF RULES for preparing formal drawings. Even if you plan to submit informal drawings, the rules should be followed as much as possible so that much of the work will already be done in the event you later need to submit formal drawings (which are required if your patent application is allowed).

Now let's summarize and discuss these rules which you'll find set out in the PTO's Rules Of Practice 81 to 85.

Summary of PTO Rules

1. **Need For Drawings:** Drawings (or only a single drawing) must be filed whenever necessary to understand the invention.

2. **Flow Charts:** Flow charts should also be included whenever useful for an understanding of the invention.

3. **Must Show Features Claimed:** The drawings must show every feature recited in the claims.

4. **Conventional Features:** Conventional features which are not essential for an understanding of the invention, but which are mentioned in the description and claims, can be shown by a graphical drawing symbol or a labeled rectangular box, e.g., a motor can be shown by an encircled "M," a CPU in a computer can be shown by a rectangle labeled "CPU," etc.

5. **Improvements:** When your invention consists of an improvement in an old machine, the drawing should show the improved portion disconnected from the old structure with only so much of the old structure as is necessary to show how your improvement fits in. For example, if you've invented a new tail light for a bicycle, show the bicycle itself with the new tail light (without detail) in one figure. Then show just the portion of the bike where the tail light is mounted in detail in another figure, together with details of the mounting hardware.

6. **Paper:** The filed drawings (xerographic copies) should be on paper which is flexible, strong, white, smooth, non-shiny, and durable. Ordinary 20# bond is acceptable. (You should do the originals on Mylar film or hard, rather than soft, bristol board; this is available in most good art supply stores. Strathmore Paper Co. makes excellent patent boards in both U.S. and A4 sizes (about $1/sheet), but you can get your sheets more economically by buying larger sheets of hard bristol board and cutting them to the proper size. (If you're using CAD, do the originals on regular bond and, since additional originals are so easy to make, send originals to the PTO.)

7. **Lines:** The main requirement for all drawings is that all lines must be crisp and perfectly black. A good photocopy on bond paper is usually used, but the lines should be crisp and sharp. A good xerographic copy from a dark-penciled Mylar film original will be accepted. Jagged slant lines from a dot matrix printer are verboten for formal drawings.

8. **White Pigment:** The use of white pigment (e.g., White Out, Liquid Paper) to cover lines is not normally acceptable, but one or two small occurrences on a sheet will usually pass muster.

9. **Uniform Size:** All drawing sheets in an application must be exactly sized in the same U.S. or in A4 size. Fig. 10-A shows these three sizes.

10. **Invisible Margins:** The margins must not contain any lines or writing (except for identifying indicia at the top—see #28 below); all writing and lines must be in the remaining "sight" (drawing area) on the sheet. Margin border lines are forbidden.

11. **Optional Holes:** U.S.-size sheets can have two 6.4 mm (.25") horizontally-centered holes at the top having their center lines spaced 17.5 mm (11/16") below the top edge and 7 cm (2.75") apart, but this is optional.

12. **Instrument Work:** All lines must be made with drafting instruments or a laser printer and must be very dense, sharp, uniformly-thick, and black. Fine or crowded lines must be avoided. Solid black areas are not permitted. Freehand work must be avoided unless necessary.

13. **Hatching:** Parts in section must be filled with slanted parallel lines (hatching) which are spaced apart sufficiently so that they can be distinguished without difficulty. Criss-cross hatching is forbidden.

14. **Shading:** Objects should be shaded with surface and edge shadings so that the light appears to come from the upper left at a 45-degree angle. Thus the shade sides of all objects (the right and bottom) should be done with heavier lines. Surface shading should be open.

15. **Scale:** The scale should be large enough to show the mechanism without crowding when the drawing is reduced 2/3 for reproduction. Detailed parts should be shown on a larger scale and spread out over two or more sheets if necessary to accomplish this, but the number of sheets should not be more than necessary.

16. **Figures:** The different views should be consecutively-numbered figures, e.g., "Fig. 1A", "Fig. 1B", "Fig. 2," etc. Each figure should be separate and unconnected with any other figure. If possible you should number the figures consecutively on consecutive sheets. However, if you want to arrange the figures in non-consecutive order to use space efficiently, that's okay, albeit less desirable.

17. **Reference Numerals:** These must be plain, legible, carefully formed, and not encircled. They should be at least 3.2 mm (1/8") high. When parts are complex, they should not be placed so close so that comprehension suffers. They should not cross or mingle with other lines. When grouped around a part, they should be placed around the part and connected by lead lines to the elements to which they refer. They should not be placed on hatched or shaded surfaces unless absolutely necessary; if then, they should be placed in a blank space in the hatching or shading. (Numerals are preferred to letters.) If a numeral refers to an entire assembly or group of connected elements, its lead line can have an arrowhead, or it can be underlined to distinguish it from the lead lines of numerals which refer to a single part.

18. **No Duplication of Reference Numerals:** The same part in different figures must always be designated by the same reference numeral. Conversely, the same reference numeral must never be used to designate different parts. Numbers with primes and letter suffixes are considered different numbers.

19. **Graphic Symbols:** These can be used for conventional parts, but must be defined in the specification. For instance, if you use an encircled "M" for a motor, the specification should say, e.g., "A motor, represented in Fig. 2 by an encircled 'M.'" Conventional symbols, such as those approved by the IEEE, ASA, etc., or from any standards or symbols book, can be used. Arrows should be used to show direction of movement, where necessary.

20. Descriptive Matter: The rules state that descriptive matter on the drawings is not permitted. I vehemently oppose this rule since the use of descriptive matter on drawings makes them far more meaningful, and since textbooks, magazine articles, etc. all use drawings with ample descriptive matter. Fortunately, this rule is usually enforced only in flagrant cases, e.g., where extremely large or wordy descriptive lettering is used. I thus regularly label almost all figures with a title (e.g., "Fig. 3—Side View Of Carburetor") and have never experienced any objection (except once, when I used an inexperienced drafter who used very large titles).

The Rules do permit (and even require) legends to be used within rectangular boxes, on flow charts, piping (plumbing) lines, or wherever else additional clarity is highly desirable. If used, the descriptive matter lettering should be as large, or larger, than the reference numerals.

21. Views: The drawings should have as many views (figures) as is necessary to show the invention. The views may be plan, sectional, exploded, elevational, or perspective; detailed larger-scale views of specific elements should be employed. Engineering views (front, side, bottom, back, etc.) should not normally be used if perspective views can adequately illustrate the invention. If exploded views are used, the separated parts of the same figure must be joined by assembly lines or embraced by a bracket. See Fig. 8-B.

A large machine or schematic or flow chart can be extended over several sheets, but they should be arranged to be easily understandable and so that the sheets can be assembled adjacent each other to show the entire machine. One figure must never be placed within another.

22. Sectional Views: The plane upon which a sectional view is taken should be illustrated in the general view by a broken line, the ends of which should be designated by numerals corresponding to the figure number of the sectional view with arrows indicating the direction in which the sectional view

is taken. E.g., suppose your Fig. 1 shows a left side front view of your carburetor and Fig. 2 shows a cross-sectional front of the back half of the carburetor on a plane vertically bisecting the carburetor into front and back halves. In this case, Fig. 1 should contain a broken vertical line spaced halfway from left to right with arrows pointing to the right at the top and bottom of this line; the arrows should each be labeled "2" to indicate the section is shown in Fig. 2; see Fig. 8-C.

23. Moving Parts: To show two positions of a moveable part, show its main position in full lines and its secondary position in broken lines, provided this can be done clearly. If not, use a separate view for the secondary position.

24. Modifications: Show modifications in separate figures, not in broken lines.

25. No Construction Lines: Construction lines, center lines, and projection lines connecting separate figures are forbidden. However, projection lines to show the assembly of parts in an exploded view in one figure are permitted; see Fig. 8-B.

26. Position Of Sheet: All views (figures) on a sheet must have the same orientation, preferably so that they can be read with the sheet upright (i.e., its short side at the top) so the examiner won't have to turn the sheets or the file to read the drawing. However, if views longer than the width of the sheet are necessary for the clearest illustration of the invention, the sheet can be turned on its side so that its short side and the appropriate top margin is on the righthand side. The orientation of any lettering on a sheet must conform with the orientation of the sheet.

27. OG Figure: One figure should be a comprehensive view of the invention for inclusion in the Official Gazette, a weekly publication of the PTO which shows the main claim and drawing figure of every patent issued that week.

28. No Extraneous Matter: No extraneous matter , i.e., matter which is not part of the claimed invention or its supporting or related structures is permitted on the drawings. However, you can (and should) place additional matter, such as a hand on a special pistol grip, if necessary to show use or an advantage of the invention. Also, you should put identifying indicia, such as your name, sheet number (e.g., "Sheet 3 of 4"), and even title of the invention, if there is space at the top of the sheet, in a centered location. The identifying indicia should not be wider than 7 cm (2.75") and should be within 19.1 mm (.75") of the top edge of the paper. See Fig. 10-A above.

29. **No Wrinkled Sheets:** The sheets should be sent to the PTO with adequate protection so that they will arrive without wrinkles or tears. You should send the sheets flat, between two pieces of corrugated cardboard within a large envelope, but they can also be rolled and sent in a mailing tube, provided they don't wrinkle. Of course the sheets must never be folded.

30. **Hidden Lines:** Parts which are hidden, but which you want to show, e.g., the inside of a computer, should be shown in broken lines. See Fig. 8-A. Reference numeral lead lines which refer to hidden parts should also be broken, in accordance with standard drafting practice. Broken lines must never be used to designate a part of the actual invention, unless to illustrate a hidden part or a moved position of a part. Indicate an alternative position of a part by phantom lines [____ _ _ ____ _ _]. Connect parts of an exploded view by projection lines [____ _ ____ _].

When your patent application arrives at the PTO, your drawings are inspected by the PTO's drawing inspectors, who are themselves draftspersons. If they find that any of your drawings are informal or in violation of any of the above rules, they will fill out and insert a drawing objection sheet in your file. A copy of this will be sent to you with your first office action (see Chapter 13). You must correct the drawings before a patent can issue by substituting new drawings; your drawings may no longer be borrowed from the PTO. Thus you should keep the originals of your drawings and send in good photocopies. Then if you have to correct the drawings, you can correct your originals and then send in new photocopies.

The most common drawing defects are listed on the drawing inspector's sheet. These are as follows:

- Lines are pale;
- Paper is poor;
- Numerals are poor;
- Lines are rough and blurred;
- Shade lines are required;
- Figures must be numbered,
- Heading space is required;
- Figures must not be connected;
- Criss-cross or double line hatching is objectionable;
- Arrowheads are used on lead lines for individual parts;
- Parts in section must be hatched;
- Solid black is objectionable;
- Figure legends are placed incorrectly (e.g., inside figure or vertically when drawing is horizontal);
- Drawing has mounted photographs;
- Drawing contains extraneous matter;
- Paper is undersized or oversized;
- Margins are too small;
- Lettering is too small;

- The sheets contain wrinkles, tears, or folds;

- Both sides of the sheet are used;

- Margin lines have been used;

- Sheets contain too many erasures;

- Sheets contain broken lines to illustrate regular parts of the invention;

- Sheets contain alterations, interlineations, or overwritings;

- Sheets contain unclear representations;

- Sheets contain freehand lines;

- Sheets contain figures on separate sheets which can't be assembled without concealing parts;

- Sheets contain reference numerals which aren't mentioned in the specification;

- Sheets contain the same reference numeral to designate different parts;

- Figures aren't separately numbered.

Although India ink on bristol board is still the preferred way to do the original drawings, you'll find it hard to work with since it tends to smear and is difficult to erase. If you do intend to use India ink, you should already be proficient in its use, or practice a good deal to attain proficiency before attempting to do your drawings. A Rapidograph or other brand of tube pen is much better than the old double-nib pen. If you need to erase a previously-drawn line, an electric eraser is best; the trick is to hold the eraser lightly and take a long time to erase the line; otherwise you'll burn or tear the paper. These problems will be non-existent if you're fortunate enough to have CAD facilities and a laser printer

D. Consider Using a Professional Patent Draftsperson

IF YOU DON'T FEEL COMPETENT to do your own drawings, you'll want to hire someone to do them for you. You can locate people who specialize in preparing patent drawings by letting your fingers do

the walking through the nearest metropolitan area yellow pages. Look under the heading "Drawing Services," which should list several patent drafters. While expensive (about $30 to $50 per hour, or $80 to $150 per sheet), these people should do the job correctly the first time in India ink on bristol board or with CAD facilities. Also, you can use a "starving artist" who's proficient in India ink, CAD, etc., and reads and understands the rules thoroughly.

E. Finaling Your Specification, Claims, and Abstract

BEFORE PUTTING THEM IN FINAL FORM, re-read your specification, claims, and abstract, to make sure they're clear, complete, and understandable. Again, make sure that the main substantive desiderata (Chapters 8 and 9) are satisfied.

You may type your specification, claims, and abstract on either U.S. or A4-size paper. If you don't think you'll be filing abroad, you can use the U.S. format, where the standards are very loose, i.e., anything in the letter-legal range will be fine, but preferably the paper should be 8.5 inches wide and 11 to 14 inches long, with at least a 1-inch left-hand margin and at least a 3/4-inch top margin, but preferably a 1.5 inch top margin. Use 1.5 or double spacing and number the sheets at the top or bottom, inside the margin

If you think you may later want to file corresponding foreign applications, you should type your application on U.S. size paper with proper margins so that if photocopied onto A4 size paper, it will have the proper A4 margins. To do this print out or type the application on letter-size or computer paper (8.5" x 11" after removal of the selvage or carrier strip). Use a 1" left margin, 6.5" line width, 1" top margins; (3" on p. 1) and a print length of about 9.5" so that the last line is almost at the bottom of the page. The sample specification in Ch. 8 (Fig. 8-E) is typed this way. Save the original for possible later use in making an A4 version for an international application. There is no typeface style requirement, and dot matrix printers are okay so long as the printout, or its photocopy, is clearly readable.

Alternatively, you can type and file your own U.S. application on A4 paper with the proper margins etc. However, I don't recommend this since, although the rest of the world uses it everywhere, it's very hard to obtain in the U.S., unless you cut it yourself or have it cut for you. If you do want to file on A4, the sheets should be 21 by 29.7 cm in size, with top margins of 8 to 9 cm on the first sheet and 2 to 4 cm thereafter, left margins of 2.5 to 4 cm, and bottom and right margins of 2 to 3 cm, with sheets numbered consecutively at the top and lines typed one-and-one-half spaces apart, i.e., 4 lines per inch. Keep the originals and file an A4 xerographic copy. As stated, the PTO doesn't care much about format, but if you later file a PCT application (see Ch. 12) these measurements will be strictly enforced.

You should start your claims and abstract on new pages. The titles should go on the first page and on the last (abstract page). Don't submit an application on easily erasable paper, or on paper which has white pigment covering any typewritten lines, since these are not considered permanent, unaltered records. If you're not a good typist, and you don't have a word processor, one solution is to type your application on easily erasable paper or regular paper, cover the errors with white pigment, type in the corrections, and then make bond paper photocopies of your typewritten original for submission to the PTO.

If, after putting your specification in final form, you find you must make a few minor changes (one or two words in up to a few places), it's okay to do so, provided you make these changes neatly in ink, in handwriting, and date and initial the margin adjacent to each change *before* you sign the application.

You don't have to file your drawings and your typewritten papers on the same size sheets: the drawings can be on A4 paper and the typewritten

pages can be on U.S.-size paper, or vice-versa. All drawing sheets must be the same size, as must all typed sheets. Never, never use both sides of a sheet, either for drawings or the specification. A neatly-typed specification will certainly make a very favorable impression on your examiner; one inventor I know went all out and did his application with beautiful assorted typefaces using a laser printer—the result was most impressive.

F. Name All True Inventors and Only True Inventors

IN SEVERAL DIFFERENT PARTS of your application, you're required to name the applicants and inventors. For example, in Form 10-1, your transmittal letter, you must list the applicants. And Form 10-2 (your Patent Application Declaration) must be signed by the inventors, e.g., see Figs. 10-B and 10-C below. These are discussed in the following sections.

As previously mentioned, while anyone can apply for a patent, the named applicant(s) must be the true inventor(s) of the invention. If you've conceived the invention (as defined by the claims) entirely on your own, of course there's no problem about who should be named co-inventor. On the other hand, if you've invented it with someone else, both of you should be named as "joint inventors." But be sure that both of you actually are joint inventors. If somebody other than you played a significant role in conceiving the invention, turn to Chapter 16, Section B for a more detailed discussion on inventorship.

G. Completing the Patent Application Declaration

EACH PATENT APPLICATION must be accompanied by a patent application declaration

(PAD), which is a written statement under oath. The form for the PAD is provided as Form 10-2, and a completed version is provided below in Fig. 10-B.

While completing the PAD is a straightforward process, you should not treat it lightly. Rather, you should read and review it very carefully before you sign. If anyone can prove that you signed the declaration knowing that any of its statements were false, your patent can be held invalid.

The title of the invention from p.1 of the specification goes in the space near the top; the name, residence, citizenship, and post-office address of each inventor (if there are more than one) go in the appropriate spaces. Your residence is your city and state. This will normally be the same as the city and state of your post-office address, unless you have a post-office box or a postal address in a city other than the one in which you actually reside. If there are more than two inventors (see Chapter 16, Section B), photocopy or type the last few lines of the form on a second page as many times as necessary and label Form 10-2 "Page 1 of 2" and the second page "Page 2 of 2." Fig. 10-B shows how the PAD is completed for two joint inventors.

Each inventor should then sign and date the appropriate spaces at the bottom of the form. Note that the PAD directs the PTO to send correspondence and calls to the first-named inventor. While every joint inventor must sign all papers which are sent to the PTO, the PTO will correspond with one inventor only. Therefore you should list the inventor who is most available (or who has best access to a photocopier) first.

Note the sentence of the PAD which states that you acknowledge a duty to disclose information of which you are aware and which is material to the examination of the application. This provision is designed to impress upon inventors their duty to disclose (to the PTO) any information which could affect the examination or validity of the application. This means you must disclose to the PTO all relevant prior art which you have uncovered; any

disadvantages of your invention of which you are aware; or any other act you think the examiner would want to be aware of when examining the application. Normally, all of this information will be provided in your Information Disclosure Statement (see Section N below).

This disclosure requirement is very important; so much so that courts have, as mentioned, held patents invalid when inventors have neglected this duty. Thus I've made it Inventor's Commandment #10.

INVENTOR'S COMMANDMENT #10.
Avoid Fraud: In addition to making a full disclosure, you should also tell the PTO, in Information Disclosure Statements, about any pertinent "prior art" or other material facts concerning your invention of which you are aware.

H. Fill Out the Small Entity Declaration If Appropriate

IF YOU'RE A "SMALL ENTITY," i.e., an independent inventor, or an independent inventor(s) who hasn't assigned or licensed, or isn't under an agreement to assign or license, the invention to a large, for-profit business (over 500 employees), you'll be entitled to pay small-entity filing, issue, and maintenance fees, which are half the large-entity filing fees. See the Fee schedule in the Appendix for the amounts of these fees. To qualify, you must complete and send in a Small Entity Declaration (SED) with your application. The SED form is Form 10-3 in the Appendix.

Declaration For Utility Patent Application

As a below-named inventor, I hereby declare that my residence, post office address, and citizenship are as stated below next to my name and that I believe that I am the original, first, and sole inventor [if only one name is listed below] or an original, first, and joint inventor [if plural names are listed below] of the subject matter which is claimed and for which a patent is sought on the invention, the specification of which is attached hereto and which has the following title:

" _FOOD CHOPPER WITH CONVOLUTE BLADE_   "

I have reviewed and understand the contents of the above-identified specification, including the claims(s) as amended by any amendment specifically referred to in the oath or declaration. I acknowledge a duty to disclose information which is material to the examination of this application in accordance with Title 37, Code of Federal Regulations, Section 1.56(a).

I hereby declare that all statements made herein of my own knowledge are true and that all statements made on information and belief are believed to be true; and further that these statements were made with the knowledge that willful false statements and the like so made are punishable by fine or imprisonment, or both, under Title 18, United States Code, Section 1001, and that such willful false statements may jeopardize the validity of the application or any patent issued thereon.

Please send correspondence and make telephone calls to the First inventor below.

Signature: Sole/First Inventor: _Mildred Goldberger_

Print Name: _MILDRED GOLDBERGER_   Date: _1988 August 9_

Residence: _Philadelphia, PA_   Citizen Of: _Hungary_

Post Office Address: _1901 Kennedy Blvd., Philadelphia, PA 19103_

Telephone: _215_   _222-2972_

Signature: Joint/Second Inventor: _Nathaniel Briskin_

Print Name: _NATHANIEL BRISKIN_   Date: _1988 August 9_

Residence: _Philadelphia, PA_   Citizen Of: _USA_

Post Office Address: _1919 Chestnut St., Philadelphia, PA 19103_

Telephone: _215_   _227-6639_

Fig. 10-B—Completed Patent Application Declaration

How to Complete Form 10-3

- Fill in the inventors' names adjacent the "Applicant" lines at the top of the form and add the title of the application where indicated.

- If you haven't sold or granted any interest in (assigned or licensed) your application to anyone (see Chapter 16), and aren't under any obligation to do so (this will be the normal case), check the box before the line reading, "there is no such person, concern, or organization."

- If you have sold or granted such an interest, or are obligated to do so, then check the line that ends in an asterisk, list the person or organization on the appropriate lines, and check the appropriate box to indicate whether this person or organization (your "assignee") is an individual, small business, or non-profit organization. I cover assignments in Chapter 16, Section E.

- Your assignee (or licensee) (if any) should complete and file a supplemental, non-inventor SED, if appropriate. Form 10-4A is for individual assignees, Form 10-4B is for small businesses, and Form 10-4C is for non-profit institutions. These forms are included in the Appendix.

- Print your names and date, and sign the main SED at the bottom where indicated. If you have more than two joint inventors (see Chapter 16, Section B), add a line at the top of the SED reading "Joint/Third Applicant: _____ _____ " [squeeze it in or retype the form] and add another three-line signature section on a second page. Label these pages "Page 1 of 2" and "Page 2 of 2," as with the PAD.

     If you don't qualify as a small entity, i.e., you (or any co-inventor) have assigned or licensed the invention, or are under an obligation to assign or license it, to a for-profit business with over 500 employees, then you should omit the SED and pay the large entity filing (and other) fees.

I. Complete the Transmittal Letter, Check, and Postcard

NOW IT'S TIME to prepare the routine paperwork necessary to actually send your patent application to the PTO. Here's how to do it.

1. Prepare a Transmittal Letter

The transmittal letter (Form 10-1) should be dated as of the date the entire patent application is mailed. Fig. 10-D below shows how it's completed.

     The names of the inventor(s), the title, the total number of pages of specification and claims, the number of sheets of drawing, and whether the drawings are formal or informal should all be indicated in the appropriate spaces on this form. The date the application was signed should also be indicated in the space provided. The application should be mailed to the PTO shortly after this date of signature.

In the United States Patent and Trademark Office

Mailed 1988   Aug. 9

Commissioner of Patents and Trademarks
Washington, District of Columbia 20231

Sir:

Please file the following enclosed patent application papers:

Applicant #1, Name: _Mildred Goldberger_

Applicant #2, Name: _Nathaniel Briskin_

Title: _Food Chopper with Convolute Blade_

(X) Specification, Claims, and Abstract: Nr. of Sheets: _12_

(X) Declaration: Date Signed: _1988 Aug. 9_

(X) Drawing(s): 3 copies each of _4_ Sheets : Formal: _____ Informal: _2_

(X) Small Entity Declaration.

(X) Assignment; please record and return; $ _7.00_ recordal fee enclosed.

(X) Check for $ _177_ for:

　　(X) $_170_ for filing fee (not more than three independent claims and twenty total claims are presented).

　　(X) $_7_ additional if assignment is enclosed for recordal.

(X) Return Receipt Postcard Addressed to Applicant #1.

Very respectfully,

Mildred Goldberger 　　　　　　　　　　 _Nathaniel Briskin_
Applicant #1 Signature 　　　　　　　　　　 Applicant #2 Signature

1901 Kennedy Blvd. 　　　　　　　　　　 _1919 Chestnut St._
Address (Send Correspondence Here) 　　　　 Address

Philadelphia, PA 19103 　　　　　　　　　 _Philadelphia, PA 19103_

EXPRESS MAIL LABEL # _B 167283123_ ; 　　**DATE OF DEPOSIT** 1988 _Aug. 10_

I hereby certify that this paper or fee is being deposited with the United States Postal Service using "Express Mail Post Office To Addressee" service under 37 CFR 1.10 on the date indicated above and is addressed to "Commissioner of Patents and Trademarks, Washington, DC 20231."

Signed: _Mildred Goldberger_

Inventor(s) _MILDRED GOLDBERGER_

NATHANIEL BRISKIN

Fig. 10-D—Completed Patent Application Transmittal Letter

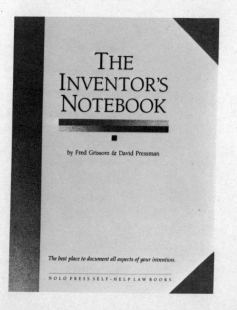

To order your copy of **The Inventor's Notebook,** fill out this order form or call our toll-free telephone lines:

1-800-992-6656 (U.S.)

1-800-445-6656 (CA outside 415 area)

Please refer to coupon code NOTE *when ordering.*

Name
_____

Street (no P.O. boxes) for UPS delivery. P.O. boxes for priority mail
_____

State, Zip
_____

(   )
_____
Daytime phone

SALES TAX (California residents)
Add your local tax:
6 1/4%, 6 3/4%, or 7 1/4%

METHOD OF PAYMENT
☐ check ☐ Visa ☐ Mastercard ☐ DiscoverCard

Acct. #
_____

Exp. Date   Signature
_____

QTY	ITEM	UNIT PRICE	TOTAL PRICE
	The Inventor's Notebook	$19.95	

SUBTOTAL	_____
SALES TAX (California only)	_____
SHIPPING AND HANDLING	_____
2ND DAY UPS	_____
TOTAL	_____

Shipping Policies
Charges: 1 book $3.00
              2-3 books $4.00
              Each additional book $ .50
We ship all orders in two business days via UPS (ground rate) or US Priority Mail. UPS cannot ship to post office boxes, so please give us a street address. Please allow 1-2 weeks for delivery.
In a Hurry?
For UPS 2nd day delivery add $3.00 (contiguous states), $6 (AK & HI) to shipping/handling charges.

NOLO PRESS • 950 PARKER STREET • BERKELEY • CA 94710   NOTE

If there are two inventors, both should sign the letter, and their respective addresses should be provided. If there are more than two inventors, retype the form with additional spaces for the additional inventors. As stated, whenever there is more than one inventor, all inventors must sign every communication to the PTO.

2. Write a Check

Write a check for the basic filing fee (large entity or small entity). This fee entitles you to file up to three independent claims and up to 20 total claims, assuming that each dependent claim refers back to only one preceding claim (dependent or independent). If you're filing more than three independent or 20 total claims (not generally recommended) indicate this on your transmittal letter and increase the fee by necessary amount. For example if you're filing 5 independent claims and 23 total claims, increase the fee by 5-3=2 times the additional independent claim fee and 23-20=3 times the additional "over 20" claim fee.

You can obtain a considerably speedier processing of your application than is usually the case by filing a Petition to Make Special along with your application (see Section P below).

If you're enclosing an assignment (see Section M below), check the "Assignment" box and the "Additional Fee" box and the assignment recordal fee.

The check should be made out to Commissioner of Patents and Trademarks for the total amount, and should be attached to the transmittal letter. Be absolutely sure you have enough money in your checking account to cover the check; if the check bounces, you'll have to pay a stiff surcharge.

3. Postcard

INVENTOR'S COMMANDMENT #11:
Postcard: Every time you send any paper(s) to the PTO, you should include a postcard addressed to you with all paper(s) listed on the back of the card.

The PTO will stamp your postcard receipt with a date and serial number and mail it back to you as soon as they open your letter. All attorneys use receipt postcards because the PTO receives many thousands of pieces of mail each day and occasionally loses some. It may be months before you receive any reply to a paper you've sent to the PTO, so you'll want to be assured it arrived safely.

Fig. 10-E (a and b) indicate how an application receipt postcard should be completed, front and back. Note that the back of the card contains the inventors' names, title of invention, number of pages of specifications, claims, and abstract, the date the Patent Application Declaration was signed, the number of sheets of drawing (and whether formal or informal), the Small Entity Declaration, and the check number and amount. Occasionally, receipt postcards get lost because of their size and inconspicuous color. I have had better results by using colored (bright red) postcards.

J.   Orderly File

I OFTEN CONSULT with "pro se" inventors (i.e., who have prepared and filed their own patent applications). Usually they bring me their "application" in the form of a sloppy file of mixed-up papers, with occasionally lost documents. You'll avoid this problem, and the serious trouble it can get you into, if you'll heed the next commandment.

INVENTOR'S COMMANDMENT # 12:
Orderly File: You should mount an identical copy of every paper you send to the PTO in a file jacket, together with every paper you receive from the PTO.

A three-part folder for (a) your application, (b) correspondence from the PTO, and (c) correspondence to the PTO is useful. Keep your references in a large envelope loose inside the folder.

K. Assembly and Mailing of Your Application—Final Checklist

CONGRATULATIONS. You're now ready to mail your patent application to the PTO, unless you want to include an assignment (Section N), an Information Disclosure Statement (Section M), and/or a Petition to Make Special (Section O). If you do want to include any of these with your application (this is optional), consult the indicated sections, complete your paper work, and then come back to this point.

Assemble in the following order and carefully check the following items, which are the third part of the checklist I started in Chapter 8:

(a) Return Postcard addressed to you with all papers listed on back. [ ]

(b) Check or money order for correct filing fee (basic fee and fee for any excess claims); adequate funds on deposit. [ ]

(c) Transmittal letter properly completed and signed. [ ]

(d) All drawing sheets present in triplicate; drawings clear, complete and understandable. Your name and sheet number are on the top front or back of each sheet of drawing and each sheet numbered.

Originals of drawings (or disc file if CAD used) kept in safe place.[ ]

(e) All pages of specification, claims, and abstract included; description of invention clear and complete and claims drafted per Chapter 9. [ ]

(f) Typing as clear and readable and 1-1/2 or double spaced. [ ]

(g) Application is prepared in formal for making proper A4 copies later if foreign filing contemplated.(optional) [ ]

(h) Top margin (above page numbers) is at least 2.5 cm on all pages. [ ]

(i) No sentences longer than about 13 words, paragraphs are not longer than about 1/2 page, and a heading is supplied for about every two pages. [ ]

(j) Claims are separated by an extra line. [ ]

(k) Claims and abstract start on new pages.[ ]

(l) Patent Application Declaration (PAD) completed, signed, and dated. [ ]

BURPING BABY DOLL
No. 7,283,406

The following received today:

Application of Mildred Goldberger and Nathaniel Briskin for "Food Chopper WithConvolute Blade," 12 sheets specification, claims, and abstract, declaration signed 1988 Aug. 9, 2 sheets informal drawing, small entity declaration, check nr. 334 for $_____:

Back of Receipt Postcard

affix
postage
here

Mildred Goldberger

1901 Kennedy Blvd

Philadelphia, PA 19103

Front of Receipt Postcard

Fig. 10-E—Completed Postcard to Accompany Patent Application

(m) Small Entity Declaration (SED) completed, signed, and dated. Additional non-inventor SED(s) included if anyone else has any interest in the application. [ ]

(n) Your Information Disclosure Statement and PTO-144a with references attached if you're filing it with your application (see Section N below). [ ]

(o) Your Petition to Make Special if you include it with your application (this speeds up the application processing—see Section P below). [ ]

(p) A completed assignment if you include it with your application (see Section O below). [ ]

(q) All documents bear original ink signatures. [ ]

(r) Envelope addressed to:

Commissioner of Patents and Trademarks Washington, DC 20231 [ ]

Mark the outside of the envelope "Patent Application."

I suggest that you file a signed photocopy and keep the original of your application so you can make copies later if you need to send them to manufacturers when you market your invention. See Chapter 11.

Staple the pages of the specification, claims, abstract and declarations together. Attach the drawings with a paper clip or other temporary device.

The papers can be transmitted in a large envelope, with one or two sheets of stiff cardboard to protect the drawing from bending, or, if they are thin enough, they can all be rolled and mailed in a mailing tube.

The application should be sent to the PTO by first-class mail. You may want to register the parcel to cover the expense of making a new drawing in case it's lost. However, registering your mail will not cover you for loss of any legal rights in case your application is lost in the mail on the way to the PTO.

L. Using Express Mail to Get an Instant Filing Date

IN ORDER TO HELP PREVENT LOSS, secure full legal rights in case of loss, and get an "instant" filing date (the date you actually mail your application), you can send it express mail. You must use "Express Mail Post Office To Addressee" service and you must indicate that you're using this service by completing and signing an Express Mail Certification (Fig. 10-D) at the bottom of your transmittal letter (Form 10-1). Your filing date will then be your date of mailing.

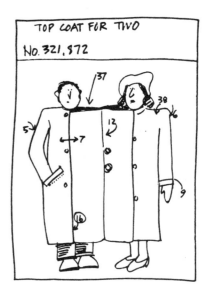

M. Receipt That Application Was Received in PTO

ABOUT TWO TO FOUR weeks after you send your application to the PTO, you'll get your postcard back, stamped to indicate the filing date of your application, and also stamped with a six-digit serial number that has been assigned to your application. Within about a week to a month after that, you should get a blue filing receipt back from the PTO indicating that your application has been officially filed in the PTO.

If for any reason your application is incomplete or deficient, it will not be officially "filed" but will be regarded as "deposited." The Application Branch of the PTO will send you a letter stating the deficiency in your application, and you should promptly remedy it. However, if you follow all the instructions in this chapter carefully, you'll get your blue filing receipt in due course.

Once you get the filing receipt, your application is officially "patent pending," and unless you want to keep your invention a trade secret, as discussed in Chapter 7 (in case your patent application is eventually disallowed), you may publish details of your invention or market it to whomever you choose without loss of any legal rights, domestic or foreign. If you manufacture anything embodying your invention, you should mark it "patent pending."

N. File the Information Disclosure Statement Within Three Months

THE PTO'S RULES impose on each patent applicant a "duty of candor and good faith" toward the PTO. This means that all inventors (and attorneys) have "a duty to disclose to the PTO information they are aware of which might influence the patent examiner in deciding on the patent application. To comply with the "prior art" part of I.C. #10, the PTO asks all applicants to submit an Information Disclosure Statement (IDS) at the time of filing the application or within the following three months. I suggest that you use the option to file afterward; this will prevent overload while preparing your basic application.

IDS forms are provided as Forms 10-5 and 10-6. A filled-in sample is set out as Figs. 10-G and 10-H.

The IDS should list all prior-art references known to the inventors that are relevant to the patentability of the application. These should include all the references you discovered in the course of your patentability search (see Chapter 6), plus any other prior art of which you're aware. In addition, the inventor must include with the IDS a copy of each cited reference (or the relevant parts thereof) and a discussion of its relevance to the invention. If you didn't make a search and aren't aware of any prior art, don't file an IDS.

You may well ask why, if the prior art you discover consists of patents, you have to send patents to the PTO; isn't this like carrying coals to Newcastle? Well, yes, but the PTO claims that it's such a large and complex organization that it would cause too many administrative difficulties and require too much examiners' time to dig out all of the patents cited by applicants; thus they want you to send them copies of your references.

You can send the IDS with your application. In this case, the names of the inventors and title of your invention are the only information you need to put at the top of the Form 10-5. If you send it in after your application is filed, you'll know the serial number, filing date, and group art unit, and can insert them. You normally won't know the examiner's name unless you've received a first "office action" from the PTO.

After the first paragraph of this form, describe the relevance of each prior-art reference of which you are aware briefly. Normally, one sentence for each reference will suffice. For example, "Smith shows a manually operated vas valve." Conclude with a final sentence stating briefly how your invention as claimed differs from the references, e.g., "None of the references show any male contraceptive device using a vas valve which is magnetically operated from outside the body, as is claimed." You needn't stress unobviousness considerations here; stress mainly the physical differences in your invention. Then add a closing salutation, for example, "Very respectfully"; type your name and address below this and sign your name. (If there are more inventors than one, each inventor should sign.)

Then, complete the Form 10-6 (PTO's Form 1449) titled "List of Prior Art Cited By Applicant" to identify the references you have discussed on your Form 10-5. Figs. 10-G and 10-H above show how these forms should be completed.

If you send in the IDS with the application, you should list it in the postcard and transmittal letter that you send with your application. If you send it in after the application is filed, you should send in a separate postcard. Again, the front of the card should be addressed to you; the back should read as in Fig. 10-I.

Information Disclosure Statement, Form 1449, and
[Number] References in   application of [name(s) of inventor(s)],
Serial No. _____,Filed _____, Title: _____
received for filing today.

Fig. 10-I — Postcard For Sending I D S

O. Assignments

AS I MENTIONED, a patent application must be filed in the name or names of the true inventor or inventors of the invention claimed in the patent application. The inventors are the applicants for the patent, and the law considers that they automatically own the invention, the patent application, and any patents that may issue on the application (Chapter 16, Section B). However, often all or part of the ownership of the invention and the patent application must be transferred to someone else, either an individual or a legal entity such as a corporation, a partnership, or an individual. To make this transfer, the inventor(s) must "assign" their interest. How this is done is covered in Chapter 16, Section E.

If you want to send an assignment you have made to the PTO for recording (highly advised), you can either send it in with the patent application or at any time afterward. If you send it in with the application, you need merely include it with the application papers, check the "Assignment" and "$_____ additional" boxes on Form 10-1, and increase your fee accordingly Be sure that the date the application was signed is included on the assignment; otherwise the PTO won't record it.

In the United States Patent and Trademark Office

Appn. Number: _06/123,456_

Appn. Filed: _1988 June 22_

Applicant(s): _Douglass, L. Clarkson_

Appn. Title: _Food Cutter_

Examiner/GAU:_____ /324

Mailed 1988 Oct 24, Mon
San Francisco, CA

Information Disclosure Statement

Commissioner of Patents and Trademarks
Washington, District of Columbia 20231

Sir:

Attached is a completed Form PTO-1449 and copies of the pertinent parts of the references cited thereon. Following are comments on these references pursuant to Rule 98:

Oot shows a fruit knife comprising a curved blade and a curved guard parallel to the blade for controlling the thickness of the rind which can be peeled by the blade.

Landers shows a knife which is similar to Oots, but in Landers the blade is straight and the spacing of the guard or gage plate is adjustable.

Gambino shows a fruit peeler with a curved frame spaced from the blade for making cuts of adjustable depth.

Wolff shows a knife with another form of adjustable slicing guide which is parallel to the blade.

Rasmussen shows a fruit peeler with a bent guide for controlling the thickness of fruit as it is being peeled.

Gillet shows a knife mounted parallel to a spacer which knife can be tilted out or in to adjust the spacing of its edge.

None of the references shows a knife for making a cut of controlled depth wherein a flat-bladed knife with an elongated sharpened edge with an outward protrusion attached to the blade which is spaced back from the edge for limiting the depth of cut which can be made by said edge when it is used to **cut** in a direction **perpendicular** to the plane of the blade, as is recited in independent claims 1 and 17, and hence their dependent claims 2 to 11 and 18 to 20.

To the contrary, all of the references show guides which are mounted generally parallel to the blade for limiting the thickness of the peel which can be cut by the blade when it is used to **peel** in a direction **parallel** to its plane.

Also, none of the references show any blade having a substantially right-angle bend parallel to its direction of elongation, as is recited in independent claim 12 and its dependent claims 13 to 16.

Very respectfully,

L. Clarkson Douglass

L. Clarkson Douglass

15 Chestnut Street
San Francisco, CA 94109
415 227-4494

[This (and all other papers sent to the PTO) should always be typed with 1.5 or double spacing.]

Fig. 10-G—Completed IDS

FORM PTO-1449
(REV. 7-80)

U.S. DEPARTMENT OF COMMERCE
PATENT AND TRADEMARK OFFICE

Sheet _1_ of _1_

ATTY. DOCKET NO.	SERIAL NO.

LIST OF PRIOR ART CITED BY APPLICANT
(Use several sheets if necessary)

APPLICANTS *Clarkson, L. D.*

FILING DATE	GROUP
1982 June 22	324

U.S. PATENT DOCUMENTS

EXAMINER INITIAL		DOCUMENT NUMBER	DATE	NAME	CLASS	SUBCLASS	FILING DATE IF APPROPRIATE
	AA	21695	1858	Oot			
	AB	602758	1898	Landers			
	AC	1478501	1923	Woodin			
	AD	1488200	1924	Gehring			
	AE	1636287	1927	Clawson			
	AF	2054480	1936	Leitshuh	30	20	
	AG	2083368	1937	Gambino	146	206	
	AH	2968867	1961	Wolff	30	284	
	AI						
	AJ						
	AK						

FOREIGN PATENT DOCUMENTS

		DOCUMENT NUMBER	DATE	COUNTRY	CLASS	SUBCLASS	TRANSLATION YES	TRANSLATION NO
	AL	69640	1949	DANMARK (Rasmussen)				
	AM	1029924	1953	FRANCE (Gillet)				

OTHER PRIOR ART (Including Author, Title, Date, Pertinent Pages, Etc.)

	AR	
	AS	

EXAMINER	DATE CONSIDERED

EXAMINER: Initial if reference considered, whether or not citation is in conformance with MPEP 609; Draw line through citation if not in conformance and not considered. Include copy of this form with next communication to applicant.

USCOMM-DC 80-3985

Fig. 10-H—Completed PTO-1449 Form to Accompany IDS

I prefer to send in assignments later, after I get the postcard receipt back, when I know and can add the serial number and filing date of the application to the assignment to make the two documents (the assignment and the application) correspond to each other more directly. In this case, you can add the serial number and filing date to the assignment in the spaces indicated. Then prepare a simple assignment recordal request letter. The heading should be as in Form 10-1. The body of the letter should merely be a simple request, e.g., "Please record and return to me the attached assignment of a one-third interest in Application Ser. Nr. [give number], Filed [give date], Title: [give title], Applicants: [give names] to [give name of assignee]; the $_____ recordal fee is enclosed."

If an assignment of a patent application has been recorded and it is referred to in the issue fee transmittal form (see Chapter 13), the PTO will print the patent with the assignee's interest indicated. However, even if you fail to indicate the assignment on the issue fee transmittal, so that the patent doesn't indicate the assignment, the assignment will still be effective if it has been recorded.

If an assignment has been made, and as a result there are two or more owners of the patent application, then the owners should consider signing a Joint Owners' Agreement (Form 16-1) for the reasons indicated in Chapter 16, Section C.

P. Petitions to Make Special

IF YOU DO NEED to have your patent issue sooner than in the normal course (1.5 to 3 years), you can have it examined ahead of its normal turn by filing a "Petition To Make Special" (PTMS— Form 10-7), together with a Supporting Declaration (SD). This can be filed with the application or at any time after. An example of a properly-completed PTMS and SD is shown below in Figs. 10-J and 10-K.

Before you get excited about filing a Petition to Make Special, however, please consider this. Unless you have a specific need for the early issuance of a patent—e.g., an infringement is occurring, you need a patent to get capital for manufacturing the invention, or the technology is rapidly becoming obsolete—most cognoscenti agree that you will not gain any advantage in having the patent issue at an early date. In fact, you're usually better off if issuance of the patent is delayed, e.g., until well after the invention is commercialized.

Why? The answer will be apparent if you remember what rights a patent gives its owner. As stated in Chapter 1, a patent gives its owner the right to exclude others from making, using, or selling the invention. Thus, unless the patentee (owner of the patent) or her licensee (person or company who's making the patented invention and paying the patentee royalties) has competition who infringes the patent, the patentee's royalties will be maximized if the 17-year monopoly clock doesn't start running until an infringement occurs, or is likely to occur.

Also, once a patent issues, the technology is made public (remember, the patent application must teach clearly how to make and use the invention), so that potential competitors can see the patent and start copying its technology and designing around it.

In the United States Patent and Trademark Office

Ser. Nr.: _06/ 123,456_
Filed: _1988 Aug. 9_
Applicant(s): _Goldberger, David_
Title: _Wind Generator Using Stratus Rotor, Etc._
Examiner/GAU: _Hayness   1654_

Mailed _1988 September 20_
At: _San Francisco, CA_

Petition to Make Special

Commissioner of Patents and Trademarks
Washington, District of Columbia 20231

Sir:

Applicant hereby respectfully petitions that the above application be made special under MPEP Sec. 708.02 for the following reason; attached is a declaration in support thereof:

I.   [ ] Manufacturer Available*;   VI.   [ ] Energy Savings Will Result;

II.   [ ] Infringement Exists*;   VII.   [ ] Recombinant DNA is involved*;

III. [ ] Applicant's Health Is Poor;   VIII. [ ] Special Procedure: Search Was Made*;

IV.   [ ] Applicant's Age is 65 Or Greater;   IX   [ ] Superconductivity is advanced.

V.   [X] Environmental Quality will Be Enhanced;

* [ ]   Also attached, since reason I., II., VII., or VIII has been checked, is the $_____ Petition Fee pursuant to Rules 102 and 17(i).

Very respectfully,

Applicant(s): _David Goldberger_

Attachment: Supporting Declaration And Fee if Indicated

C/o: _David Goldberger_
119 Walnut St.
San Francisco, CA 94123
Tel.: _(415) 733-0362_

Fig. 10-J—Completed Petition to Make Special

In the United States Patent and Trademark Office

Appn. Number: __06/123,456__
Filing Date: __1988 Aug. 9__
Applicant(s): __Goldberger, David__
Appn. Title: __Wind Generator Using Stratus Rotor, Etc.__
Examiner: __Hayness / GAU 654__

Mailed __1988 Dec. 11, Wed.__
At: __San Francisco, CA__

Declaration in Support of Accompanying Petition to Make Special

Reason V — Enhancement of Environmental Quality

In support of the accompanying Petition to Make Special, applicant declares as follows:

1. I am the applicant in the above-identified patent application.

2. The invention of the above application will materially enhance the quality of the environment of humankind by contributing to the restoration or maintenance of the basic life-sustaining natural elements of air and water in the manner described below.

3. Specifically, the invention of the above application is an improved electrical power generator employing wind energy. It provides a more efficient wind power generator than heretofore available because it uses a highly-efficient stratus rotor in combination with a Loopis vane, thereby intercepting an average of 25% more of the wind energy passing therethrough than prior-art conventional fan-blade wind turbines, as described in full detail on pages 3 to 5 of the specification.

4. By more efficiently using wind power, it enables the installed cost per average kilowatt of generated wind power on a yearly basis to be materially lowered. This will make wind power generators more economical, cost-effective, and attractive to inventors, individual power consumers, and power companies. As a result, more utilization of wind power generation will occur, causing less dependence on and less utilization of conventional power planets using fossil-fuel sources such as coal and oil, or nuclear fission, thereby resulting in less air and water pollution due to reduced effluents into the air and waterways from such conventional power plants. Thus thermal and other pollution of such air and waterways will be reduced so that air and water quality will be maintained and will actually be restored due to natural self-purification.

5. I further declare that all statements made herein of my own knowledge are true and that all statements made upon information and belief are believed to be true, and further that these statements were made with the knowledge that willful false statements and the like so made are punishable by fine or imprisonment, or both, under Section 1001 of Title 18 of the United States Code, and that such willful false statements may jeopardize the validity of the application and any patent issuing therefrom.

Very respectfully,

David Goldberger

David Goldberger, Applicant

1919 Chestnut Street
Philadelphia, PA 19103
215 L07-6639

[This (and all other papers sent to the PTO) should always be typed with 1.5 or double spacing]

Fig. 10-K—Completed Declaration to Accompany PTMS

Lastly, most potential licensees (companies whom you'd like to license under your patent) would prefer to sign the license while the patent application is still pending and hence kept in secrecy in order to get an edge on the competition. As stated in the next chapter, you should try to license your patent application as soon as it's filed and not wait until the patent issues.

As you'll note on the PTMS, an application can be made special, and hence examined ahead of turn, for any of nine reasons. Those reasons marked with an asterisk (*—numbers 1, 2, 7, and 8) will require a petition fee (for large or small entities)—see Fee Schedule. If you use any of the other reasons (numbers 3 to 6 and 9) you won't have to pay any fee since these are "favorable public policy" (non-mercenary) reasons. Here are the nine reasons:

1. Manufacturer Available: A manufacturer is available, i.e., a person or company exists which will manufacture the invention provided the patent application is allowed or a patent issues.*

2. Infringement Exists: Someone is making, using, or selling the invention covered by the patent application and you need a patent to sue the infringer or get the infringer to pay you royalties.*

3. Applicant's Health Is Poor: You're in such poor health that your normal lifespan is likely to be shortened and you want to get the fruits of your invention before you depart this life.

4. Applicant's Age Is 65 Or Greater.

5. Environmental Quality Will Be Enhanced.: Your invention conserves natural resources and/or keeps the air, water, or landscape pristine. See Fig. 10-K.

6. Energy Savings Will Result: The invention provides a way to use energy more efficiently, thereby also conserving natural resources.

7. Recombinant DNA Is Involved.: Public policy favors the full and rapid exploitation of recombinant deoxyribonucleic acid.

8. A Search Was Made: If you've made a search and submitted an Information Disclosure Statement—as you're supposed to do anyway (see Sec. N above)—you can get the case made special, since the examiner's task is made easier by your search.

9. Superconductivity Is Advanced: Public policy favors the exploitation of this phenomenon.

The supporting declaration which accompanies the PTMS should be in the format of Fig. 10-K with the introductory paragraph, paragraph 1, and the last paragraph left intact. The remaining paragraphs must give detailed facts (MPEP 708.02) in support of the reason for the petition. Here are some suggestions:

- If reason 2 is applicable (infringement exists), you should state in your Supporting Declaration (SD) that you've made a rigorous comparison of the claims of your application with the infringer's device and find that the claims read on such device. You can even attach a two-column table, listing the elements of one of your claims as separate paragraphs in the left column and explaining how each element "reads on" the infringing device in corresponding paragraphs in the right column.

- If reason 4 (senior citizen) is applicable, you need merely state that you're over 65 and give your birthdate.

- If reasons 7 or 9 (DNA or superconductivity) is involved, refer to your application and tell how it involves DNA or superconductivity.

- If reason 8 is applicable, state that an IDS has been filed or is enclosed, and state where (e.g.,class and subclass) and by whom the search was made.

- If any of the other reasons is involved, give detailed facts or reasoning in support of your main reason, as I have done in Fig. 10-K. Don't hesitate to attach photocopies of letters, advertisements, etc. to your SD if they are relevant.

If you file your PTMS with the application, you should refer to it in your transmittal letter and your postcard receipt. In this case, you won't be able to include the PTO's filing data on the PTMS. If you file it later, add the application filing data to the

PTMS, as I have done in Fig. 10-J. As always, don't forget the postcard receipt.

If your PTMS is accepted, you'll receive a letter from the PTO stating that your petition has been granted and the examiner in charge of your application has been instructed to examine it ahead of turn.

You should then receive an official action (see Chapter 13) in a month or so. If your PTMS isn't accepted, you'll also receive a letter telling you why. Usually the rejection will be because your facts and reasons in the Supporting Declaration aren't detailed enough. In this case, file a revised Declaration beefing up your facts and reasons.

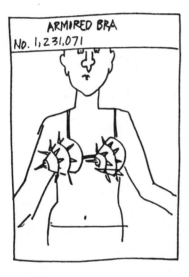

ARMORED BRA
No. 1,231,071

Q. Filing A Design Patent Application

AS I'VE INDICATED IN CHAPTER 1, Section B a design patent covers the ornamental external appearance rather than the internal structure, function, composition, or state of an invention. You may file both a design-patent application and a separate utility patent application on the same device, but of course, they should not cover the same feature of the device. The utility patent application

should cover only the structure (or a method) which makes the device or invention function or operate, while the design-patent application should cover an entirely separate "invention," namely, the ornamental (aesthetic) external (non-functional) appearance of something. For example, you can file a utility-patent application on the circuitry, keyboard mechanism, or connector structure of a computer, and a design patent application on the shape of the computer's case.

You'll be relieved to know that design patent applications are very easy to prepare. A design-patent application consists simply of the following:

- A preamble, specification, and claim,
- The drawing(s),
- A PAD (Form 10-2),
- A SED (Form 10-3),
- The fee (see Fee Schedule in Appendix), and
- The receipt postcard.

A completed preamble, specification, and claim for a design application is shown in Fig. 10-L.

If you believe that your invention has a unique ornamental appearance that is significantly different from anything heretofore designed, you can file a design-patent application on it.

Although not 100% kosher, some inventors file a design patent application on the external appearance of a utility invention which is unpatentable in the utility sense, and which has unfinalized or trivial novelty in the design sense. They do this mainly to be able to truthfully and legally state for a few years that the invention is "patent pending."

The first step in completing a design-patent application is to prepare drawings in the same format as for a regular patent application (see Section A above). However, the drawings for the design-patent application should show only the exterior appearance of your invention; no interior parts or workings should be shown Usually only one embodiment of a

Design Patent Application—Preamble, Specification, and Claim

Commissioner of Patents and Trademarks
Washington, District of Columbia 20231

Sir:

Preamble:

The petitioner(s) whose signature(s) appear on the declaration attached respectfully request that Letters Patent be granted to such petitioner(s) for the new and original design set forth in the following specification. The filing fee of $ _10_ , a patent application declaration, a small entity declaration, and a return receipt postcard are attached.

Specification:

The undersigned has (have) invented a new, original, and ornamental design entitled
" _Dress Hanger with Bosom Bosses_   "
of which the following is a specification. Reference is made to the accompanying drawings which form a part hereof, the figures of which are described as follows:

Fig. 1 is a ___perspective_____ view.
Fig. 2 is a ___side_____ view.

Claim: I (We) Claim:
The ornamental design for a __Dress Hanger with Bosom Bosses_____ as shown.

Fig. 10-L—Completed Design Patent Application—Preamble, Specification, and Claim

design is permitted. But it's important to remember that drawings of your design-patent application should show all of the details of the external surface of your design. A company I once worked for had an important design patent on a TV set held invalid because the design patent's drawings failed to show the rear side of the TV set.

Once you've made your drawings (in formal or informal form), fill out Form 10-8 from the Appendix as indicated in Fig. 10-L above. The title of your design can be very simple and need not be specifically directed toward your invention. For example, "Bicycle" is sufficient. Each view of the drawing should be separately indicated. For example, "Fig. 1 is front perspective view, Fig. 2 is side view," etc.

Note that the design-patent application has one claim only, and to write that claim you need merely fill in the blank on Form 10-8 with the title of your design. Fill out the PAD (Form 10-2) and SED (Form 10-3) as instructed in Sections G and H above.

The design-patent application with the declarations, drawings, and receipt postcard should be sent to the PTO in the same manner as your regular patent application. Be sure to keep an identical copy of your design-patent application and the drawings. The filing fee is indicated in the Fee Schedule in the Appendix (large/small entity). No transmittal letter is needed since Form 10-8 inherently provides a transmittal letter.

You'll receive your receipt postcard back in a week or two and you'll receive a blue filing receipt a week a month or so thereafter. If you're aware of any prior art, don't forget to file an Information Disclosure Statement (Part L above) within three months of your filing date.

Chapter 11

How to Market Your Invention

IN THIS CHAPTER I make an important detour from the central task covered by this book—obtaining a valid and effective patent on your invention. The reason for this sudden turn is simple. In the usual course of events, you'll have an interval (six months to two years) after you file your patent application before you need to either consider foreign filing or reply to an office action from the PTO. I strongly recommend that you use this interval to get your invention out on the market. This advice is so important that I've put it in an Inventor's Commandment reads as follows:

INVENTOR'S COMMANDMENT #13:
You should try to market your invention as soon as you can after filing your patent application; don't wait until your patent issues. You should favor companies who are close to you and small in size, and who already make and sell items close to yours.

If you already know how your invention will be marketed, or you work for a corporation that plans to handle this task, you can skip this chapter and continue reading about patent protection. In this regard, Chapter 12 deals with obtaining patent protection in other countries and Chapter 13 with getting the U.S. PTO to grant your patent.

It's now time for you, as an independent inventor to consider how best to get your invention out to the public. "Out to the public"? you ask. Shouldn't you keep your invention, and the fact that you've filed the application, secret? The answer is, "No." In fact, once you file a patent application on your invention, you may show it to whomever you think might be interested in buying or licensing it without risk of having someone scoop you on your patent.

This is because it would be very difficult for someone to steal your invention when you're the first to file a patent application on it. A patent thief

would have to file another application (the filing date would necessarily be later than yours due to the preparation time), get into a patent contest (called an "interference"—see Chapter 13, Section K) with you, and be able to win it. It's unlikely that this will happen because the thief's filing date would be later than yours, making the thief a "junior party" with a large burden of proof. You would also be able to prove that the thief "derived" the invention from you if you keep records of those to whom you reveal your invention Moreover, the thief would have to commit perjury (a serious felony) by falsely signing the Patent Application Declaration (Chapter 10) Of course, if you plan to maintain the invention as a trade secret, you should take the proper precautions (Chapter 1, Section P). At any rate, inventions are seldom stolen, in their early stages, before they're proven in practice.

Your next question might be, why try to sell or license your invention before a patent has been issued? While there are advantages to selling an already-patented invention, generally it's best to try to sell or license your invention as soon as possible after filing your patent application. This is because prospective corporate purchasers of your invention will want time to get a "head start" on the competition and to have the 17-year monopoly coincide with the time the product's actually on the market. The lack of prestige that a pending patent has as compared to an already-issued patent can be compensated for by a favorable search report showing that there's no strong prior art, i.e., that a patent is likely to issue on your invention.

A. Perseverance Is Essential

AS PAUL SHERMAN, NY Asst. Attorney General, said in his excellent article, "Idea Promoter Control: The Time Has Come" (Journ. Pat. Off. Soc., 1978 April, p. 261), "It is a failing of our system that there are no recognized avenues for amateur inventors to have their ideas evaluated and presented to manufacturers..." Even if you get a patent, it will almost certainly be totally worthless unless it protects a commercially-exploited invention. In fact, millions of patents have issued on inventions which were never successfully commercialized. None of these patents ever yielded a nickel to their owners. Thus, to get your invention into commercial production, you'll have to persevere. There's no magic solution to the invention marketing process. As noted toy inventor Paul Brown says, "You almost have to be obsessed with your invention to get it going." Or put another way, Emerson's famous adage about building a better mousetrap would have been better written, "If you build a better mousetrap, you'll still have to beat a path to many doors to get it sold." This brings us to the next Inventor's Commandment.

INVENTOR'S COMMANDMENT #14:
If you want your invention to be successful, you should persevere with commercial exploitation with all the energy which you can devote to it.

B. Overview of Alternative Ways to Profit from Your Invention

AS YOU CAN SEE from the chart of Fig. 11-A, there are seven main ways or routes for the independent inventor to get an invention into the marketplace and profit from it—Routes 1 to 7. These choices involve increasing difficulty and work for you. I recommend that most inventors use Route #3.

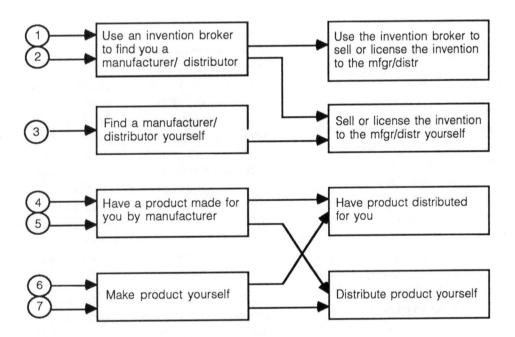

Fig. 11-A—Alternative Ways to Profit from Your Invention

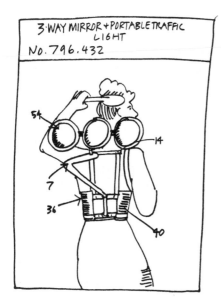

1. Route 1: Using A Contingent-Fee Intermediary

Starting at the top, Route 1 involves getting a contingent-fee invention broker or intermediary to find a suitable manufacturer/ distributor for you and then using the broker to represent you in the sale or license of your invention.

Also termed "invention developers," "invention marketers," "invention promoters" and the like, these contingent-fee brokers are firms that will represent you and try to market your invention by selling or licensing it for a percentage of your rights, the "contingent-fee basis." Most of the invention brokers are reputable and honest, and provide a legitimate service for a fair fee.

Obviously, Route 1 is the easiest possible path since the broker will do all of the work for you. However, it's neither that difficult to find suitable manufacturer/distributors (Section F below) if they exist, nor to present your invention to them once you locate them (Section I below). Thus, I recommend that you consider handling this task yourself. No one can sell an invention as sincerely and with as much enthusiasm and conviction as you, the true inventor. Also, you'll get 100% of the benefits and won't have to share the fruit of your creativity with a salesperson.

If you do use an invention broker, you should be concerned about two main possibilities for harm:

1. Loss of your invention rights through theft or communication to a thief; and

2. Loss of time and hence other opportunities.

The first possibility isn't great because you've already got a patent application on file. However, the second possibility is very real, and you should accordingly verify the efficacy of any promoter beforehand Unfortunately, about the only sure-fire way to do this is by word-of-mouth. Check with a patent attorney (Chapter 6, Section F) or an inventors' organization (Chapter 2, Section F) if your own associates are unable to provide you with a lead.

Once you're satisfied with the promoter's honesty and references, you should next investigate the contract you are offered by the promoter to be sure you don't lose time needlessly. Thus, the contract should specify that the promoter will perform substantial services, such as identifying the prospective manufacturers, preparing an invention presentation or demonstration, building and testing the invention, negotiating a license or sales agreement for you, etc. And most important, the agreement should set a time limit for the promoter to succeed—that is, get you a firm offer to buy, license, or get your invention on the market in product form. I feel that a year is reasonable, 18 months about the maximum you should ever consider. Make sure that if the promoter fails to succeed in the allotted time, all of your rights will be returned to you, together with all of the promoter's research, presentation documents, models, etc.

Here's a list of some invention brokers and developers who will work on a contingent-fee basis. If there are no results, there is no cost to you. I haven't made a detailed investigation of these or any other agencies; it's up to you to do this before you sign. Don't confine your efforts to this list, however;

there are many other reliable, honest contingent-fee promoters around the U.S. who are listed in local telephone and business directories or who are known to local patent attorneys and inventors' organizations.

Arthur D. Little, Inc.
(Invention Management Group)
Acorn Park
Cambridge, Massachusetts 02140
(617) 864-5770

Battelle Development Corporation
505 King Avenue
Columbus, Ohio 43201
(614) 424-6424

Control Data Technotec, Inc.
8100 34th Avenue South
Minneapolis, Minnesota 55440
(612) 853-4405

Patentvestments International
1050 Connecticut Ave., N. W.
Suite 300
Washington D. C.
20036

Product Resources International
1414 Avenue of the Americas
New York, N.Y. 10019
(212) 682-2966

REFAC Technology Development Corporation
122 E. 42nd Street
New York, New York 10017
(212) 687-4741

Watertight Imports
2973 Harbor Blvd., Suite 249
Costa Mesa, CA 92629-3989
(714) 548-1998

In addition, many universities now have an invention marketing department which exists primarily to market the technology developed in the university's research labs, but also take ideas from outsider on a contingent-fee basis. Check with your local college.

Don't Use a Fee-based Invention Intermediary

On the other hand, there are other firms which I'll call fee-based invention intermediaries. You should generally avoid like the plague. These companies or organizations run ads in newspapers, magazines, etc., stating something like "Inventions and Ideas Wanted!" They will commonly require you to pay a relatively small fee (say $200 to $600) for an "evaluation" (which almost always is glowingly positive). Then they'll ask for a relatively large fee, in return for which they will do some "market research" and try to sell your invention or have it manufactured. They sometimes also take a percentage share (for example, 20 percent) of your invention.

The large fee charged by invention promoters is $1,000 to $2,000, and often a lot more.It is especially high in relation to the services rendered. Generally, they will do little more than write a brief blurb describing your invention and send it to prospective manufacturers in the appropriate fields. For this reason I recommend you not use an invention promoter unless you find one that can establish a successful track record, i.e., a record of bringing its clients a significant percentage in royalties in relation to the fees it charged them. Most can't. Most of them, if asked for a track record will balk, get evasive, say track records aren't relevant, will demonstrate tremendous expertise in evasion, etc. They'll also tell you that you won't be able to do it yourself, that they have specialized expertise and files, etc. However, no special expertise is needed to market an invention; a sincere approach to a suitable company from an independent inventor (using techniques of this Chapter) will be far more effective than any "professional's" effort. I would look elsewhere.

2. Route 2: Partial Use of an Intermediary

Route 2 (a seldom-used path) is the same as Route 1, except that here you use a broker to find prospects and then you take over and do the selling. Contingent-fee brokers won't accept this type of arrangement since they'll want to control the sales negotiations. However, if you feel that your strong suit is in presenting and selling, and that sales research is for someone else, you can pay a broker (either contingent-type or fee-based) to research possible purchasers. Then go out and present your invention yourself.

3. Route 3: Finding a Manufacturer and Distributor Yourself

Route 3 is the path I most favor and which most independent inventors use. Here you do your own research and selling. If you succeed, you'll get 100% of the rewards and you'll control the whole process, yet you won't be bothered with manufacturing or distributing.

4. Route 4: Having Your Invention Manufactured and Distributed for You

Route 4 is a viable alternative for some relatively uncomplicated products. Here you have your invention manufactured for you—a Far-Eastern manufacturer will usually be cheapest— and then use U.S. distributors to sell the product. Of course, you have the headaches of supervising a manufacturing operation, including quality control, red tape associated with importing, etc. But, if you succeed, you'll keep much of the manufacturing profit for yourself.

5. Route 5: You Distribute

In Route 5, you handle distribution as well as supervising manufacturing. More profit, but more headaches and work.

6. Route 6: You Manufacture

In Route 6, you really get into it; you have to do the manufacturing yourself, with all of its headaches (see Section J), but you'll get a lion's share of the profits, if there are any.

7. Route 7: You Manufacture and Distribute

Last, and most difficult, in Route 7 you do it all yourself—manufacturing and distributing. While you get all of the profits, you'll have all of the headaches, as explained in Section J.

Because, as I said, Route 3 makes the most sense for most independent inventors, I devote the bulk of this chapter to finding a manufacturer/distributor to build and market your patent. If you want to pursue the possibility of manufacturing and distributing your invention, I've included an overview of potential resources in Section J below to help you do this.

C. Be Ready to Demonstrate a Working Model of Your Invention to Potential Customers

ASSUMING THAT YOU CHOOSE Route 3, the best way to get a manufacturer or others to "buy" your invention is to demonstrate an actual working model. Pictures and diagrams may convey an idea and get a message across, but the working model is the thing that will make believers out of most people and show them that your invention is real and fulfillable, and not just chicken scratchings on paper. So, if you haven't made a model before, do your best to make one now, even if it has to be made of cardboard or wood. One essential is to make your model or prototype as simple as possible. Simplicity enhances reliability, decreases cost, decreases weight, and facilitates salability, both to a manufacturer and to the public.

If you're not handy, hopefully you can afford to have a professional model maker or artisan make you one, or you may have a handy friend or relative. Where can you find model makers? Ask your local inventors' organization (Chapter 2, Section F). If that fails, an obvious place is in your nearest metropolitan area yellow pages under "Model Makers." Also try "Machine Shops" and "Plastics—Fabricating, Finishing, and Decorating."

In addition, your local college or community college may have a design and industry department which may be able to refer you to a model maker. If you live near an industrial plant which employs machinists or model makers, perhaps you can get one of these employees to do a moonlighting job for you—put a notice on the plant's bulletin board, call, or ask around.

If you do use a model maker and you disclose critical dimensions, materials, suppliers, or other information you consider to be proprietary (i.e., maintained by you as a trade secret), it is best to have the model maker sign your Consultant's Work Agreement (Form 4-3) before you turn over your drawings or other papers. Follow the instructions in Chapter 4 to fill out this form. I also suggest that you also add a confidentiality legend to any drawings or descriptions you turn over to your model maker. Such a legend, which can be made in rubber-stamp form or can be typed on the drawings, can read as follows:

This drawing or description contains proprietary information of [your name] and is loaned for use only in evaluating or building an invention of [your name] and must be returned upon demand. By acceptance hereof, recipient agrees to all of the above conditions. © 198_ [your name].

After you've made a working model, you should take at least one good photograph of it. The photograph should be of professional quality. Or, if you are not a good photographer, have a professional do it, and order several views if necessary. Have at least fifty glossy prints made of the photo, possibly with several views on one sheet. Then write a descriptive blurb about your invention, stating the title or the trademark, what it is, how it works, its main advantages and selling points, plus your name, address, telephone number and the legend "Patent Pending." Don't get too bogged down in detail, however. In other words, make your write-up snappy and convincing. Then have it typed or printed and have at least fifty copies made to go with the photographs.

D. Finding Prospective Manufacturers/Distributors

THE NEXT STEP is to compile an initial listing of manufacturers who you believe could manufacture and distribute your invention profitably. (You should keep your marketing notes, papers, and, correspondence in a separate file from your patent application (legal) file.) Your initial list should comprise all the manufacturers who meet the following three criteria:

1. They're geographically close to you;

2. They already manufacture the same or a closely-related product; and

3. They're not too large.

Nearby or local manufacturers who already work in your field are best. If they manufacture your invention, you can monitor the progress and consult with them frequently. Obviously it's a big help to deal with a company which has experience with devices similar to yours. They already know how to sell in your field, are aware of competitive pricing policies, can make your invention part of their existing product line, which allows them to keep sales costs low, and presumably want new models related to their existing products in order to keep ahead of the competition. If the manufacturer is not in a closely-allied line, both the seller and the product will be on trial, so why start with two strikes against you?

The reasons for avoiding giant manufacturers are these:

1. Smaller manufacturers are more dependent on outside designers. In other words, most don't have a strong inbred prejudice against inventions they did not invent themselves(see the "NIH" Syndrome in Sec. E. below);

2. You can contact the decision-makers or the owners of the company directly, or more easily;

3. Decisions are made more rapidly because the bureaucracies are smaller; and,

4. You are less likely to be required to sign a waiver form (see "The Waiver," in Section F below).

Obviously, you shouldn't use companies that are so small that they don't have enough money to finance the manufacture of your invention or the marketing of it adequately. Companies with sales of about 5 to 50 million dollars are best.

To find companies meeting the above criteria, start by first considering people you know. Which one of them is likely to have contacts in the field of your interest? Put them to work for you and you may be amazed that with a few phone calls you can get just the introduction you need. If this doesn't work, try looking in your appropriate local stores for manufacturers of closely-allied products which are already on the shelves. You'll know for sure that these companies have a successful distribution and sales system or operation. Also, check the library for books listing local manufacturers (such as the California Manufacturers Register) and check national resources such as the *Thomas Register* or *Duns Million Dollar Directory*. In addition, check the ads in pertinent trade and hobby magazines. Lastly, stock advisory services, such as *Value Line Investment Survey*, *Standard & Poor's*, and *Moody's*, supply excellent information about companies. Get the names of the company presidents, vice presidents, directors of engineering, marketing, etc. Find out all you can about each company you select; know its products, sales corporate history, factory location(s), etc.

If your invention is in the gadget category and you believe it would appeal to the affluent, your first choice might be Hammacher Schlemmer, a specialty store and mail-order house at 147 East 57th Street, New York, NY 10022. This outfit develops and sells a wide variety of gadget exotica, both through its catalogs and over the counter. The company receives about 3,000 ideas for inventions each year, accepts about 50 to 75 of these, and arranges to have them produced by manufacturers. Many items that Hammacher Schlemmer financed and had manufactured, or first sold as strictly luxury gadgets, have become commonplace in American homes. For example, the steam iron, the electric razor, the pressure cooker, the blender, the humidifier, the electric can opener, the high-intensity lamp, the microwave oven, and the automatic-drip coffee maker were first introduced by this unique and innovative firm. Many other "gadget exotica" mail order forms exist, such as *The Sharper Image*, *JS&A*, etc., but these firms don't develop or manufacture any products. Also trade fairs or shows, e.g., The Gift Show, are good places for you to wander about, looking for prospective manufactures. Talk to their people who run the exhibits to get a feel for the companies, whom to contact, what their attitude to outside inventions is, etc.

If you can't find any U.S. companies, try foreign ones. Sadly, there has been a recent trend in many

U.S. firms to become complacent or tight. They've refused to undertake new ventures which foreign firms have jumped at. So, if chauvinism fails, try abroad, using the techniques above to find suitable prospects.

MILKMAID'S PORTA-STOOL
No. 359.921

E. The "NIH" Syndrome

BEFORE PRESENTING YOUR INVENTION to any manufacturer, two possible impediments should be kept in mind: the NIH (Not Invented Here) syndrome and the common insistence that you give up many of your legal rights by signing a waiver (Section F below). Generally, the larger the manufacturer, the greater the chances of encountering one or both of these impediments.

The NIH syndrome is an unwritten attitude which handicaps inventors who submit their ideas to a company, no matter how meritorious such ideas may be. Put simply, many companies have a bias against any outside invention because it was "not invented here." This attitude prevails primarily because of jealousy. The job of the corporate engineering department is to create new and profitable products for their company. If an engineering department were to recommend an outside invention, it would almost be a tacit admission that the department had failed to do its job in solving a problem and coming up with the solution the outside inventor has found.

How can you overcome the NIH syndrome? First, realize that it's more likely to exist in larger companies, or companies with extensive engineering departments. Second, when forced to deal with engineering departments or any department in a company where the NIH syndrome may be present, always remember that the more your invention appears to be a logical extension of ideas already developed within the company, the better your chances of acceptance will be.

F. The Waiver and Precautions in Signing It

MOST COMPANIES with legal advice will require you to sign their agreement (called a "waiver"), under which you give up a number of important rights that you would otherwise possess under the law. The reason for this waiver is that many companies have been sued by inventors claiming violation of an implied confidentiality agreement, or an implied agreement to pay if the invention is used. Even though the company's own inventor may have come up with the invention independently of the outside inventor, many companies have lost these suits or were forced to compromise because of the uncertainties and expenses of litigation.

The waiver itself usually requires you to give up all your rights, except those which you may have under the patent laws. Specifically, the waiver typically asks you to agree that:

1. The company has no obligation to pay you if they use your idea;

2. The company isn't bound to keep your idea in confidence;

3. The company has no obligation to return any paper you submit; and

4. The company has no obligation whatever to you, except under the patent laws.

Many companies add many other minor provisions, which are not significant enough to discuss here. The effect of the waiver is that you have no rights whatever against the company if they use your invention, except to sue them for patent infringement if and when you get a patent.

The usual procedure, if you send a letter mentioning your idea to the company, is for the company to route your letter to the patent or legal department, which will send you a form letter back stating their policy and asking you to sign the waiver before they can review your idea. Once you do so, the patent or legal department will approve your submission for review and send it to the appropriate engineering manager of the company.

Since you may not get a patent, since the company may use a variation of your idea that may not be covered by any patent you do get, and since you would like to have the company keep your submission in confidence, it's best to avoid signing any waiver if at all possible. Ergo, you should, at least initially, concentrate on smaller companies. The smaller the company, the less likely they are to make you sign a waiver. In fact, the best sort of relation you can have with a company to which you submit your ideas is to have them sign an agreement that you have drafted. Many small companies actually want to review outside inventions and are willing to sign a proprietary-submission agreement.

If the company is willing, or if you can swing it (say, by touting the commercial potential of your invention), have the company sign a Proprietary Submission Agreement such as the following:

Sample Proprietary Submission Agreement

X Company agrees to review an invention from [your name] for a new and improved [describe invention], to keep such invention and all papers received in confidence, to return all papers submitted upon request, and to pay [your name] a reasonable sum and royalty to be settled by future negotiation or arbitration if it uses or adopts such invention.

If a company won't sign the above agreement, you can make it a bit more palatable by eliminating the last clause regarding the payment of a reasonable fee and royalty. Even with the last clause eliminated, you're in a very good position if you've gotten them to sign. If the company still refuses to sign your agreement, you can add an exculpatory clause such as follows:

The forgoing shall not obligate X Company with respect to any information which X Company can document (a) Known to it prior to receipt from me, either directly or indirectly, or (b) Which is now or hereafter becomes part of the public domain from a source other than X Company.

Lastly, if you can't get them to sign even this, you're still in a pretty good position legally if you can get them to review your invention without any agreement being signed by either side.

If all else fails and you do have to sign a waiver before the company will look at your invention (that's what will usually happen), it's not all that bad, since you do, at least, have a pending patent application. And most companies are far more afraid of you suing them (for taking your invention) than they are interested in stealing your invention. Now you can understand why I emphasized the need to file your patent application before submitting your invention to any company. If you sign the waiver, your position won't be seriously jeopardized if your patent issues. However, if you're submitting an invention to a company without having first filed a patent application (Block B of the Invention Decision Chart from Chapter 7, Section C), it's very important that you try to get the company to sign a

Proprietary Materials Agreement (Form 3-1, Appendix) or, failing that, try to submit it without signing their waiver.

If you do have to sign a waiver, it's best to make sure the company is a reliable and fair one. Also, it's important to insist, by means of a separate letter, that the company make its decision within a given time, say six months, or else return all of your papers to you. This is because many companies, especially large ones, can take forever to make a decision if you let them, which may interfere with your efforts to market the invention to others.

To the extent you are uncertain about whether signing a waiver is a good idea under the circumstances, a consultation with a patent attorney might be wise. On the other hand, don't let the waiver prevent you from showing your invention to a reputable manufacturer that promises to give you a decision in a reasonable time. As long as your patent is pending and eventually issues, you'll have reasonably strong rights.

G. The Best Way to Present Your Invention to a Manufacturer

THE BEST AND MOST EFFECTIVE way to sell your invention to a manufacturer is personally to visit the decision-maker in the company you elect and demonstrate a working model or prototype of your invention (or present drawings of it if you have no working model). To accomplish this, write a brief, personal, friendly, and sincere letter to the president of the company, saying that you have a very valuable invention you believe would be profitable for the company's business and that you would like to make an appointment when convenient to provide a brief demonstration. You can disclose the general area of your idea, but don't disclose its essence until you can present it properly. Keep the initiative by stating that you will call in a few days. Follow through accord-

ingly. Here's an example:

Mr. Orville Billyer
President, Billyer Saw Co.
[etc.]

Dear Mr. Billyer:

I'm employed as an insurance agent, but in my spare time I like to tinker. While building a gun rack, I thought of and have perfected a new type of saw fence which I believe can be produced at 60% of the cost of your A-4 model, yet which can be adjusted in substantially less time with greater accuracy. For this reason, I believe that my fence, for which I've applied for a patent, can be a very profitable addition to your line. I'll call you in a few days to arrange a demonstration of my invention for you in your plant.

Most sincerely,

_____

Marjorie Morgenstern

When you come to the demonstration, be prepared! Set up your presentation well in advance. Practice it on friends. Tell the advantages of your invention first, how it works, how it will be profitable for their business, why it will sell, etc. Make sure your model works. Also, prepare appropriate and attractive written materials and photos for later study by the decision-maker.

In your presentation and written material, it's wise to cover the "Three F's"—Form, Fit, and Function. Form is the appearance of your invention. Stress how it has (or can have) an attractive, enticing appearance.

Demonstrate how your invention mates with other products, or with the environment in which it is to be used. If your invention is a highly functional device, such as a saw fence, show and tell how it fits onto a sawing machine. If it's a clock, show (or present attractive pictures showing) how it looks attractive on a desk or coffee table.

Function is what your invention does, how it works, what results it attains. Demonstrate and discuss its function and its advantage here. Mention all of the advantages from your Positive And Negative Evaluation (Form 4-1, Appendix).

In addition, be prepared to discuss cost of manufacture, profit, retail price, competition, possible product liability, product life, etc. Review all of the positive and negative factors from the list in Chapter 4 to be sure you've covered all possible considerations.

During the verbal part of your presentation, it's wise to use diagrams and charts, but keep your model, written materials, and photos hidden from view. Otherwise, the people you're trying to sell to will be looking at these instead of listening to you. Then, at a dramatic moment, bring out your model and demonstrate how it works. Don't apologize if your model is a crude or unattractive prototype, but radiate enough confidence in yourself and your invention that they will overlook any lack of "cosmetics." If you can't bring or show them your model for any reason, a videotape filmstrip or slide presentation which shows the three F's will be a viable, though less desirable way, to show the invention.

If possible, make them think that the invention is basically their idea. You can do this by praising their related product line and then showing how your idea compliments theirs, or by enthusiastically endorsing any reasonable suggestion they make for your idea.

At the end of your verbal presentation, produce your written materials and pictures for study (either then and there or at a later time). If they're interested in the invention, be prepared to state your terms and conditions—see Chapter 16, Section G. If they're really serious and ask for it, you can show them your patent application without your claims, but only with the understanding that it won't be copied and will be returned to you. You shouldn't offer the claims or prior art from your search unless you're asked.

If you've done your best and still get a rejection, don't accept it blindly and walk away with your tail between your legs, but turn it into an asset for next time. Talk to the executives about it and learn exactly why they decided not to accept your idea so that in the future you'll be better prepared to answer and overcome the disadvantage which blocked your initial acceptance.

Assuming the company is interested, you shouldn't blindly or automatically accept it as your patron. Rather, you should evaluate the company to which you're demonstrating your invention just as they're evaluating you and your invention. For example, if the company seems to lack energy or vision, don't go with them. After all, you're risking a lot, too, when you sign up with a company. If the company doesn't promote your invention enthusiastically and correctly, it can fail in the market, even if it's the greatest thing to come down the pike in 20 years.

H. Presenting Your Invention by Correspondence

THE OTHER WAY to present your invention is by correspondence. Because letters are easy to file and forget, and because any salesman will tell you a personal presentation is a thousand times more likely to make a sale, I strongly advise against submitting an invention to a manufacturer by correspondence if you can avoid it. Try your utmost to arrange a personal demonstration with a working model as described in the previous section. Nevertheless, if you do have to resort to correspondence, don't let your efforts slacken.

Your letter should always be addressed to a specific individual. Find the president's name from the directories mentioned in Section D above. If you receive an expression of interest from the company, you will probably be faced with the waiver question. My comments in the previous discussion cover how to handle this problem. Before you send a model, get an advance written commitment from the company that they'll return it within a given time. You should send your model by certified, insured mail, return receipt requested, and make follow-up phone calls as appropriate.

I. Making an Agreement to Sell Your Invention

IF YOU SELL YOUR INVENTION to a manufacturer/distributor, the next step is to sign an agreement of some sort with the manufacturer. The question thus arises, what will be the terms of the agreement, exactly what will you sell them, and for how much? There are many possibilities. These are covered in Chapter 16, which deals with ownership and transfers of patents.

J. Manufacturing and/or Distributing the Invention Yourself

FOR REASONS STATED EARLIER, manufacturing and/or distributing a product embodying your invention—unless you already have manufacturing experience, a plant, and/or distribution facilities—is very difficult. Besides, you can spend your time more effectively selling your invention or patent application, rather than dealing with manufacturing and product-marketing problems.

If you do plan to manufacture and/or distribute your invention yourself (Routes 6 or 7), I strongly suggest that you learn about the subject thoroughly beforehand so you will know what is involved and which pitfalls to avoid. The best place to obtain literature and reading material is your local SBA (Small Business Administration) office, which has scads of literature and aids available to apprise you of the problem and pitfalls. They even have a service which allows you to obtain the advice of an experienced executive free; ask for a "Counseling Request from Small Business Firm" form. Nolo Press publishes an excellent book, *Start-Up Money: How to Finance Your Small Business,* which tells potential businesspeople how to assess the costs of a proposed business, how to draft a business plan, and how to obtain sufficient start-up money.

1. Financing the Manufacture Of Your Invention

Financing any manufacturing venture of your own is a separate and formidable problem. If you have an untried and unsold product, most banks will not loan you the money to go ahead. However, if you can get orders from various local firms, the bank may loan you the money. Thus a local test marketing effort on a limited scale may be desirable.

For obtaining money to finance untried products, a money lender who's willing to take more risk in needed. Such a person is usually termed a "venture capitalist" (VC). A VC will loan you money in exchange for shares or a portion of your enterprise. The *Venture Capital Monthly*, the *Guide to Venture Capital Sources*, and the *Directory of the Natural Venture Capital Association* (listed in resources section at the back of this book) should aid you here. Also the Venture Capital Hotline, tel. 800 237-2380, will provide you with a list of suitable VCs for a fee (about $75). However, VCs won't loan you money on the same terms a bank would. Because of the higher risks they take, they demand a much larger return—namely a piece of the action. Also, they'll want to monitor your company and exercise some degree of control, usually by putting their people on your board of directors. A thorough discussion of the pros and cons of working with venture capitalists can be found in the Nolo Press book, *Start-Up Money*, mentioned earlier.

A recent development in the VC field is the "Incubator VC." This is a VC which provides several different inventors with offices, labs, and/or a manufacturing area in a special building, called an "innovation center." Also the VC may provide technical, financial, and marketing consultation, as well as other services, until each nurtured enterprise is ready to leave the "nest." The sources in the preceding paragraph, as well as inventors' organizations (Ch. 2, Sec F) will give you the names of Incubator VCs. One of the largest is Genexus International, Inc., 200 N. Main St., Suite 200, Salt Lake City, UT 84103, tel (801) 328-1504. They have several innovation centers around the U.S.

2. Prepare a Business Plan

To obtain venture capital to start a business based on your invention, you'll have to prepare a business plan—a presentation which tells all about your invention, the market for it, and how you plan to use the money. You can get an excellent booklet from a national accounting firm gratis which will tell you how to prepare your business plan. This is *Raising Venture Capital—An Entrepreneur's Guidebook*, available free from any office of Deloitte, Haskins & Sells. This practical guide tells how to write and present your business plan. Again, *Start-up Money* is also recommended for this purpose.

3. Distribution Through Mail Order

Mail order is often an easy way for an individual to distribute an invention, whether the inventor makes it or has it made. An excellent guide is *How I Made $1,000,000 In Mail Order*, by E. Joseph Cossman (Prentice-Hall). Once your mail order operation starts bringing in some cash, you can branch out and try to get some local, then regional, then state, and then (hopefully) national distributors who handle lines similar to yours.

There are two principal ways to contact your potential customers:

1. Magazine/media advertising; and

2. Direct mail advertising.

If you're interested in the latter, order the *Dunhill Marketing Guide to Mailing Lists* from Dunhill International List Company, Inc., 444 Park Avenue South, New York NY 10016.

You can also try to use a mail-order distributor. Many mail-order houses will, if you send them a production sample and they like it and feel you can meet their demand, buy your production. They'll put in their own ads, manufacture and distribute their own catalog, and thus are valuable intermediaries for many garage-shop manufacturers. Walter Drake & Sons, Colorado Springs, CO 80940, is one of the largest, but you can obtain the names of many others by looking for ads in *Redbook, House Beautiful, Better Homes and Gardens, Apartment Life, Sunset, Holiday,* etc. These mail-order firms are always looking for new gadgets, and most of their products come from small firms. While many of them will purchase quantities of your product outright, some will want to take them on consignment, which means they do not pay you until and unless they sell it themselves.

4. Distribution Through Government

If your invention is or can be used in a product which the federal government might purchase, write to the General Services Administration, Federal Supply Service, 1734 New York Avenue, N.W., Washington, D.C. 20406, telling them that you're offering a product which you feel the government can use. They'll send you appropriate forms and instructions. Also, don't neglect your corresponding state and local purchasing agencies.

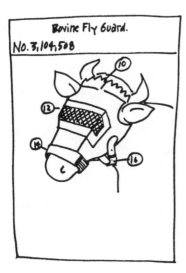

5. Publicity

Publicity will sometimes be of great aid to you before you get your invention into production, and is invaluable once it's on the market. Assuming it's not yet on the market and you're either looking for a manufacturer, distributor, or thinking of manufacturing or distributing it yourself, publicity can cut both ways. As stated, many manufacturers like to get a secret head start on their competition and thus won't be too interested if your invention has already been disseminated to the public.

If you're going to make and sell it yourself, I believe you should wait until you've got the product out before you try to publicize it. Why? The public's memory span is short, so they'll be likely to forget about your product by the time you get it on sale. My advice is to not seek publicity until a product with your invention is almost or actually on the market, unless you've tried unsuccessfully, after substantial efforts, to get it on the market.

Assuming you're ready for publicity, one way to get it (at a price) is to hire a public-relations or marketing research firm to promote your invention for you. There are many reputable firms who can come up with many creative and valuable ideas for a fee. However, since the cost of public-relations services is very high, I don't recommend it unless you can bear the cost without difficulty.

Many magazines will feature new ideas free if you send them a clear, understandable, professional-quality photo or drawing of your invention, plus a brief, clear, and understandable description of it. They may even write a full-length feature about your invention if they think it's interesting enough. Suitable magazines are *Popular Science, Mechanics Illustrated, Popular Electronics, Better Homes and Gardens, Pageant, Parade, Playboy, This Week, True Story, Jet, Outdoor Life, House and Garden, House Beautiful, Outdoor Living, Changing Times, McCalls, Apartment Life, Argosy* and *Sunset.* You can obtain the addresses of those you think are relevant from *Ulrich's International Periodicals Directory* in your local library.

The magazine *Advertising Age* has a feature called "Idea Marketplace" in each issue in which they publicize new inventions gratis. Write to them at

Crown Communications, Inc., 740 Rush Street, Chicago IL 60611, sending a picture and brief description of your invention. Thomas Publications, 1 Pennsylvania Plaza, New York NY 10001 has a bimonthly called Technology Mart which offers a similar service.

Many other trade magazines, such as *Dental Product Reports*, have a similar column. Review the trade magazines in the field of your invention for ideas.

Nolo also publishes an excellent book, *Marketing Without Advertising*, by Phillips and Raspberry; its title is self-explanatory.

Other sources of publicity are exhibits, trade fairs, and business shows. You pay a relatively small fee to the sponsor of the show, in return for which you're given a table or booth, or equivalent space to demonstrate your invention at the fair or show. Naturally, your exhibit should be attractive, interesting, and it is preferable to have a working model or very good literature available in connection with your invention. There are exhibition-service companies that will prepare a display exhibit for you for a fee.

The Invention Trade Center (part of Inventors Workshop International—see Chapter 16), 121 North Fir Street, Ventura, CA 93001, will exhibit your invention to prospective manufacturers for six months for a fee. The ITC even has a separate security area for unpatented inventions; visitors must sign a keep-confidential agreement before they can enter. Also, several of the Contingent-Fee Invention Brokers listed above have exhibition areas.

Don't overlook the media (radio, TV, newspapers, and magazines) as an excellent source of publicity. Many local radio and TV stations feature programs in which new inventors can demonstrate or discuss their inventions. One of the best ways to get media publicity is to dream up or pull a stunt. For example, if you've invented a new bicycle drive mechanism, you might enter and win a local bike race, or sponsor some type of contest (which you can win!).

Chapter 12

Going Abroad

A. Introduction

BY NOW YOU'VE GOTTEN your U.S. application on file and have taken steps to have your invention manufactured and distributed in anticipation of receiving a patent. Your next step will be either to file in one or more other countries (this chapter) or to deal with the first substantive response by the U.S. PTO (called an "Office Action") to your application (Chapter 13).

If you've already received your first Office Action from the US PTO, you'll have a pretty good idea of the patentability of your invention, and consequently your chance of getting foreign patents abroad. If you want to help determine your chances of getting foreign patents, see Ch. 10, Sec. P to see how to get your US application examined earlier.

Why file your patent application in other countries? Simply because a US patent will give you a monopoly only in the US. If you think your invention is important enough to be sold or manufactured in large quantities in any other countries, so that you'll want to create a monopoly there, you'll have to go through the considerable effort and expense of foreign filing in order eventually to get a patent in each desired foreign country. Otherwise, anyone in the foreign country will be able to make, use, and sell your invention with impunity. However, they won't be able to bring it into the US, if you have a US patent, without infringing your US patent.

This chapter doesn't give you the full, detailed instructions necessary to file abroad. That would take another book. Instead, my mission is to alert you to the basic procedures for accomplishing filing in other countries so that you won't lose your opportunity to do so through lack of information. However, once you decide to foreign file, you'll probably need some professional guidance, notwithstanding the availability of other resources (discussed in Section K of this chapter) that will answer most of your questions.

The single most important point you can learn from this chapter is presented in the following inventor's commandment:

INVENTOR'S COMMANDMENT #15:
Foreign Filing: Do any desired filing of your invention abroad in Convention (industrialized) Countries within one year after your U.S. filing date, but only file abroad if you believe that your invention has very strong commercial value there. (File in non-Convention (non-industrialized) Countries before the invention is publicly known.) In any case, do not send your invention abroad until you get a foreign filing license or until six months after your US filing date.

Prior to discussing the ins and outs of foreign filing, it's important that you familiarize yourself with several important treaties and arrangements. Let's wade in.

B. The Convention and the One-Year Foreign Filing Rule

THE MOST IMPORTANT THING to know about foreign filing rules is that many of them are governed by the International Convention for the Protection of Industrial Property. Most people in the patent field simply call it the "Paris Convention," or simply "the Convention." The majority of nations of the world, except those listed in Section E below, are parties to this international treaty, which was entered into in Paris in 1883 and has been revised many times since. Generally, the Paris Convention governs almost all reciprocal patent filing rights.

For the purpose of this chapter, there's only one thing you need to know about the all-important Paris Convention: If you file a patent application in any one member (Paris Convention) country (such as the US), and you file a corresponding application in any other member country or countries (such as the UK, Japan, the EPO, Australia, the USSR, etc.), within one year of your original filing date, your applications in the other countries will be entitled to the filing date of your original application for purposes of overcoming prior art. You do have to claim "priority" of your original application. If you fail to file any foreign convention applications within the one-year period, you can still file after the one-year period in Convention jurisdictions (including the Patent Cooperation Treaty (PCT) and European Patent Office (EPO)), provided you haven't sold or published your invention yet. However, any such late application won't get the benefit of your original U.S. filing date, so any relevant prior art which has been published in the meantime can be applied against your applications. Put differently, once the one-year rule is missed, your application becomes a non-Convention application, even in Convention countries.

C. European Patent Office/Europäisches Patentamt/Office européen des brevets (EPO)

THE EUROPEAN PATENT OFFICE (EPO) is a separate and vast tri-lingual patent office in Munich, across the Isar from the famous Deutsches Museum. The EPO grew out of the earlier formation of the European Economic Community (EEC, also known as the Common Market) and the economic integration that resulted. Member nations of the EEC are also members of a treaty known as the European Patent Convention (EPC). Under the EPC you can make one patent filing in the EPO, whose main branch is at Ehrhardstrasse 27, D-8000, München 2, West Germany. If this filing matures into a European patent which, it will, when registered in whatever individual member countries you've selected, cover your invention in these countries. And since the EPC is in turn considered the same as a single country (a jurisdiction) under the Paris Convention, your effective EPO filing date will be the same as your original U.S. filing date, so long as you comply with the one-year foreign filing rule. In other words, filing in the EPO allows you to kill many birds with one stone.

Once your application is on file, the EPO will subject it to a rigorous examination, including an opposition publication (see Chapter 13) 18 months after filing. Even though you'll have to work through a European agent, patent prosecution before the EPO is generally smoother than the PTO because the examiners are better trained (all speak and write three languages fluently) and because they actually take the initiative and suggest how to write your claims to get them allowed. If your application is allowed, you'll be granted a European patent which lasts for 20 years from your filing date (provided you pay maintenance fees in the member countries you've selected). Your patent will be valid automatically in each member country of the EPC that you've designated in your application, provided that you register it in and file translations in each country and appoint an agent there.

Here's a list of the EPO countries as of 1988:

Austria	Netherlands
Belgium	Spain
France	Sweden
Germany (Fed. Republic)	Switzerland/
Greece	Liechtenstein
Italy	United Kingdom
Luxembourg	

NECK SHOWER.
No. 821.716

D. The Patent Cooperation Treaty (PCT)

THE PCT IS ANOTHER IMPORTANT TREATY to which most industrial countries are a party. Under the Patent Cooperation Treaty (PCT), which was entered into in 1978, you can file in the U.S. and then make a single international filing within the one-year period; this can cover all of the PCT countries, including the European Patent Office (EPO). Then you must file separate applications in each PCT country or the EPO where you desire coverage. These separate filings, which must be translated for non-English speaking countries, must be made within 20 months after your U.S. filing date, or eight months after your PCT application is filed. Moreover, if you elect a newly-accepted part (Chapter II) of the PCT by 19 months after your US filing date, you can wait up to 30 months after your US filing date to make these separate filings. Thus, except for the single international filing, the PCT affords you an eight or eighteen-month extension of the one-year rule for PCT countries or the EPO.

Also, you can file your first application under the PCT and then file in any PCT countries (including the US) within 20 or 30 months from your PCT filing date. Also, since the PCT is a member of the Paris Convention, if you file with the PCT first, you can file in any non-PCT Convention country within one year from your PCT filing date. You can either file a "regular" PCT application, in which case you'll receive a "search report," but will have no opportunity to prosecute the application or get claims allowed, or you can elect Chapter II of the PCT, in which case you will receive an examination and can prosecute and amend your claims and receive a formal indication of allowability (or rejection!). Here's a list of the PCT countries and jurisdictions as of 1987 March:

AIPO#	Korea, Rep. of
Australia	Luxembourg
Austria*	Madagascar
Barbados	Malawi
Belgium*	Monaco
Brazil	Netherlands*
Bulgaria	Norway
Canada	Romania
Denmark	Spain
European Patent Office	Sri Lanka
Finland	Sudan
France*	Sweden*
Germany, Fed. Rep.*	Switzerland/
Hungary	Liechtenstein[+*]
Italy*	United Kingdom*
Japan	USA
Korea, Dem. Ppls. Rep. of[+]	USSR

#African Intellectual Property Organization: Common patent system for Central African Republic, Senegal, Cameroon, Chad, Togo, Gabon, Congo, Mauritania, Mali, & Benin.

*Also EPO members

+These countries have not accepted Chapter II of the PCT, although Switzerland/ Liechtenstein can be accessed via the EPO under Chapter II.

E. Non-Convention Countries

THERE ARE A SIGNIFICANT NUMBER of countries (generally non-industrial) that aren't parties to the Paris Convention. These include:

Bangladesh	Jamaica
Chile	Jordan
China, Republic of (Taiwan)	Kuwait
	Liberia
Colombia	Malaya
Costa Rica	Nicaragua
Ecuador	Pakistan
El Salvador	Panama
Ethiopia	Paraguay
Gambia	Peru
Ghana	Philippines
Guatemala	Sierra Leone
Honduras	Singapore
Hong Kong	Venezuela
India	Zaire
Iraq	

If you want to file in any of these countries, generally you may do so at any time before you make your invention publicly known, i.e., before you or any patent office makes a publication of or you make a commercial use of your invention. Remember, however, that you aren't permitted to file your invention abroad until six months after your filing date, unless your filing receipt says, "Foreign License Granted [date]." (See Section G below.)

Filing isn't common in these countries, but if you do want to file in any of them, you may do so at any time, provided:

a. Your invention hasn't yet become publicly known, either by your publication, by patenting, by public sale, or by normal publication in the course of prosecution in a foreign jurisdiction (the PCT and the EPO publish 18 months after filing), and

b. You've been given a foreign-filing license on your U.S. filing receipt (see Section G below), or six months has elapsed from your U.S. filing date

I won't discuss filing in non-Convention countries further, except to note that if you do wish to file in any, you should do so in exactly the same manner as you would for an individual filing in a Convention country (see Section J below), except that you won't need a certified copy of your U.S. application.

F. Never Wait Until the End of Any Period

AS STATED, YOU HAVE ONE YEAR after you file your U.S. application to file foreign Convention patent applications (and be entitled to your U.S. filing date) in the PCT, the EPO, or any country that's a member of the Paris Convention. You also have eight months (eighteen months under Chapter II) after you file a PCT application to file in the individual PCT countries, including the EPO. You have one year if you file under the PCT first, to file in non-PCT Convention countries or 20 months (30 months under Chapter II) to file in the PCT countries, respectively. However, you should never wait until the end of any of these periods. You should normally make your decision and start to take action about three or four months before the end of the period. This is to give you and the foreign agents time to prepare (or have prepared) the necessary correspondence and translations and to order a certified copy, if needed, of your US application.

G. The Early Foreign Filing License or Mandatory Six-Month Delay

NORMALLY, the blue official filing receipt (Chapter 10) which you get after filing your U.S. application gives you express permission from the PTO to file abroad, and such permission usually will be printed on your filing receipt, e.g., as follows: "Foreign Filing License Granted 1988 Aug 9." However, if your filing receipt fails to include a foreign filing license (only inventions with possible military applications won't include the license), you aren't allowed to foreign file on your invention until six months following your U.S. filing date. What's the reason for this? To give the U.S. government a chance to review your application for possible classification on national security grounds. You probably won't be affected by any of this, as most applications get the foreign filing license immediately and, in any case, there is usually no good reason to file before six months after your U.S. filing. If your situation is different, however, and your filing receipt doesn't include a license, see a patent lawyer (Chapter 6, Section F). If your invention does have military applications, not only will fail to get a foreign filing license on your filing receipt, but after you receive the receipt, you may receive a Secrecy Order from the PTO. This will order you to deep your invention secret until it's declassified, which often takes 12 years. Your patent can't issue till then, but the Government may compensate you if they use your invention in the meantime. You can foreign file an application which is under a secrecy order, but it's complicated; see a patent lawyer who has experience in this area.

H. Don't File Abroad Unless Your Invention Has Very Good Prospects in That Country

BECAUSE PATENT PROSECUTION and practice in other countries is relatively complicated and very expensive, you should file applications only in those countries where:

- A significant market for products embodying the inventions is *very* likely to exist; or

- Where *significant* commercial production of your invention is *very* likely to occur; or

- You've got a foreign licensee (someone who's paying you money for your inventions and know-how).

It's been my experience that far too many inventors like to file abroad because they're in love with their invention and feel it will capture the world. Unfortunately, this almost never happens. Almost all inventors who do file abroad never recoup their investment, i.e., they usually waste many thousands of dollars in fees and hardly ever derive any royalties, let alone enough royalties to cover their costs. Thus, as a rule of thumb I suggest that you file in another country only if you feel that you're:

1. Very likely to sell at least $300,000 worth of your invention there, if you're selling it yourself; or

2. Very likely to earn at least $20,000 in royalties from sales of your invention there by others; or

3. Associated with a licensee or sales representative there who contracts to pay you royalties with a substantial advance or guarantee, or who will pay for your foreign filing in that country.

Note that even if an infringement occurs in a country where you didn't file, it still wouldn't have paid to file unless the infringement is substantial enough to justify the expense of filing, getting the patent, and the uncertainties of licensing and litigation.

Remember: The U.S., with its approximately 300 million people, provides a huge marketplace which should be a more-than-adequate market from which to make your fortune, especially if its your first invention. In comparison, most foreign countries are relatively insignificant. E.g. Switzerland is smaller in size than San Bernardino County in California and smaller in population than Los Angeles County; Canada has fewer people than California.

I. The Patent Laws of Other Countries Are Different

DESPITE THE PARIS CONVENTION and other treaties covering patent applications, and except for Canada, whose patent laws and practice are practically identical to ours, almost all countries differ from the U.S. in their substantive patent laws and practices. Some of the main differences are as follows:

- Patents expire 15 years (Italy) to 20 years (Israel, U.K.) from the filing date, rather than 17 years from the issue date. In Japan it's 20 years from filing or 15 years from publication, whichever's later.

- Many smaller countries (e.g., Belgium, Portugal) don't conduct novelty examinations, but instead simply issue a patent on every application filed and leave it up to the courts (in the event of an infringement) to determine whether the invention was novel and unobvious.

- Some jurisdictions (the EPO, France, West Germany, Italy, Australia, the Netherlands) require the payment of annual maintenance fees while the application is pending.[1]

- Almost all foreign countries (e.g., Canada, W. Germany, U.K., Japan) require that patented inventions be "worked" (put into commercial use) within a set number of years. Otherwise, the government may force you to license others to work it (called "compulsory licensing") at government-set fees.

DOG TOOTH BRUSH
No. 3,002,103

---

[1]If you file in these countries (except Australia) via the EPO, no fee is due until the third year after filing.

J. The Ways To File Abroad

UNTIL SEVERAL YEARS AGO, there was only one way to foreign file, namely, to file a separate Convention application in each country in which you wished to file. As this was a cumbersome and expensive process, many of the countries got together to simplify things. Now there are three basic approaches to filing abroad in Convention countries. For filing in non-Paris Convention countries, see Section E above. You may end up using different approaches for different countries, or the same approach for all. The chart below, Fig. 12-A, summarizes these alternatives. In essence, they are:

Route A: File in U.S.; then file in individual countries or jurisdictions (including the EPO) under the Paris Convention within one year.

Route B: File in U.S.; then, within one year, under the Paris Convention file:

1. A PCT application to cover the PCT countries and jurisdictions (including the EPO) and file

individually in the PCT jurisdictions by eight or eighteen months after your PCT filing, and

2. File in whatever non-PCT Convention jurisdictions you desire.

Route C: File under the PCT; then

1. Within one year, under the Paris Convention, file in non-PCT jurisdictions, and

2. Within 20 or 30 months, under the PCT, file in PCT jurisdictions (including the U.S. and EPO).

Let's discuss each of these alternatives in more detail.

1. Route A: Paris Convention/EPO

Here you file in the U.S. first and then go abroad solely via the Paris Convention. This is basically the same as the old way of filing in individual countries under the Paris Convention, except that now, instead of filing in the individual countries of Europe, you can cover most of them with one application filed in the EPO. Let's discuss the Paris Convention first.

As stater this treaty provides an international norm which determines when patent applications must be filed in other countries (the one-year rule). However, it doesn't affect the various patent application procedures in place in each country. If you file under the Paris Convention, either in individual countries or through the EPO, the process typically will involve:

- Preparation of a separate application in the language of the country (you can file in English in the EPO and in the English-speaking countries);

- A separate patent search by that country's patent office, or by the EPO;

- Separate prosecution before that country's patent office or the EPO; and

- Payment of initial fees to file the application and often annual fees (called "maintenance fees") to continue the processing of the application to its conclusion.

- Laying open of your application to the public, generally 18 months after filing.

- Publishing your application for opposition after it's allowed but before it can issue. (An opposition is a procedure in which any member of the public can object to the issuance of a patent, generally by citing additional prior art to the patent office concerned. The US currently has no opposition procedure.)

To file individually in any country or the EPO, you'll need to have a foreign patent agent in each country you select prepare an appropriate application. The easiest way to do this is to send the agent a copy of your U.S. application and ask what else is needed. The requirements vary from country to country, and the EPO, but special drawings in each country's format will always be needed. You can have your foreign agent prepare these, or you can have these prepared yourself at lesser cost by the same companies that make drawings for U.S. divisional applications—see discussion of "Divisional Applications" in Chapter 14, Section B. Also, the agent will send you a power of attorney form which you'll have to sign and sometimes get notarized, certified by your county clerk, and legalized by the consulate of the country to which the form is being sent. Also you'll generally need a certified copy of your U.S. application; this can be obtained from the PTO—See Fee Schedule In Appendix. The cost for filing a foreign application in each individual country is about $700 to $2500, depending on the country, on the length of your application, and whether a translation is required.

If you wish to correspond directly with the foreign patent agents yourself, you'll first have to get the name of a patent agent in each country. See Section K below.

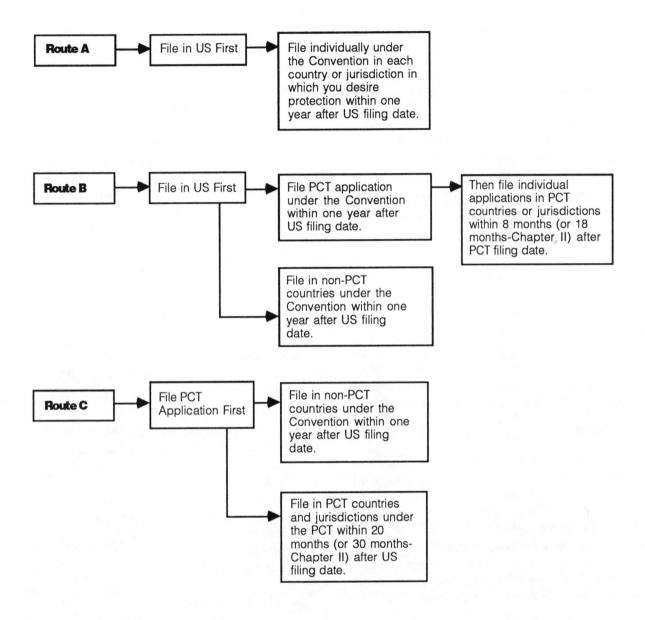

Fig. 12-A—Foreign Filing Choices

European Patent Office

If you wish to file in several European countries, the best way to do this, instead of filing a separate application in each country, is to file a Convention application in the European Patent Office (EPO), designating those countries you desire to cover. An EPO filing, while expensive, is generally considered cheaper than separate filings if:

1. Two or more non-English-speaking countries are involved (e.g., it's cheaper to file in the EPO than to file separate applications in France and Germany); or

2. If the UK and more than one non-English country is involved. Conversely, it's cheaper to file separate applications in the UK and Germany, for instance, than to go through the EPO.

As mentioned, to file a Convention application in the EPO you'll have to go through a European patent agent, unless you have an address in one of the EPO countries, in which case you can do it yourself. Correspondence with the EPO must be in English if your application is based on your U.S. case.

Including the agent's fee, expect to spend a stiff fee (see Fee Schedule) to get your application on file and examined in about six countries. Additional large fees will be incurred for prosecution (getting your application approved once it's filed) and issuance. Then, translations and individual agents for the respective countries you designate will have to be arranged. For more information, write to the EPO, ask for a copy of *How to Get a European Patent* (address in section C above).

2. Route B: Convention /PCT /EPO

Here you file in the U.S. first and then go abroad via the PCT, insofar as possible. Here's how it works for U.S. inventors:

- A PCT request form and a separate "international application" is filed with the U.S. PTO. The application designates the PCT member countries or jurisdictions (e.g., the EPO) in which protection is desired;

- The request and application are forwarded to the "International Searching Authority" (a branch of the PTO) where an "international search report" is prepared if you've elected a "regular" PCT application, or a full examination if you've elected Chapter II. In either case the examination will generally be done by the same examiner who will handle your U.S. application;

- If you've elected the regular procedure, copies of the search report and application are then forwarded to the countries designated in the application;

- If you elect Chapter II (by 19 months after your US filing date), you'll prosecute the application similar to the way you do with a US application and you'll ultimately get your claims allowed or rejected.

- If you wish to prosecute the application in the individual countries, you must provide a translation (except in the EPO) and must pay any fees that are required. While a separate prosecution is required in each country, it's commonly made easier by the fact that the PCT member countries generally rely on the international search is generally relied on by the PCT member countries.

a. Preparation of an International Application

If you think you'll want to file an international application under the PCT, prepare your original U.S. application and drawings in the new international format. The main differences between the PCT and U.S. national formats (both of which are now acceptable for U.S. applications) are the drawing size and margins, location of page numbers, and spacing between typed lines. These differences are detailed in Chapter 10.

b. How to Begin

Assuming your application is in the proper format, the first step in filing a PCT application is to obtain a "Request" (Form PCT/RO/101) and transmittal letter (Form PTO1382) from Box PCT, Patent and Trademark Office, Washington, DC 20231. Complete the forms (full instructions will be attached), requesting the PTO to prepare a certified copy of your U.S. application for use with your PCT application, and attach a check payable to the Commissioner of Patents and Trademarks for the international application filing fees as computed on the Request form.

c. PCT Fees

The fee for a certified copy of your U.S. application is listed in the Fee Schedule in the Appendix. The PCT fees vary frequently due to exchange rate fluctuations. They're composed of several parts as follows:

- Transmittal Fee;
- Search Fee, (a) if you haven't already filed in the U.S.; (b) if you've already filed in the U.S.; and (c) if you want to use the EPO as your searching authority;
- International Fee;
- Country Designation Fees, (the EPO counts as one country);
- Examination Fee (Chapter II) (a) if the U.S. PTO is the searching authority; (b) if you've used the EPO to make the search.
- Handling fee (Ch II).

A common course of action is to file a PCT application after filing a U.S. application, designating the EPO and Japan with an EPO search. You should designate the EPO as your searching authority if you intend to file there since they generally do a better search than the U.S. PTO and you'll save money and time in the EPO later.

If you want to elect Chapter II, i.e., have the PCT application examined in the U.S. PTO, and you've used the EPO to make the search, the cost will be substantially more.

d. How to File

Mail the Transmittal Letter, Request, copy of your application and drawings (both on A4 size), and check to: Box PCT, Patent and Trademark Office, Washington, DC 20231, which, as mentioned, is a designated receiving office for the International Bureau. Like Convention applications, the international (PCT) application should be filed within one year of your U.S. filing date. I advise filing it at least a month before the anniversary of your US filing date so you'll have time to correct any serious deficiencies.

e. What Happens to Your International Application?

You'll receive a filing receipt and separate serial number for your international application, and the application will eventually be transmitted for filing to the countries (including the EPO) you've designated on your request form. If you're make any errors in your PCT application, the PCT Dept. of the US PTO will give you a month to correct them If you've elected a regular PCT application within 20 months after your U.S. filing date (about eight months after your PCT filing date), you'll need to file separate national or regional translated applications.[2] If you've elected Chapter II, you'll have to file the separate national and regional applications by 30 months after your U.S. filing date.

---

[2]An application to the EPO doesn't need to be translated until after allowance, when you'll have to register and translate it in each country you've designated. (No translation is needed for the UK of course.) As stated, you (or your EPO agent) will also need to engage the services of independent patent agents in the respective countries to assist you in representing you in their countries. See Section C above.

As mentioned, each of the separate countries will rely to a great extent on the international search (or allowance if Chapter II was elected), they'll receive from the International Bureau (in most cases this will be the EPO search or an adoption of the U.S. search), so one advantage of the PCT approach is that you'll save much of what used to be the agonizing, extremely expensive job of separately and fully prosecuting an application in each country in which you elected to file.

However, if you file through the PCT, don't forget that you'll also have to file individual Convention applications in each non-PCT Convention country you desire protection in (such as Canada and Mexico) within one year of your U.S. filing date. These individual Convention filings will be identical to those of Route A above.

3. Route C: PCT/Convention/EPO

Here you file in the PCT first and then go to the U.S. and the PCT countries (including the EPO) via the PCT and to non-PCT Convention countries via the Convention.

If you know for certain, before you file anywhere, that you'll want to file in the U.S. and at least one other foreign PCT country, then you can save some fees and effort by filing a PCT application first, before you file in the U.S. In your PCT application you must designate the U.S. and any foreign PCT countries (including the EPO) you desire. Then, within one year of your PCT filing date, you should file Convention applications, based upon your PCT application, in any non-PCT countries, such as Mexico and Canada, you desire.

Within 20 months of your PCT filing date (30 months if you elected Chapter II), file separately under the PCT in each country or jurisdiction you've designated in your PCT application, including the U.S. and the EPO. Your U.S. application should be identical to a "regular" U.S. application, except that you should add the following sentence to the PAD (Form 10-2) to get the benefit of your PCT filing date: "I hereby claim foreign priority benefits under 35 USC 119 of the PCT patent application, Ser. Nr. _____, Filed 198_____." Then order (from the PTO) a certified copy of your PCT application and file this within a few months after your U.S. filing date.

Whether you're filing in a PCT or non-PCT jurisdiction based upon a PCT filing, your foreign patent agents will tell you what you'll need to file PCT-based applications in their countries; allow at least two months before the 20-or 30-month deadline to give them (and you) time to prepare the applications and translations, if necessary.

Which route should you take? That depends on when you know you're going to foreign file and in which countries you do decide to file.

If you're sure you want to foreign file before you file anywhere, you should take Route C. Filing in the PCT first gives certain economies and saves work later on.

If you're not sure you'll want to foreign file before you file anywhere (the usual case), use Route A or B.

If, after about 10 months, you decide to file in three or more PCT-member countries or jurisdictions use Route B, the most common choice. File a PCT application to get an eight- or eighteen-month delay in the PCT countries and jurisdictions, such as Japan and the EPO. However, you'll have to file ordinary Convention applications within the one-year deadline in the non-PCT countries, such as Canada and Mexico.

If, after about 10 months, you want to file in one or two PCT-member countries, use Route A (foreign file Convention cases directly wherever you desire, including Japan and the EPO).

If, after about 10 months you aren't sure, file a PCT application just to get the eight or eighteen month delay in the PCT-member countries. For the non-PCT countries, you'll have to file now if you want to obtain the benefit of your US filing date, but you can postpone your decision till later (provided your invention hasn't yet been made publicly available), although you won't get the benefit of your US filing date and thus will be at the mercy of any additional prior art that has been published since your US filing date.

If you do file a PCT application, it's strongly advisable to elect Chapter II since you'll be able to do most of your foreign prosecution before the U.S. PTO. Although the cost for a Chapter II case is much more, you'll save this in foreign agent's fees and you'll buy more time in case you change your mind.

K. Resources to Assist in Foreign Filing

THERE ARE A NUMBER of resources to assist you in foreign filing your patent application. Let's look at them separately.

1. Foreign Patent Agents

As I've mentioned, if you desire to file abroad you'll almost certainly need to find a foreign patent agent who's familiar with patent prosecution in the countries where you desire protection. Your best bet is find one through a U.S. patent attorney (see Chapter 6, Section F), as most are associated with one or more patent agents in other major countries.

If you don't know a U.S. patent attorney or someone who's familiar with foreign patent agents, there are several other ways to obtain the names. One is to look in the telephone directory of the city where the patent office of the foreign country is located. Most large libraries have foreign telephone directories. Another simple way is to inquire at the consulate of the country; most foreign countries have consulates in major U.S. cities and should have a list of patent agents.

A third possibility is to hire a local patent attorney to do the work for you, although this involves an intermediary's costs. Because of the complicated nature of foreign filing, many patent attorneys even use their own intermediaries, namely specialized patent-law firms in New York, Chicago, or Los Angeles, which handle foreign filing exclusively.

A fourth possibility is to hire a British firm of patent agents to do all your foreign filing. The reason for this is that they speak fairly good English and they're familiar with foreign filing. This would be especially appropriate if you're filing with the EPO, but most German agents in Munich, although not as fluent in English, have the compensating advantage of their physical proximity to the EPO. Finally, you can look in the Martindale-Hubbell Law Directory (in any law library) which lists some foreign patent agents in each country.

Whichever way you get your foreign patent agent, be careful, since some foreign patent agents, like some U.S. patent attorneys and agents, aren't competent.

2. Written Materials

As you've gathered by now, filing abroad can become very complicated. If you want to learn more, including the laws of each country, see *Patents Throughout the World* by Greene (Clark Boardman). This book is revised annually, so be sure you have the most recent version. Also, you can call the consulate of any country to get information on their patent laws. For more information on how to utilize the PCT, a brochure "The PCT Applicant's Guide," is available free from the PCT Department of the U.S. PTO, and a comprehensive book, *The PCT Applicant's Guide,* is available from World Intellectual Property Organization, Post Office Box 18, 1211 Geneva 20, SWITZERLAND.

Bonne chance et au revoir!

Chapter 13

Getting the PTO to Deliver

THERE'S AN OLD SAYING in the law: *You can sue the bishop of Boston for bastardy*. This means that you can file a lawsuit against anyone for anything. Whether you can prove your case and win is, of course, a very different matter.

Similarly, anyone can file a patent application on anything. But getting the Patent and Trademark Office (PTO) to issue you a patent is, of course, a very different matter.

This chapter tells you how to get the PTO to deliver, assuming your invention meets the standards of patentability (Chapter 5). This material is sure to seem confusing the first time you read it. A little familiarity with the process, however, should do a world of good when it comes to your understanding. So please relax and read this chapter a couple of times before you swing into action.

A. What Happens After Your Patent Application Is Filed?

IT WILL BE HELPFUL to review exactly what will occur after your patent application is filed.

1. Receipt Postcard

After sending your patent application to the PTO, you'll receive your receipt postcard back in about two to four weeks. It will be stamped with a date and a six-digit number, e.g., US Patent & TM Office, 22 May 1988; 407893 The date is the "deposit" date (date of receipt), and the number is the serial number (sometimes called "application number") of your application.

2. Official Filing Receipt

About a month later (if you followed my instructions in Chapter 10) you should receive an official filing receipt. This is a blue sheet containing the following:

- The name(s) of the inventor(s);
- The title of your patent application;
- The examining group to which your application has been assigned;
- The filing date and serial number of your application; (The serial number on your official filing receipt will be the same as that on your receipt postcard, except for the addition of a two-digit prefix, e.g., "06/407,893." Use this eight-digit serial number from now on.)
- The number of claims (total and independent);
- The filing fee you paid;
- Your name and address;
- The words "Small Entity" if you filed an SED (Form 10-3); and
- The words "Foreign Filing License Granted [date]" if the invention hasn't been militarily classified (most won't be). This means that you can foreign file at any time, rather than waiting six months. However, for reasons stated in Chapter 12, you still should wait until approximately eight months have passed before considering filing abroad.

Check all of this information carefully; it's what's entered into the PTO's data-processing system about your application. If the filing receipt has any errors, write a brief letter. The caption should be as in Form 13-1, but substitute "Request For Corrected Filing Receipt" instead of "Amendment" and point out the errors and request a new filing receipt.

Once you receive the blue official filing-receipt sheet, your patent application is officially pending and you may label your invention and any descriptive literature, "Patent Pending," or "Patent Applied For." They have the same legal meaning.

If for any reason your application hasn't been filed properly (e.g., you forgot the SED, your check bounced, you forgot to sign the PAD (Form 10-2), etc.) you won't get the blue filing receipt. Instead the Application Branch of the PTO will send you a deficiency notice telling you what's needed and what surcharge (fine) you'll have to pay for the error of your ways. Once you comply with the deficiency notice (they usually give you a month), you'll get your blue filing receipt a few weeks later.

3. Patent Pending Status

What does it mean to say "patent pending?" Many people believe that a person who copies an invention on which a patent is pending is liable for infringement. This isn't true. You have no monopoly or rights until your patent actually issues. In other words, a manufactured article covered by a pending patent application can be freely copied by anyone. However, even though "patent pending" status can't be used to preclude others from copying the invention, it's not common for them to do so. This is because if a patent later issues, the infringer takes the chance that you'll use your patent to enforce your monopoly, i.e., stop any further production and marketing. In this case the money already spent on expensive tooling will have been mostly wasted. (If you're willing to license the infringer under your patent, the infringer's tooling outlay will be worthwhile, but few infringers will be willing to take this chance.)

Finally, it's a criminal offense to use the words "patent applied for" or "patent pending" in any advertising when there's no active, applicable patent application on file.

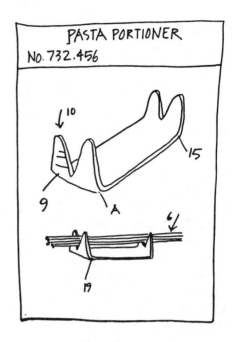

4. Send In Your Information Disclosure Statement (IDS)

If you haven't done so already, after receiving your official filing receipt you should send in your Information Disclosure Statement as discussed in Chapter 10, Section N, together with copies of the references you listed on the form. Remember that the PTO likes the IDS to be filed within three months of the application's filing date. Don't forget to print your serial number and filing date on the forms.

5. First Office Action

Usually about six months to two years after the filing date, you'll receive a communication from the PTO known as a "first office action (OA)," sometimes called an "official letter." It consists of forms and a letter from the examiner in charge of your application, describing what is wrong with your application and why it cannot yet be allowed.

(Rarely will an application be allowed in the first OA.) Specifically, the OA may:

- Reject claims,
- List defects in the specification and/or drawings,
- Cite prior art which the examiner believes either (a) shows your invention is not novel, or (b) obvious; and/or
- Raise various other objections.

To find out approximately when you'll receive the first OA from the PTO, you can look in a recent issue of the *Official Gazette* (OG). One of the initial pages will be headed "Patent Examining Groups." Fig. 13-A is a copy of such a page. Look for the actual examining group to which your application is assigned, and in the column headed "Actual Filing Date of Oldest New Case Awaiting Action," note the date opposite your examining group. You can expect to receive the first OA in about the interval between this date and your filing date.

6. Response to First Office Action

The OA itself will specify an interval, usually three months from the date the OA was mailed, within which you must file a response. Your response must take whatever action is necessary to overcome the objections and rejections listed in the OA. The response you file is technically called an "amendment" (assuming it contains any changes) and the entire process of correspondence (office actions and amendments) to and from the PTO is known as "patent application prosecution," although no one is "prosecuted" in the usual sense. I show you how to draft your response in Section F below.

7. Second/Final Office Action

About two to six months after you file your first amendment, you'll receive a second OA from the PTO; this will usually be designated a "final" OA by the PTO. A final OA is supposed to end the prosecution stage before the examiner. However, as we'll see later, this is far from true. In other words, a "final action" is rarely final. Again you have three months to reply.

8. Notice of Allowance

Assuming you submit what is necessary to get your application in condition for allowance, you'll be sent a Notice of Allowance, indicating that all of your claims are allowed and that an issue fee is due within three months. (Sometimes you'll get a "Notice of Allowability" before or with the formal allowance; this merely states that your claims are all allowed and the Notice of Allowance will be sent soon.)

9. Issue Fee

After you pay the issue fee (see Fee Schedule in Appendix), you'll receive a receipt from the PTO, indicating the issue date and number of your patent. When you pay the issue fee, you can also order ten or more printed copies of your patent at the usual charge.

10. Receipt of Official Patent Deed

Shortly after the date your patent issues, you'll receive your official letters patent or deed from the PTO, plus any printed copies of the patent which you've ordered.

PATENT EXAMINING CORPS

RENE D. TEGTMEYER, Assistant Commissioner

WILLIAM FELDMAN, Deputy Assistant Commissioner

CONDITION OF PATENT APPLICATIONS AS OF JULY 31, 1976

PATENT EXAMINING GROUPS	Actual Filing Date of Oldest New Case Awaiting Action

CHEMICAL EXAMINING GROUPS

GENERAL CHEMISTRY AND PETROLEUM CHEMISTRY, GROUP 110—S. N. ZAHARNA, Director........... 1-2-76
Inorganic Compounds· Inorganic Compositions; Organo-Metal and Organo-Metalloid Chemistry; Metallurgy; Metal Stock; Electro Chemistry: Batteries; Hydrocarbons; Mineral Oil Technology; Lubricating Compositions; Gaseous Compositions; Fuel and Igniting Devices.

GENERAL ORGANIC CHEMISTRY, GROUP 120—A. L. LEAVITT, Director........... 1-2-76
Heterocyclic, Amides; Alkaloids; Azo; Sulfur: Misc. Esters; Carbohydrates; Herbicides; Poisons; Medicines; Cosmetics; Steroids; Oxo and Oxy; Quinones; Acids; Carboxylic Acid Esters; Acid Anhydrides; Acid Halides.

HIGH POLYMER CHEMISTRY, PLASTICS AND MOLDING, GROUP 140—A. P. KENT, Director........... 10-1-75
Synthetic Resins; Rubber; Proteins; Macromolecular Carbohydrates; Mixed Synthetic Resin Compositions; Synthetic Resins With Natural Polymers and Resins; Natural Resins; Reclaiming; Pore-Forming; Compositions (Part) e.g.: Coating; Molding; Ink; Adhesive and Abrading Compositions; Molding, Shaping, and Treating Processes.

COATING AND LAMINATING, BLEACHING, DYEING AND PHOTOGRAPHY, GROUP 160—R. FRIEDMAN, Director. 9-12-75
Coating; Processes and Misc. Products; Laminating Methods and Apparatus; Stock Materials; Adhesive Bonding; Special Chemical Manufactures; Special Utility Compositions; Bleaching; Dyeing and Photography.

SPECIALIZED CHEMICAL INDUSTRIES AND CHEMICAL ENGINEERING, GROUP 170—H. S. VINCENT, Director... 1-2-76
Fertilizers; Foods; Fermentation; Analytical Chemistry; Reactors; Sugar and Starch; Paper Making; Glass Manufacture; Gas; Heating and Illuminating; Cleaning Processes; Liquid Purification; Distillation; Preserving; Liquid, Gas, and Solid Separation; Gas and Liquid Contact Apparatus; Refrigeration; Concentrative Evaporators; Mineral Oils Apparatus; Misc. Physical Processes.

ELECTRICAL EXAMINING GROUPS

INDUSTRIAL ELECTRONICS, PHYSICS AND RELATED ELEMENTS, GROUP 210—W. L. CARLSON, Director.... 5-21-75
Generation and Utilization; General Applications; Conversion and Distribution; Heating and Related Art Conductors; Switches; Photography; Motion Pictures; Illumination; Horology; Acoustics; Recorders; Weighing Scales.

SPECIAL LAWS ADMINISTRATION, GROUP 220—C. D. QUARFORTH, Director........... 11-6-75
Ordnance, Firearms and Ammunition; Radar, Underwater Signalling, Directional Radio, Torpedoes, Seismic Exploring, Radio-Active Batteries; Nuclear Reactors, Powder Metallurgy, Rocket Fuels; Radio-Active Material.

INFORMATION TRANSMISSION, STORAGE AND RETRIEVAL, GROUP 230—J. F. COUCH, Director........... 11-3-75
Communications; Multiplexing Techniques; Facsimile; Data Processing, Computation and Conversion; Storage Devices and Related Arts.

RECEPTACLES, SANITATION AND CLEANING, WINDING, AND MEASURING, GROUP 240—N. ANSHER, Director.. 1-30-76
Receptacles; Joint Packing; Conduits; Plumbing Fixtures; Textile Spinning; Food; Agitating; Cleaning; Pressing; Geometrical Instruments; Sound Recording; Winding and Reeling; Measuring and Testing; Indicating.

ELECTRONIC COMPONENT SYSTEMS AND DEVICES, GROUP 250—L. FORMAN, Director........... 12-3-75
Semi-Conductor and Space Discharge Systems and Devices; Electronic Component Circuits; Wave Transmission Lines and Networks; Optics; Radiant Energy; Measuring.

DESIGNS, GROUP 290—C. D. QUARFORTH, Director........... 4-30-75
Industrial Arts; Household, Personal and Fine Arts.

MECHANICAL EXAMINING GROUPS

HANDLING AND TRANSPORTING MEDIA, GROUP 310—D. J. STOCKING, Director........... 11-3-75
Conveyors; Hoists; Elevators; Article Handling Implements; Store Service; Sheet and Web Feeding; Dispensing; Fluid Sprinkling; Fire Extinguishers; Coin Handling; Check Controlled Apparatus; Classifying and Assorting Solids; Boats; Ships; Aeronautics; Motor and Land Vehicles and Appurtenances; Brakes; Railways and Railway Equipment.

MATERIAL SHAPING, ARTICLE MANUFACTURING, TOOLS, GROUP 320—S. S. MATTHEWS, Director........... 2-18-76
Manufacturing Processes, Assembling, Combined Machines, Special Article Making; Metal Deforming; Sheet Metal and Wire Working; Metal Fusion—Bonding, Metal Founding; Metallurgical Apparatus; Plastics Working Apparatus; Plastic Block and Earthenware Apparatus; Machine Tools for Shaping or Dividing; Work and Tool Holders, Woodworking; Tools; Cutlery; Jacks.

AMUSEMENT, HUSBANDRY, PERSONAL TREATMENT, INFORMATION, GROUP 330—G. M. FORLENZA, Director. 12-18-75
Amusement and Exercising Devices; Projectors; Animal and Plant Husbandry; Butchering; Earth Working and Excavating; Fishing, etc.; Tobacco; Artificial Body Members; Dentistry; Jewelry; Surgery; Toiletry; Printing; Typewriters; Stationery; Information Dissemination.

HEAT, POWER, AND FLUID ENGINEERING, GROUP 340—B. R. GAY, Director........... 7-9-75
Power Plants; Combustion Engines; Fluid Motors; Reaction Motors; Pumps; Rotary Engines and Pumps; Heat Generation and Exchange; Refrigeration; Ventilation; Drying; Temperature and Humidity Regulation; Machine Elements; Couplings; Gearing; Bearings; Clutches; Power Transmission; Fluid Handling and Control; Lubrication.

GENERAL CONSTRUCTIONS, TEXTILES AND MINING, GROUP 350—M. M. NEWMAN, Director........... 1-2-76
Joints; Fasteners; Rod, Pipe and Electrical Connectors; Miscellaneous Hardware; Locks; Building Structures; Closure Operators; Bridges; Closures; Earth Engineering; Drilling; Mining; Furniture; Supports; Cabinet Structures; Centrifugal Separations; Coating; Textiles; Apparel and Shoes; Sewing Machines.

Expiration of patents: The patents within the range of numbers indicated below expire during August 1976, except those which may have expired earlier due to shortened terms under the provisions of Public Law 690, 79th Congress, approved August 8, 1946 (60 Stat. 940) and Public Law 619, 83rd Congress, approved August 23, 1954 (68 Stat. 764), or which may have had their terms curtailed by disclaimer under the provisions of 35 U.S.C. 253. Other patents, issued after the dates of the range of numbers indicated below, may have expired before the full term of 17 years for the same reasons, or have lapsed under the provisions of 35 U.S.C. 151.

Patents........... Numbers 2,897,500 to 2,901,748 inclusive
Plant Patents........... Numbers 1,856 to 1,860 inclusive

1285

Fig. 13-A—Page of Official Gazette Indicating Status of Pending Applications in Examining Groups

B. General Considerations During Patent Prosecution

Patent application prosecution is generally more difficult than the preparation of the initial application. Although you've probably gained sufficient experience in the patent field to adequately handle major portions of the prosecution phase yourself, you may wish to consult with a patent attorney if you feel you're in over your head.

Assuming you handle the prosecution phase pretty much on your own, I recommend that you keep the following general considerations in mind.

1. The PTO Can Write Claims for You

As I mentioned in Chapter 9 (claims drafting), you can have the PTO write a claim for you if you wish. Then you can either accept this claim or amend it if you think you can get it past the examiner.

2. Consultation With a Patent Professional Might Be Wise

You might wish to consult with a patent expert at this point of the proceedings. Paying $200-$400 (if you use a "discount" patent attorney—see Chapter 16) to have an expert amend your claims (which is usually what's required) may prove to be relatively cheap in the long run if you can afford the expense now. As you review the following, often dense, material, remember that expert outside help is available.

3. Intervals Are Approximate

Except for official periods, such as the three-month period for response to an OA or to pay the issue fee, the dates and times I've given in this chapter are only approximate and are gleaned from recent experience. They can vary quite widely, depending on conditions in the PTO at the time you file your patent application. If you don't receive any communication from the PTO for a long time, say over 1.5 years after you file your application, or over 6 mos. after you file an amendment, you should check a recent *Official Gazette*, make a call, or send a letter to the examiner or examining group to determine the status of your case.

4. You'll Be Given an Opportunity to Correct Technical Errors

Don't worry too much about minor technical errors (except for dates—see next consideration) when dealing with the PTO. If you make one, you'll be given an opportunity to correct it. The PTO has so many rules and regulations that even patent attorneys who deal with it all the time can't remember them all. Also, the PTO is flexible in giving do-it-yourself applicants opportunities to correct errors that don't affect the substance of the application.

5. Dates Are Crucial

Every OA which you receive from the PTO will specify an interval by which you must reply to the OA. If you fail to reply in the time the PTO allots you, the penalty is draconian; your application will go abandoned, although it can be revived at a price. See Section Q below. Thus, you should note the due date for every OA promptly on your calendar and heed it carefully. If you're not the type who can faithfully heed due dates, you must do something about this; e.g., by hiring a methodical friend to bug you, or by even turning the whole job of prosecution over to a patent attorney.

6. Situations Not Covered

If any situation occurs that isn't covered in this book, and you can't find the answer by looking in the *Rules of Practice* or *Manual of Patent Examining Procedure* (see Section 6 below for how to obtain these), call the PTO, consult an attorney or agent, or use common sense and do what you would expect to be the logical thing to do in such a situation.

For example, if after you've filed your patent application you find a reference that considerably narrows what you thought your invention to be, bring it to the attention of the PTO by way of another (supplemental) IDS and submit an amendment substituting narrower claims that avoid the reference. Remember that you have a continuing duty to disclose all material information about your invention to the PTO, see Form 10-2. If you discover that an embodiment of your invention doesn't work, delete it from your application (see Section E below for how to do this). If you discover a new embodiment of your invention that supersedes the present embodiments, file a continuation-in-part application (see Chapter 14).

If you change your address, you should send an appropriate letter (caption as in Form, 13-1 but headed "Change Of Applicant's Address") to the PTO.

If the examiner cites a reference against your application that is later than the filing date of your application, obviously the examiner made an error (this happens occasionally) and you should call or write to bring it to the examiner's attention so that a new office action can be issued. If the PTO fails to send you a copy of a reference which it has cited against you (this happens often), send an appropriate paper (captioned as in Form 13-1) headed "Request For Copy of Missing Reference" to the PTO.

Finally, as a wise person said, "Don't be afraid to ask dumb questions: they're easier to handle than dumb mistakes."

7. Standards of Patentability Vary

While I've tried to give the proper standards of patentability in this book (see Chapter 5), what actually happens when your application is examined will vary depending upon the personality, whims, and current emotions of the examiner assigned to handle it. Most examiners adhere to the basic standards of patentability outlined here and are competent, knowledgeable, and commonly helpful when it comes to telling you what to do to put the case in condition for allowance. Unfortunately, many are very new and inexperienced, new to the US and unfamiliar with English, and/or incompetent, which can sometimes lead them to make arbitrary, irrational rulings and deny patents that should be granted or vice-versa. Services have deteriorated everywhere in recent years, but especially in the PTO.

The solution to the problem with a tough or experienced examiner is to, first, be persistent, go to the PTO (or have a patent attorney) to interview your examiner, and if necessary, appeal. Examiner's don't like to write answers to appeal briefs since these take a lot of time, they have to have an appeal conference with another examiner, and it looks bad on their record if they have many appeals or get reversed often. Appealing is thus a powerful weapon against a tough examiner. The problem with an easy examiner is that your allowed application might not stand up in court (should this ever become necessary). Accordingly, if you believe that your examiner is not rigorous enough (for instance, all your claims are allowed in the first office action), make especially sure yourself that at least some of your claims are clearly patentable, in the sense that they will withstand a court challenge (see Chapter 15).

It may help to know that examiners themselves have to contend with two opposing forces: on the one hand they're expected to dispose of (allow or get the applicant to abandon) a certain number of cases, but on the other hand they're subject to a quality review program to make sure they're not too lenient.

Note that even if you have a great invention which is quite patentable, but you haven't claimed it properly, many US PTO examiners, unlike their counterparts in the European Patent Offices won't volunteer help or constructive suggestions or try to assist you. They'll simply reject your claims or make a requirement and leave it to you to figure out how to do what's necessary to remedy the situation.

8. Dealing with the PTO Can Be Frustrating and Unfair

Dealing with the PTO, as with any other government agency, can sometimes be a very difficult, time-consuming experience. For example, I once filed an application for an inventor whose last name was "Loe." The filing receipt came back with the name "Lee." After several letters and calls with no response, a "corrected" filing receipt arrived with the name spelled "Leo." After a few more calls and much frustration, a correct filing receipt finally arrived. All I can tell you is to be philosophical, scrupulously check your correspondence with the PTO to make sure they get it right, and persist in correcting errors when they occur.

As far as the unfairness goes—there are many situations when you deal with the PTO (and the IRS) where you'll find an inherent unfairness due to non-reciprocity. E.g., while you have to reply to an OA when the PTO tells you to, they can reply to your amendment whenever they get around to it. While you have to make your claims and specification clear, grammatical, and free of spelling errors,

you'll often find that the correspondence you receive from the PTO doesn't meet these standards. While you have to pay a stiff fine if you forget to sign your check or make some other inadvertent error, the PTO never is liable, no matter how negligent they are. There's nothing you can do about this unfairness except, again, to be philosophical and resign yourself to accept the rules of the game before you play.

One inventor was so frustrated that he sued his Examiner and the PTO for negligence. The judge said, "This is the sad tale of an inventor frustrated by the bureaucratic mindset and Byzantine workings of the PTO." While he won in trial court, the appellate court reversed, holding that examiners are not legally responsible for their actions.

9. PTO Rule Books

During patent prosecution, you may need to refer to the PTO's Rules of Practice. These can be obtained from any government bookstore in a paperbound book entitled *Code of Federal Regulations—37—Patents, Trademarks, and Copyrights* (see the bibliography at the back of this book). As stated in the Introduction, the PTO's patent rules are given the prefix number (1.), distinguishing them from trademark rules (2.) and copyright rules (3.). For example, Patent Rule 111, referred to later in this chapter, is officially identified as Title 37 of the Code of Federal Regulations, Section 1.111, or in legal citation form, 37 C.F.R. 1.111.

In addition to the Rules of Practice, an aid that's very useful during prosecution is the *Manual of Patent Examining Procedure* (MPEP), which is often referred to as the "examiner's bible," and which covers almost any situation you can encounter in patent prosecution. It is an expensive, large, loose-leaf volume; it can be found in many public libraries and in most Patent Depository Libraries.

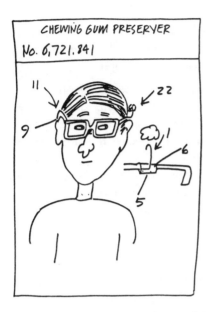

CHEWING GUM PRESERVER
No. 6,721.841

10. Never Make Negative Statements on the Record

When dealing with the PTO, you should never, never say or write anything which derogates your invention, and you should never admit that any prior-art reference shows (demonstrates) any feature of your invention. Admittedly, this advice may be very difficult to follow in some situations, but it's extremely important that you comply with it. Why? Any negative admission by you in your correspondence with the PTO will be put into your official file (called your "file wrapper") and can be used against your patent if it ever becomes involved in litigation. Thus, if you always anticipate that your patent may later be involved in litigation, you'll do a much better job in the prosecution phase. This is so important that I've made it an inventor's commandment:

INVENTOR'S COMMANDMENT #16:
You should never admit or state anything negative about your invention on the record since anything negative which you admit may be used against you later by an adversary.[1]

11. Be Available to Answer Office Actions

As mentioned, you'll normally be required to respond to a PTO office action within three months. If an OA is sent while you're away or unavailable and you fail to reply to it, your application will, as stated in Subsection 5 above, be considered abandoned. Thus, if you'll be unavailable for an extended period while your application is pending, you should empower a patent attorney to handle it for you or arrange to have your mail forwarded by a reliable friend or relative. Unfortunately a layperson isn't licensed to act for you.

12. Consider Foreign Filing

About eight to ten months after you file your patent application, you should consider whether you want to file for protection in other countries, as stated in Inventor's Commandment #15. There are international conventions or agreements among all countries that entitle you to the benefit of your U.S.

---

[1]You do have a duty to disclose all information known to you that bears on the patentability of your invention (see Chapter 10, Section N), but do not have to (and shouldn't) admit or state anything negative about your invention, even if your disclosure is tantamount to a smoking gun.

filing date on any foreign applications you file within one year after you file your U.S. application. Refer back to Chapter 12 for how to file for a patent in other countries.

13. You Can Call and Visit Your Examiner

If you have any questions about your application, or any reference which is cited against it, you are permitted to call, and/or make an appointment with and visit the examiner in charge of your application. Your examiner's telephone number will be listed on official letters which you receive from the PTO. However, usually only one, or at most two, applicant-initiated interviews are permitted. So save this privilege for when you really need it. If you have an interview, you must summarize its substance (unless the examiner does so) in the next amendment. An interview is often a very valuable way to get a difficult case allowed since communication is greatly enhanced when you and the examiner can discuss your differences and reach an understanding via the give and take and multiple feedback loops an interview permits. Also it's harder to say "no" directly to a person face-to-face. However I recommend that you try to avoid calling or interviewing any examiner on Fridays since, like most of us, they're likely to be less attentive then.

14. No New Matter Can Be Added to Your Application

As I've repeatedly stressed, once your application is filed, you cannot add any "new matter" to it. New matter consists of any technical information that was not present in your application as originally filed. If you do want to add any new developments to your application, consider a special type of supplementary application (termed a "continuation-in-part application" and covered in Chapter 14) or, if your improvement is really significant, an independent, subsequent patent application.

New matter should be distinguished from prior art that may be discovered after an application has been filed. You are obligated to inform the PTO about any newly-discovered, relevant prior art.—see Subpart 6 above. Such prior art doesn't affect the nature of your invention, but rather provides the PTO with more information by which to judge your invention for patentability.

15. Official Dates Are When the PTO Receives Your Submission

Every paper which you send to or receive from the PTO has an official date. This is the date on which it was mailed from or received by the PTO. You should put your actual date of mailing on anything you send to the PTO, but the date of the PTO's "Received" stamp on your paper will be the "official" date of the paper. If you send in your application by Express Mail with an EM Certification (see Chapter 10), the PTO will stamp it as of the date you express mailed it, even though they receive it one to three days later. **Fax now available:** Amendments, petitions, appeals, and elections (but not applications, PCT papers, fees, or drawings) can also be filed by fax at (703) 557-9564. Faxed papers should include, "I certify I have transmitted this paper by fax on [date]." Keep your signed original and your machine's record of successful transmission.

16. Know Whose Court the Ball Is In

To use an analogy drawn from the game of tennis, during patent prosecution, the "ball" (burden of action) will always be in either your court or the PTO's. If you just sent in your case, the ball will be in the PTO's court until they return your postcard, send you an official filing receipt, and send you a first office action. It doesn't go back to your court until that first OA. Once they send the first OA, you have the ball and must usually take action within three months. Once you file an amendment, the PTO has the ball again, etc. You should always know the status of your case, i.e., whose court the ball is in.

17. Re-Read Appropriate Chapters

When you respond to an OA, you should go back and reread the chapter that covers the issue you need to address. For example, if a claim is disallowed for prolixity, reread Chapter 9 (drafting claims). If the application is disallowed on prior art grounds, reread Chapter 5. If your specification or drawings aren't in proper form, reread Chapters 8 and 10.

ANTI-SNAKE BITE PANTS
No. 5049

C. A Sample Office Action

NOW THAT YOU HAVE AN OVERVIEW of the patent application prosecution process, it's time to get more concrete. Figures 13-B to 13-D below are reproductions of pages 76.2, 76.3, and 78 of the *Manual of Patent Examining Procedure* (MPEP); they show a sample OA in an imaginary patent

application. A study of this example will enable you to deal with your first OA far more effectively. It has been purposely written to include almost every possible objection and rejection; an actual OA is never this complicated and OA's are now typed and quote applicable statutes. First let's look at Fig. 13-B.

At the top of the OA, the examiner's name (Callaghan) and his examining section (Art Unit 353) are given. Art Unit 353 is part of Examining Group 350. Before that, in the large brackets, are the filing date, serial number and inventor's name (John A. Novel). To the right is the date the OA was mailed; this is its official date.

Below the address of the attorney (John C. Able), the first box that is checked indicates: "This application has been examined," denoting that this is the first OA in application No. 999,999. If it had been a second and non-final OA, the second box, "Responsive to communication filed on [date]," would have been checked; had it been a final OA, the third box, "This action is made final," would have been checked.

The next paragraph indicates that the period for response will expire in three months and that failure to respond will cause the application to be abandoned. Since the OA was mailed April 19, 1976, the period for response expires July 19, 1976. If the last date of the period falls on a Saturday, Sunday, or holiday, the period for response expires on the next business day.

Under "Part I," the check at box 1 indicates that one attachment, a "Notice of References Cited" is part of the OA. A typical Notice of References Cited is shown in Fig. 13-D below.

Under "Part II—Summary of Action," the examiner has checked various boxes to indicate what action he has taken with the application. Of the eleven claims that are pending, he has rejected claims 1 to 8, objected to claims 9 to 11, approved a drawing request, and acknowledged a claim for "priority" of an earlier corresponding foreign

PAPER NO. __2__

U.S. DEPARTMENT OF COMMERCE
Patent and Trademark Office
Address : COMMISSIONER OF PATENTS AND TRADEMARKS
Washington, D.C. 20231

T.F. Callaghan   Art Unit 353

[04/11/75   999,999
John A. Novel   ]

MAILED

MAILED:
APR 19 1976

GROUP 350

John C. Able
1234 Jefferson Davis Highway
Arlington, Virginia 22202

THIS IS A COMMUNICATION FROM THE EXAMINER
IN CHARGE OF YOUR APPLICATION.

COMMISSIONER OF
PATENTS AND TRADEMARKS

☒ This application has been examined.

☐ Responsive to communication filed on _____.

☐ This action is made final.

A SHORTENED STATUTORY PERIOD FOR RESPONSE TO THIS ACTION IS SET TO EXPIRE ____3____ MONTH(S)

_____ DAYS FROM THE DATE OF THIS LETTER.

FAILURE TO RESPOND WITHIN THE PERIOD FOR RESPONSE WILL CAUSE THE APPLICATION TO BECOME ABANDONED.
35 U.S.C. 133

PART I   THE FOLLOWING ATTACHMENT(S) ARE PART OF THIS ACTION:

1. ☒ Notice of References Cited, Form PTO–892.   2. ☐ Notice of Informal Patent Drawing, PTO–948.

3. ☐ Notice of Informal Patent Application,   4. ☐
Form PTO–152

PART II   SUMMARY OF ACTION

1. ☒ Claims __1–11_____ are pending in the application.

Of the above, claims _____ are withdrawn from consideration.

2. ☐ Claims _____ have been cancelled.

3. ☐ Claims _____ are allowed.

4. ☒ Claims __1–8_____ are rejected.

5. ☒ Claims __9–11_____ are objected to.

6. ☐ Claims _____ are subject to restriction or election requirement.

7. ☐ The formal drawings filed on _____ are acceptable.

8. ☒ The drawing correction request filed on _March 1, 1976_ has been   ☒ approved.
                                                                          ☐ disapproved.

9. ☒ Acknowledgement is made of the claim for priority under 35 U.S.C. 119. The certified copy has
   ☐ been received.
   ☐ not been received.   ☒ been filed in parent application:
                                    serial no. _888,888_ filed on _12-5-72_.

10. ☐ Since this application appears to be in condition for allowance except for formal matters, prosecution as to the
merits is closed in accordance with the practice under Ex parte Quayle, 1935 C.D. 11; 453 OG. 213.

11. ☐ Other

Form PTOL–326 (rev. 11–75)

Fig. 13-B

application, which was filed in a U.S. parent application of the present application.

Now it's time to look at Fig. 13-C.

In Part III of the Office Action, the examiner gives his specific reasons for rejecting or objecting to the claims. In line 1 of Part III he has rejected claims 1, 3 and 4 under Section 102 of the patent laws (see Chapter 5) on reference A.

The examiner has not made any comment, which indicates that he feels that claims 1, 3, and 4 are completely and fully anticipated by reference A in a manner which is too obvious to require explanation (see Chapter 5, Section E on "novelty").

In line 2, he has rejected claims 2 and 5 under Section 102 on reference B or reference C, and he has added a brief comment to indicate where in each reference certain features of these claims are found. The virgule or slash (/) between B and C represents "or," as indicated by the key at the bottom of the form.

In line 3, he has rejected claims 6 and 7 under Section 103 of the Patent laws on reference D in view of (symbol: v) references E an F, and he has stated specifically why these references can be combined to render the subject matter of claims 6 and 7 obvious (see Chapter, 5 Section F for a discussion of obviousness).

In line 4, he has rejected claims 6 and 7 under Section 112, second paragraph, because he feels the word "aperture" in these claims doesn't properly describe the invention.

In line 5, he has rejected claim 8 under Section 103 with an explanatory comment.

In line 6, he has stated that claims 9 to 11 are objected to because they are dependent on a rejected claim but will be allowed if rewritten in independent form (see Chapter 9 for discussion of dependent and independent claims). This means that he considers that claims 9 to 11 contain allowable subject matter, and that if they are rewritten so that they include all

the limitations of their parent or independent claims, they will be allowed.

In line 7, the examiner indicates that claim 6 would also be allowed if amended (narrowed) to recite a certain feature of the invention.

In line 8, the examiner states that an additional reference (G) has been cited (but not used in any rejection) to show another feature of the invention.

The examiner has signed the office action at the bottom and has also listed his telephone number above his official name stamp.

Finally, we turn to Fig. 13-D.

The Notice of References Cited lists eleven U.S. patents, five foreign patents, and four magazine or text references. (Most OA's will cite only a few references.) All these references will be attached to the OA, except those checked in the column marked with the asterisk(*), which were furnished in a prior office action, a prior related application, or were furnished by you in your Information Disclosure Statement. To save time, the references are referred to in the OA by the letters A to U. The "Document Number" column generally lists patent numbers, except that reference D is a defensive publication (see Subsection 10 below), reference E is a plant patent, references H and I are design patents, and reference J is a reissue patent.

The date column indicates the date the patent issued, or the document was published. If this date is later than your filing date, the reference is not a good reference against your application, unless it is a U.S. patent filed before your application. In the latter instance, the examiner is supposed to indicate the filing date of the patent reference in the last column, which he has done for reference H.

When the PTO cites patents, some inventors react in various illogical ways, as indicated by the following Common Misconceptions:

FORM PTO-1142
(3-75)

U.S. DEPARTMENT OF COMMERCE
Patent and Trademark Office

PART III

SERIAL NUMBER 999,999

GROUP ART UNIT 353

NOTIFICATION OF REJECTION(S) AND/OR OBJECTION(S) (35 USC 132)

	CLAIMS (1)	REASONS FOR REJECTION (2)	REFERENCES * (3)	INFORMATION IDENTIFICATION AND COMMENTS (4)
1	1,3,4	35 U.S.C. 102	A	
2	2,5	35 U.S.C. 102	B/C	Axle assemblies of each fixed to tubular members (Fig. 2 of B, Fig. 4 of C).
3	6,7	35 U.S.C. 103	D v E+F	Obvious to extend auxiliary wheels of D (Fig.1) laterally as in E (p. 2, ls. 1-6). Also, obvious to provide vertically adjustable wheels in D as shown by F (Fig. 3).
4	6,7	35 U.S.C. 112, 2nd paragraph	—	"Aperture" is misdescriptive in defining a sleeve within a frame member.
5	8	35 U.S.C. 103	A v E	Obvious to extend auxiliary wheels of A (Fig.1) laterally as in E (p. 2, ls. 1-6).
6	9-11	—	—	Objected to — depend from rejected claim; will be allowed if rewritten in independent form.
7				Claim 6 would be allowed if amended to recite the specific hydraulic wheel-moving arrangement.
8				G cited to show an analogous hydraulic wheel-moving mechanism.

* Capital letters representing references are identified on accompanying Form PTO-892.
The symbol "v" between letters represents - in view of -.
The symbol "+" or "&" between letters represents - and -.
A slash "/" between letters represents the alternative - or -.

NOTE: Sections 100, 101, 102, 103, and 112 of the Patent Statute (Title 35 of the United States Code) are reproduced on the back of this sheet.

EXAMINER

TEL. NO. (703) -587 - 3070

Thomas F. Callaghan
Thomas F. Callaghan
Primary Examiner
Art Unit 353

-2-

Fig. 13-C

TO SEPARATE, HOLD TOP AND BOTTOM EDGES, SNAP—APART AND DISCARD CARBON

FORM PTO—892 (REV. 9-75)	U.S. DEPARTMENT OF COMMERCE PATENT AND TRADEMARK OFFICE	SERIAL NO. 999,998	GROUP ART UNIT 425	ATTACHMENT TO PAPER NUMBER 3
NOTICE OF REFERENCES CITED		APPLICANT (S) STRUCK et al.		

U.S. PATENT DOCUMENTS

		DOCUMENT NO.	DATE	NAME	CLASS	SUB-CLASS	FILING DATE IF APPROPRIATE
	A	2717874	9-1955	VERAIN	21	102 R X	
X	B	2572144	10-1951	HEALY	340	71 X	
	C	2137376	11-1938	ALTORFER	21	DIG. 2	
	D	T881002	12-1970	JONES	96	1.6	
	E	P.P. 2400	5-1964	BOERNER	Plant	20	
	F	B207272	1-1975	DAVIDSON	75	1	
	G	1671843	5-1928	SCOTT	15	104.01 R	
	H	D23840 4	1-1976	OWENS	D6	5	11-13-1972
	I	DRe24841	6-1960	ROCHÉ	D8	189	
	J	Re18406	4-1932	MARINSKY	24	205.16 C	
X	K	3035319	5-1962	WOLFF	24	274 W8 X	

FOREIGN PATENT DOCUMENTS

		DOCUMENT NO.	DATE	COUNTRY	NAME	CLASS	SUB-CLASS	PERTINENT SHTS. DWG	PP. SPEC.
	L	136113	1-1950	AUSTRALIA	PAPER PRODUCTS	24	134 QA		
	M	Add.34622	11-1934	FRANCE	LORENZ	26	15 R	1	4-7
	N	19421	of 1913	UNITED KINGDOM	CROSSE	26	51.5		
X	O	1345890	7-1963	GERMANY	MUTHER	19	6		
	P	683125	3-1964	CANADA	FISHBURNE	100	216	1-5	1-19
	Q								

OTHER REFERENCES (Including Author, Title, Date, Pertinent Pages, Etc.)

R	Chemical Abstracts, Vol. 75, No. 20, Nov. 15, 1971, p. 163, abstract no. 120718k, Shetulov, D.I., "Surface Effects During Metal Fatigue", copy in Group 120 Library.
S	(s00840001) Winslow, C.E.A., Fresh Air and Ventilation, E.P. Dutton, N.Y., 1926, p. 97-112, TH 7653 W5, 315-22.
T	Ballistic Missile & Aerospace Technology, Vol. 3, Academic Press, N.Y., 1964, TL 78759, p. 199, 250-108.
U	Carbowax & Polyethylene Glycols, Carbide Chemical Corporation, 1946, p. 5, copy in Group 120 Library.

EXAMINER Richard Stone	DATE 4-10-76	

* A copy of this reference is not being furnished with this office action.
(See Manual of Patent Examining Procedure, section 707.05 (a).)

Fig. 13-D

▲

Common Misconception: *The PTO can't cite foreign or non-English patents or other publications against a U.S. patent application.*

Fact: *As indicated in Chapter 5, any publication, including patents, from anywhere in the world, in any language, is valid prior art against your patent application, provided it was published before your filing date, or before your earliest provable date of invention, up to one year before your filing date.*

▲

Common Misconception: *A foreign patent which shows or claims your invention will prevent you from making the invention in the U.S.*

Fact: *A patent of any country is enforceable only within the geographical area of that country and has no effect elsewhere. Thus for example, a French patent is enforceable only in France and has no force or effect in the U.S., except as a prior-art reference.*

▲

Common Misconception: *If an examiner cites an in-force U.S. patent against your application, this means that your invention, if manufactured, sold, or used, would infringe this patent.*

Fact: *The only way you can tell if your invention would infringe any patent is to compare the patent's claims against your invention. Most cited in-force patents would not be infringed by your invention since their claims are directed to a different invention. Again, examiners hardly ever read claims of patents they cite and the PTO is never concerned with infringements.*

▲

Common Misconception: *If an examiner cites a very old reference against your application, it is not as good a reference as an in-force patent or a very recent reference.*

Fact: *The age of a reference is totally irrelevant, so long as its date is earlier than your filing date or your earliest provable date of invention (see Chapter 5).*

D. What to Do When You Receive an Office Action

WHEN YOU RECEIVE AN OA, don't panic or be intimidated. It's common for examiners to reject all claims, even if the rejections are not valid. This type of rejection is termed a "shotgun" or "shoot-from-the-hip" rejection. Although they shouldn't do so, examiners sometimes do this because of the pressure of work, and sometimes to force you to more clearly state the essence of your invention and its true distinguishing features. You'll find that even if all claims are rejected, if you approach your OA in a calm, rational, and methodical manner as outlined below, you shouldn't have too much difficulty in ultimately getting your patent if your invention meets the legal tests for patentability.

1. Write Your Due Date on Your Calendar and Mount the OA In Your File

After you get your office action, write the due date of your response right on it, and also on your calendar so you don't forget it. You should actually write the date thrice on your calendar: once on the date it's actually due, once one week before it's due, and once one month before it's due. If the due date falls on a weekend or holiday, your due date is the next business day. Also mount the O.A. in your file (see Ch. 10, Part J) so you won't lose it.

2. Check the References and Review Your Application

Check all your references carefully to make sure you've received all the correct references as listed in the Notice of References Cited. If there's any discrepancy, call or write the examiner at once. This call will not count as an interview (which are limited to two.)

Next, read the OA carefully and make a detailed written summary of it so that you'll have it impressed in your mind. After that, reread your application, noting all grammatical and other errors in the specification, claims, and drawings you would like to correct or improve. Also if you really want to do a bang-up job, if any significant new reference(s) has been cited by the PTO, you can make a note to add a brief discussion of it to the prior art part of your specification; don't forget to "knock" it as usual.

3. Read and Analyze Each Cited Reference

Next, read every cited reference (except the claims of patent references) completely and carefully. Make sure that you take enough time to completely understand the reference and how it works. However, as I mentioned in Chapter 6 in connection with conducting a patentability search, don't bother reading the claims of any patent cited as a reference. Why not? Because the patent has not been cited for what it claims, but rather for what it shows about the prior art. The claims generally only repeat parts of the specification and are not directly relevant to the patent prosecution process since they are only used to

determine whether infringement exists. If you think of cited patents as magazine articles, you'll avoid this "claims trap" which most laypersons fall into.

Write a brief summary of each reference, preferably on the reference itself, even if it has an adequate abstract, in order to familiarize yourself with it in your own words.

Suppose a cited reference is a publication with a date less than one year before your filing date, or is a U.S. patent which has:

- A filing date preceding your filing date; and
- An issue date on or after your filing date; or
- An issue date up to one year before your filing date.

In any of these cases, you can "swear behind" the reference under Patent Rule 131 and thereby eliminate it from consideration. To do this, you must submit a declaration containing facts and attached copies of documents showing that you built and tested the invention, or conceived the invention, and were thereafter diligent in building and testing it, before the effective date of the reference (see Chapter 5, Section E).[2] See MPEP 715 for full details.

4. Make a Comparison Chart

Next, I find it helpful to make a comparison chart showing every feature of your invention across the top of the chart and listing the references down the left-hand side of the chart, as in Fig. 13-E.

---

[2]You can't swear behind a cited patent which claims the same invention as yours: the only way you can overcome such a patent is to get into interference with it and win "priority"; see Chapter 15.

Features of my Invention

References	Pivot arm	Bracket at end of arm	Bracket has screw tightener
A	X	X	
B	X		X

Fig. 13-E

Be sure to break up your invention so that all possible features of it, even those not already claimed, are covered and listed across the top of the chart. Remember that a feature can be the combination of two known separate features. Then indicate, by checking the appropriate boxes, those features of your invention which are not shown by each reference. This chart, if done correctly and completely, will be of tremendous aid in drafting your response to the first OA.

5. Compare Your Broadest Claim with the Cited References for Novelty

If the examiner applies any references under Section 102, you'll need to deal with the novelty question. However, if the reference is said to apply under Section 103 (obviousness), the examiner is tacitly admitting that you've made it past Sec. 102, i.e., your claimed structure is novel. Therefore, you won't have to go through the analysis in this section. Instead, turn to Section 6.

First, reread your broadest claim to see which features it recites. Remember, only positively-recited physical structure or acts count. Then consider whether these physical features distinguish your invention from each reference cited against this claim. Don't pay any attention to the advantages of your invention, your statements of function, or your whereby clauses. Only focus on the physical features, including those that are in the form of a means clause followed by a function.

Example: "A lever having a threaded end with a counterbalance thereon" is a proper physical recitation which can distinguish your invention from the prior art. The phrase "means for counterbalancing" is a means clause followed by a function and is equivalent to a physical recitation. But "said lever counterbalancing said arm" is a mere statement of result or function and can't be used to distinguish the prior art.

If only one reference has been cited against your broadest claim, consider whether your claim distinguishes over this reference under Section 102 (i.e., whether your claimed structure is novel—see Chapter 5, Section E). That is, are there any features recited in the claim which are not shown in the reference being cited against it? If not, the claim is "fully met" or anticipated by this reference and will have to be narrowed.

Remember that the examiner is entitled to interpret any claim in any reasonable way against you. I.e., if a claim, or any word in a claim, has two reasonable interpretations, the examiner is entitled to take the one least favorable to you when determining if your claim has novel physical structure under Sec. 102. E.g., suppose your invention uses a clamp which is halfway between two ends of a rod and a reference shows a clamp near one end of a rod. If your claim recites that the clamp is "intermediate" the ends of the rod, this won't distinguish over the reference since "intermediate" means "between" as well as "in the middle". The remedy? Recite that your claim is "substantially in the middle" of the rod in order to distinguish over the reference under Sec. 102 (but not necessarily under Sec. 103).

Suppose the physical features of your claim are all shown in a prior-art reference, but the prior art reference's physical features are used for a different purpose than yours. E.g., you claim "a depression in a wall plate for holding a clock" and the prior art shows a large oil drip pan under a milling machine; this pan literally constitutes "a plate with a depression." Thus your claim literally "reads on" the prior art, but your claimed elements are directed to a different purpose than the elements of the prior-art reference. Unfortunately the rejection is valid: you'll have to narrow the claim, or consider claiming your structure as a "new use" invention—see Chapters 5 and 8.

6. Analyze Novel Features for Unobviousness

If the claim recites (or has been amended to recite) novel features, consider whether these are unobvious over the reference cited against it. All possible reasons for arguing unobviousness are listed in Fig. 13-F, Part I.

If you consider the features of your invention obvious, you'll have to narrow the claim, either by adding more features (refer to Fig. 13-E above) or by reciting the existing features more narrowly.

7. If References Are Cited In Combination Against Your Broadest Claim

If two or more references have been cited in combination against your broadest claim, refer to Fig. 13-F, Part II to see whether the examiner has a point.

Note: You should especially consider reasons 20 and 21, i.e., ask yourself whether it is proper to combine these references in the manner that the examiner has done. Also note that when you use any of the reasons of Fig. 13-F, you should not merely state the applicable reason, but also supporting facts that pertain to your invention.

8. Does the Combination Disclose Subject Matter of Your Broadest Claim?

Assuming that the references are combined (whether or not they can be), ask yourself if the combination discloses the subject matter of your claim (reason 27). If not, are the distinctions in your claim patentable under Section 103 (reasons 1 to 23 and 28-32)? Also ask yourself whether there are any other errors in the examiner's logic or reasoning.

9. If Your Claims are Rejected Under Section 112 of the Patent Laws

If your claim has been rejected under Section 112, a very common occurrence, even for patent attorneys, the examiner feels that the language of your claim is not clear or proper, and you should try to work out alternative language which will satisfy this objection. Also, you can ask the examiner to write clear claims for you; see Section F(2)(i) below.

Part I - General

1. **Unexpected Results:** The results achieved by the invention are new, unexpected, superior, disproportionate, unsuggested, unusual, critical and/or surprising.

2. **Assumed Unworkability:** Up to now those skilled in the art thought or found that the techniques used in the invention were unworkable or presented an insuperable barrier.

3. **Assumed Insolubility:** Up to now those skilled in the art thought or found the problem solved by the invention was insoluble, i.e., the invention converts failure into success. The failures of prior-art workers indicate that a solution was not obvious.

4. **Commercial Success:** The invention has attained commercial success. (Prove this by a declaration with supporting documents.)

5. **Unrecognized Problem:** The problem solved by the invention was never before even recognized.

6. **Crowed Art:** The invention is classified in a crowded art; therefore a small step forward should be regarded as significant.

7. **Omission of Element:** An element of the prior art has been omitted or a prior-art version has been made simpler without loss of capability.

8. **Unsuggested Modification:** The prior art lacks any suggestion that the reference should be modified in a manner required to meet the claims.

9. **Unappreciated Advantage:** Up to now those skilled in the art never appreciated the advantage of the invention, although it is inherent.

10. **Inoperative References:** The prior-art references which were relied upon are inoperative.

11. **Poor References:** The prior-art reference(s) are vague, foreign, conflicting or very old and therefore are weak and should be construed narrowly.

12. **Ancient Suggestion:** Although the invention may possibly have been suggested by the prior art, the suggestion is many years old, was never implemented, and produced greatly inferior results.

13. **Lack of Implementation:** If the invention were in fact obvious, because of its advantages, those skilled in the art surely would have implemented it by now. I.e., the fact that those skilled in the art have not implemented the invention, despite its great advantages, indicates that it is not obvious.

14. **Misunderstood Reference:** The reference does to teach what the Examiner relies upon it as supposedly teaching.

15. **Long-Felt & Unsolved Need:** The invention solves a long-felt, long-existing, but unsolved need.

16. **Commercial Acquiescence:** The invention has been licensed.

17. **Professional Recognition:** The invention has been given an award or recognized in a professional publication.

18. **Competitive Recognition:** The invention has been copied by an infringer; moreover the infringer has made laudatory statements about it.

19. **Contrarian Invention:** The invention is contrary to the teachings of the prior art. I.e., the prior art teaches away from the concepts of the present invention.

20. **Strained Interpretation:** The examiner has made a strained interpretation of the reference which could be made only by hindsight.

21. **Paper Patent:** The reference is a "paper patent," i.e., it was never implemented or commercialized and therefore should be construed narrowly.

22. **New Principle of Operation:** The invention utilizes a new principle of operation.

23. **Inability of Competitors:** Competitors were unable to copy the invention until they were able to learn its details via a publication or reverse engineering a commercial model; this indicates unobviousness.

Part II - Combination of References Applied

24. **Synergism:** The whole (i.e., the result achieved by the invention) is greater than the sum of its parts (i.e., the respective results of the individual references).

25. **Unsuggested Combination:** The prior art references do not contain any suggestion that they be combined, or that they be combined in the manner suggested.

26. **Impossible to Combine:** Those skilled in the art would find it physically impossible to combine the references in the manner suggested.

27. **Claimed Features Lacking:** Even if combined, the references would not meet the claims, or alternatively, it would be necessary to make modifications, not taught in the prior art, in order to combine the references in the manner suggested.

28. **Inoperative Combination:** If combined, the references would produce an inoperative combination.

29. **Multiplicity of References:** The fact that a large number of references must be combined to meet the invention is evidence of unobviousness.

30. **References Teach Away:** The references themselves teach away from the suggested combination.

31. **Multiplicity of Steps Required:** The combination suggested requires a series of separate, awkward combinative steps which are too involved to be considered obvious.

32. **Reference Is From Different Field:** One reference is from a very different technical field than that of the invention, i.e., it's "non-analogous art."

Fig. 13-F—Arguments Against Obviousness Rejections

10. What to Do If You Disagree with the Examiner

If you believe your broadest claim is patentable over the prior art and that there is a serious flaw in the examiner's logic, it is theoretically permissible to leave the claim as it is and argue its patentability in your response. While it's sometimes desirable to do this to emphasize the rightness of your position, I generally advise you not to do this since it's difficult psychologically for the examiner to back down. In other words, it's easier to get the examiner to change directions slightly than to make an about turn.[3] Thus, to save the examiner's ego, it's best to try to make some amendment to the claim, even if it's insignificant. *Treat all persons you deal with as if they had a sign around their neck reading, 'Make Me Feel Important.'*—Mary Kay Ash.

---

[3]If you do file a reply to an OA without changing the specification or claims, your reply is technically not an "amendment," so call it a "response."

11. Making Amendments Without Narrowing Scope of Claim

It's usually possible to make amendments to a claim which don't narrow its scope. For example, you can recite that a member, which of necessity must be elongated, *is* elongated. By doing this, you have amended the claim without narrowing your scope of protection. Also, in the electronic field, you can state that a circuit is energized by a direct-current source. For almost any claim you can add a whereby clause to the claim stating the function of the mechanism of the claim, and you can add a longer preamble stating in more detail (but not in narrower language) the environment of your invention. The important thing is to add some words to the claim(s), even if you already believe they distinguish under Sections 102 and 103, in order to show that you're meeting the examiner part way.

12. Amending Your Claim When You Agree With the Examiner

If you believe your broadest claim isn't patentable as written, and you agree with all or part of the examiner's rejection, you'll have to narrow the claim by adding physical or structural limitations, or by narrowing the limitations already present, in the manner outlined in Chapter 9.

Here are some suggestions on how to approach the amendment of your claims:

a.   Look for the physical feature(s) in Fig. 13-E that constitute the essence of why your invention can be distinguished from the prior art. Then try to put this essence into your claim. Note that you should amend the main claim so as to distinguish physically over the references under Sec. 102. However the physical distinctions should be significant enough to define structure which is also unobvious under Sec. 103. However save your actual reasons as to why the physical distinctions are unobvious for your remarks or for a "whereby" clause at the end of the claim.

b. Don't make your main claim narrower than necessary. Often the limitation you are looking for can be found in one or more dependent claims.

c. Show your invention and the cited references to friends or associates; often they can readily spot the distinguishing essence of your invention.

d. After you've narrowed your main independent claim so that it distinguishes over the prior art cited by the patent examiner, and you feel the distinguishing features are patentable under Section 103 (i.e., they're unobvious), do the same for all your other independent claims.

e. If you've changed any independent claims, change your dependent claims so that they completely and correctly correspond in language and numbering with your main claim. If you incorporate a limitation from a dependent claim into your main claim, cancel the dependent claim. This is because the dependent claim will no longer be able to add anything to narrow the independent claim. You may also think of other, narrower dependent claims to replace those which you've cancelled; refer to the comparison chart to be sure you've claimed every feature.

f. You should write the narrowest possible claims you're willing to accept since it will be difficult to amend again if your amended claim is rejected this time around. See Section J on final office actions.

g. Be sure all of the less-important specific features of your invention are covered in your amended dependent claims.

h. Try to distinguish by adding quantitative or relative, rather than qualitative, recitations to your claims since these carry far more weight. E.g., say "a rod at least one meter long" or "a rod which is longer than said post" not "a rod of great length."

13. Plan an Outline of Your Response

Indicate on a copy of your application, or on separate sheets, the amendments you intend to make to your specification, your claims, your drawing, and your remarks. The "remarks" section of your amendment (as shown in Fig. 13-G below) should consist of:

* A brief summary of all your amendments;

* A review of the rejections made by the examiner;

* A review of the references cited by the examiner;

* A summary of how you changed the claim, quoting your changes;

* A statement of your claimed distinctions under Sections 102 and 103 if one reference was cited, together with arguments from Part 1 of Fig. 13-F;

* A statement of why the references can't be combined, followed by comments regarding Sections 102 and 103 if more than one reference was cited, together with arguments from Part 2 of Fig. 13-F;

* A request for reconsideration of the examiner's position;

* A discussion of dependent and other main claims you have;

* A discussion of any technical (Section 112) rejections;

* Any request for aid you may wish to make under MPEP 707.07(j) requesting the examiner to write claims; and

* A conclusion.

See Section G below for specifics on drafting your remarks.

At this point, read Fig. 13-G, a sample successful amendment from an actual case (now a patent) to see the format customarily used. Continue to refer to Fig. 13-G throughout the next four sections of this chapter.

Ser. Nr. 07/088,691 (Koppe)   Amendment A, contd.

P. 9, l. 9, change "(26 in" to --26 (--.

P. 11, last line, delete "4, , ".

P. 12, l. 12, delete "the".

P. 13, l. 21, delete "use of".

P. 13, l. 22, after "tages" insert --in--.

CLAIMS:

Cancel the claims of record (1 to 23) and substitute new claims 24 to 46 as follows:

24. In a bag closure of the type comprising a flat body of material having two major sides which face in opposite directions, a lead-in notch beginning at one edge of said body of material and extending into said body, a gripping aperture in said body which is adjacent to and which communicates with said notch, and a layer of paper laminated to one side of said body,

the improvement wherein said flat body of material is made of a flexible plastic of the type which can be repeatedly bent and straightened without fracture, whereby said closure can be bent so that it can easily be removed from bag without damaging said bag, and thereafter can be straightened so that it can be re-used as a closure on said bag, and whereby any bending of said closure will cause said paper layer to tear or

---

IN THE US PATENT AND TM OFFICE

Appn. Number: 07/088,691
Filing Date: 1987-8-24
Applicant(s): Koppe, Lou W.
Appn. Title: Paper-Laminated Pliable Closure For Plastic Bags

Examiner : V. N. Sakran/GAU 357
Disk Nr./File: PN1:Koppe8.412

Mailed 1988 Apr 12
San Francisco, CA

AMENDMENT A

Commissioner of Patents and Trademarks
Washington, District of Columbia 20231

Sir:

In response to the Office letter mailed 01/12/88, kindly amend the above application as follows:

SPECIFICATION:

P. 3, l. 11, change "4 ," to --4,694,542--.

P. 5, l. 3, change "4, (1987)" to --mentioned above--.

P. 8, l. 5, change "(20 in" to --20 (--.

P. 8, l. 7, change "hyphens" to --abbreviated "PET"; hyphens--.

Fig. 13-G—Sample "Regular" Amendment

Ser. Nr. 07/088,691 (Koppe)   Amendment A, contd.   3

be distorted so as to leave an indication that said closure was bent and possibly removed and replaced.

25. The closure of claim 24 wherein said body of material is composed of polyethyleneterephthalate.

26. The closure of claim 24 wherein said body of material is elongated and has a longitudinal groove on said one side of said body and extends the full length of said one side, from said gripping aperture to the opposite edge.

27. The closure of claim 26 wherein said groove is formed into and along the full length of said layer of paper.

28. The closure of claim 24 wherein said body of material is elongated and has a longitudinal groove which is on the side of said body opposite to said one side thereof and extends the full length of said one side, from said gripping aperture to the opposite edge.

29. The closure of claim 24 wherein said body of material is elongated and has two longitudinal grooves which are on opposite sides of said body and extend the full lengths of said sides, from said gripping aperture to the opposite edge.

30. The closure of claim 29 wherein the groove on said one side of said body is formed into and along the full length of said layer of paper.

31. The closure of claim 24 wherein said body has paper layers laminated to both of said sides, respectively, of said body of

Ser. Nr. 07/088,691 (Koppe)   Amendment A, contd.   4

material.

32. The closure of claim 31 wherein a groove is on one side of said body of material and extends the full length of said one side, from said gripping aperture to the opposite edge.

33. The closure of claim 32 wherein two grooves are formed on opposite sides of said body, said grooves extending the full lengths of said sides, from said gripping aperture to the opposite edge.

34. The closure of claim 33 wherein said grooves are rolled into and along the full lengths of said layers of paper, respectively.

35. The closure of claim 24 wherein said layer of paper is colored.

36. The closure of claim 24 wherein said body is elongated and has a longitudinal through-hole parallel to and between said major sides thereof.

37. A bag closure of the type comprising a flat body of material having a lead-in notch on one edge thereof, a gripping aperture adjacent to and communicating with said notch, and a layer of paper laminated on one of its sides, characterized in that said flat body of material is made of a flexible plastic of the type which can be repeatedly bent and straightened without fracture, whereby said closure can be

Fig. 13-G—Sample "Regular" Amendment

Ser. Nr. 07/088,691 (Koppe)   Amendment A, contd.   5

bent so that it can easily be removed from bag without damaging said bag, and thereafter can be straightened so that it can be re-used on said bag as a closure therefor, and whereby such bending of said closure will cause said paper layer to tear or distort so as to leave an indication that said closure was bent and possibly removed and replaced.

38. The closure of claim 37 wherein said body of material is composed of polyethyleneterephthalate.

39. The closure of claim 37 wherein said body is elongated and has a longitudinal groove which is on said one side of said body and which extends the full length of said one side, from said gripping aperture to the opposite edge.

40. The closure of claim 37 wherein said body is elongated and has a longitudinal groove which is on the side of said body opposite to said one side thereof and extends the full length of said one side, from said gripping aperture to the opposite edge.

41. The closure of claim 37 wherein said body is elongated and has two longitudinal grooves which are on opposite sides of said body and extend the full lengths of said sides, from said gripping aperture to the opposite edge.

42. The closure of claim 37 wherein said body has a paper lamination on both of said sides.

43. The closure of claim 42 wherein a groove is on one side of said body and extends the full length of said one side,

---

Ser. Nr. 07/088,691 (Koppe)   Amendment A, contd.   6

from said gripping aperture to the opposite edge.

44. The closure of claim 42 wherein two grooves, on opposite sides of said body, extend the full lengths of said sides, from said gripping aperture to the opposite edge.

45. The closure of claim 37 wherein said layer of paper is colored.

46. The closure of claim 37 wherein said body is elongated and has a longitudinal through-hole.

REMARKS—General

1. Several editorial corrections have been made in the specification, including updating the missing patent number.

2. The objection to the drawings has been noted; new drawings will be filed after allowance.

3. Enclosed is an Information Disclosure Statement listing the patents mentioned in the introductory portion of the present specification. Applicant regrets this omission.

4. The claims of record have all been rewritten and replaced with new claims 24 to 46 in order to define the invention more particularly over the cited references. These claims are all submitted to be patentable over the cited references because (1) they recite novel structure and thus distinguish physically over every reference (Sec. 102), and (2) the physical distinctions effect new and unexpected

Fig. 13-G—Sample "Regular" Amendment

Ser. Nr. 07/088,691 (Koppe)   Amendment A, contd.   7

results, thereby indicating that the physical distinction are unobvious under Sec. 103.

The Claims All Distinguish Over The References Under Sec. 102

5. The two independent claims, and hence all claims, distinguish over the references under Sec. 102 because they recite a bag closure made of a body of material having a lead-in notch, a communicating gripping aperture, and a layer of paper laminated to one side thereof, characterized in that the body of material is made of a flexible plastic which can be repeatedly bent and straightened without fracture, thereby providing a closure which can be bent and straightened so that it can be removed and re-used, yet will still provide an indication that it was removed.

6. The cited and relied-upon Arnold et al. (Arnold) patent shows a closure tab which is made of a "rigid" (e.g., col. 2, l. 36) substrate. Arnold gives, as examples of a suitable material for his substrate, polystyrene, polyvinyl chloride, and polyamide. All of these materials are brittle or very rigid and cannot be bent and straightened even once without fracture, much less repeatedly.

7. The cited and relied-upon Parmenter patent shows a bag closure but does not give any details of the type of plastic used. However Parmenter's closure uses an integral or "living" hinge. Therefore it probably is made of a soft, flexible plastic. However Parmenter does not teach, show, or suggest that any paper layer can be applied to one or both sides of his tab, as applicant's claims now recite.

8. The cited and relied-upon Mesek et al. (Mesek) patent shows an

Ser. Nr. 07/088,691 (Koppe)   Amendment A, contd.   8

absorbent bed pad. The Fig 2 embodiment comprises a polyethylene layer 12 with a fibrous layer 16 attached along bead lines 14. Mesek's Fig 8 embodiment is said to be similar, except that an additional layer 213' of a short-fiber enriched material with a paper-like skin 218 in added. The base layer in Fig 8 is layer 214r, which does not appear to be mentioned, but which appears similar to layer 14 of Fig 2. Mesek does not appear to have any paper-like layers laminated to both sides of his base layer and of course does not have the lead in notch or the communicating holding aperture, as recited in the independent claims.

9. The other cited but not relied-upon patents are also deficient in one or more of the above-discussed physical features of the independent claims.

10. Since the independent claims both recite features which are not present in any reference, applicant submits that these claims, and hence all of the dependent claims, clearly recite novel physical features which distinguish over any and all references under Sec. 102.

The Novel Physical Features Of The Claims Provide New And Unexpected Results And Hence Should Be Considered Unobvious, Making The Claims Patentable Under Sec. 103

11. Applicant submits that the above-recited novel features in the independent claims, and hence in all claims, provide new and unexpected results and hence should be considered unobvious, making the claims patentable under Sec. 103.

12. Specifically, by making the substrate or body of the closure of a

Fig. 13-G—Sample "Regular" Amendment

Ser. Nr. 07/088,691 (Koppe)   Amendment A, contd.   9

flexible plastic which can be repeatedly bent and straightened without fracture, cracking, or tearing, the closure can be repeatedly bent to be removed from its bag, and then replaced and straightened, without damage to itself or the bag. However when the closure is bent and removed and then straightened and replaced, the paper layer on the closure will provide a tell-tale tear or indication that the closure was bent and straightened and probably removed. Thus even though the closure is easily reuseable, it inherently will provide an indication when it has been removed. Thus any adulteration, poisoning, or fraud (due to tag exchanges by dishonest consumers) can easily be detected by the consumer or checkout cashier.

13. None of the prior-art closures can provide these new and unexpected results:

Arnold's tab, being made of rigid, frangible plastic, can't be removed without either (a) damaging the bag, or (b) bending the tab (whereupon it will fracture and can't be re-used). In contrast, applicant's device is bendable so that it can easily be removed and replaced repeatedly without damaging the bag.

Parmenter's closure has no paper layer so that it can't be easily labeled and lacks all of the advantages listed in the present specification. Moreover Parmenter's closure, having no paper layer, can be straightened after it is bent and removed without providing any indication that it was removed. Thus Parmenter's closure can't be used to prevent adulteration, fraud, or poisoning, as can applicant's.

Mesek doesn't show any bag closure whatever.

Ser. Nr. 07/088,691 (Koppe)   Amendment A, contd.   10

14. Since the novel above physical features of applicant's device provide these new and unexpected results over any reference, applicant submits that these new results indicate unobviousness and hence patentability. Accordingly applicant respectfully requests reconsideration and allowance of the present application with the above new claims.

Additional Reasons Militate In Favor Of Unobviousness

15. In addition to the above new and unexpected results, applicant submits that additional reasons militate in favor of patentability, as follows:

16. Unrecognized Problem: Up to now, insofar as applicant is aware, the art contained no indication of the desirability of providing a reusable plastic bag closure which also could indicate, in a permanent manner, that it had been removed and replaced. The discovery of this problem, as well as the concomitant ability to protect against adulteration, fraud, or poisoning, as well as provide reusability, is submitted to be an important one, worthy of patent protection.

17. Crowded Art: The present invention is in a crowded art (note all of the references on bag closures which are cited in the introductory portion of the present specification). It is well recognized that in a crowded art, even a small step forward is worthy of patent protection. While the present invention is submitted to be far more than a small one, nevertheless this factor militates in applicant's favor.

18. Long-Felt But Unsolved Need: The present invention solves a long-

Fig. 13-G—Sample "Regular" Amendment

Ser. Nr. 07/088,691 (Koppe)   Amendment A, contd.   11

existing but unsolved (and unrecognized) need and therefore is submitted to be worthy of patent protection. Specifically, although bag closures of the plastic type have been inuse for many years, they had numerous inherent disadvantages, as stated in the prior-art section of the present specification. Users suffered from the inability to re-use the closures and still have some indication that a closure had been used or the container had been opened. The present invention provides both of these features, thereby solving a long-felt need in this area.

19. Unsuggested Combination: The need for the prior art references themselves to suggest that they can be combined is well-known. E.g., as was stated in In re Sermaker, 217 U.S.P.Q. 1, 6 (CAFC 1983):

"[P]rior art references in combination do not make an invention obvious unless something in the prior art references would suggest the advantage to be derived from combining their teachings."

20. The suggestion to combine the references should come from the prior art, rather than from applicant. As was forcefully stated in Orthopedic Equipment Co. Inc. v. United States, 217 U.S.P.Q. 193, 199 (CAFC 1983):

"It is wrong to use the patent in suit [here the patent application] as a guide through the maze of prior art references, combining the right references in the right way to achieve the result of the claims in suit [here the claims at issue]. Monday morning quarterbacking is quite improper when resolving the question of nonobviousness in a court of law [here the PTO]."

Ser. Nr. 07/088,691 (Koppe)   Amendment A, contd.   12

21. In the present case, the rejection of certain claims uses the Mesek patent, which relates to an absorbent bed pad, a field far removed from plastic bag closures. There would be no reason for one skilled in the art to combine disparate references such as Mesek and either of the other relied-upon references. And there is no suggestion in the references themselves that they be combined. Thus applicant submits that any combination of Mesek with the two other references is an improper one, absent any showing in the references themselves that they can or should be combined.

The Dependent Claims Are A-fortiori Patentable

22. The dependent claims add additional novel features and thus are submitted to be, a-fortiori, patentable. For example, claim 25 recites that the substrate is made of PET, a plastic with a memory which retains its bend, thereby facilitating re-installation on a bag; none of the references show any closures made of PET. Claim 26 recites the groove on both sides. None of the references show this feature. The provision of a groove on both sides facilitates bending and tearing of the laminated paper layers. Claim 27 recites that the groove is in the layer of paper; this also facilitates bending and tearing. Claim 31 recites the paper layers on both sides; this insures full tearing because at least one side will be subject to a convex bend regardless of the way the substrate is bent. Claim 36 recites the through hole in the plastic substrate; this weakens the substrate and facilitates a sharp, paper-tearing bend.

Fig. 13-G—Sample "Regular" Amendment

Ser. Nr. 07/088,691 (Koppe)   Amendment A, contd.   13

The Cited But Non-Applied References

23. These subsidiary references have been studied, but are submitted to be less relevant than the relied upon references.

Request For Constructive Assistance

24. The undersigned has made a dilligent effort to amend the claims of this application so that they define novel structure (closure is made of a flexible, repeatedly-bendable material) which is also submitted to render the claimed structure unobvious because it produces new and unexpected results (repeatedly re-usable closure which provides tell-tale indication if removed). If, for any reason the claims of this application are not believed to be in full condition for allowance, applicant respectfully requests the constructive assistance and suggestions of the Examiner in drafting one or more acceptable claims pursuant to MPEP 707.07(j) or in making constructive suggestions pursuant to MPEP 706.03(d) in order that this application can be placed in allowable condition as soon as possible and without the need for further proceedings.

Very respectfully,

Lou W. Koppe

Applicant Pro Se

P. O. Box 567
Athabasca, Alberta, CANADA TOG OBO
(403) 776-3960

I hereby certify that this correspondence is being deposited with the United States Postal Service as first class mail in an envelope addressed to: Commissioner of Patents and Trademarks, Washington, D.C. 20231, on 1988-4-12
(Date of Deposit)
L. W. Koppe
Name of applicant, assignee, or Registered Rep.
L W Koppe   88-4-12
Signature   Date

Fig. 13-G—Sample "Regular" Amendment

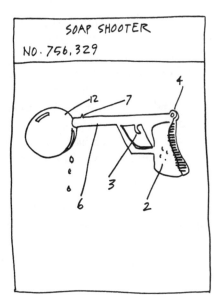

SOAP SHOOTER
NO. 756,329

E. The Format of Amending the Specification and Claims

FORM 13-1 provides the initial part of your amendment.

Fill in the serial number, filing date, your name, title of your application, and the examiner's name and examining unit or group art unit. The date on which you actually mail the amendment goes after "date," and the date of the office letter goes at the space indicated in the first paragraph. Put an appropriate letter (A, B, etc.), after "Amendment" to indicate which amendment it is (your first, second, etc.) Then immediately after the "In response to..." sentence make all the desired changes to your specification and claims, in the manner indicated below, before starting your remarks.

1. Changes to Specification

If you're going to make any changes to the specification, provide the heading, "SPECIFI-

CATION:" below the sentence printed on Form 13-1. Then indicate the specific places in your application where you want to make amendments and the actual amendments you wish to make. Use quotes to indicate existing words and dashes (in typing a dash is made of two hyphens, thus: —) to sandwich words you wish to add. For example:

Page 1, line 3, change "member" (2d occur.) to —lever—.

Page 5, lines 12 to 14, change "member 14... pivot 23" to —lever 14 is connected by way of arm 22 to bearing 23—.

Page 12, line 21, after "screw 18" insert —in contact with arm 22—.

Page 14, lines 12 to 13, delete "member 14... pivot 23".

When your amendment is received, the clerk of the examining group will make each change in red ink in handwriting on the official copy of your application in the manner you direct. Thus you should insure that there is no ambiguity in your amendments. See the example above for how to change the word "member" where it occurs twice on a line.

Be sure that your amendments to the specification don't contain any "new matter"—see part B-14 above.

2. Amendments to Claims

If you want to amend your claims, provide the heading:

"CLAIMS:" and then indicate specifically the claim changes you desire. There are three ways to amend any claim of a patent application:

a. By word cancellation and/or insert;

b. By claim cancellation and substitution; and

c. By rewriting the claim with brackets and underscoring.

Let's look at each of these in more detail.

a. The Word-Cancellation-And/Or-Insert Method

The word-cancellation-and/or-insert method can be used only if you are cancelling words and/or are adding no more than five words to your claim. Claim amendments made by this method are done in exactly the same manner as specification amendments. For example:

Claim 1, line 5, change "said elongated member" to —a lever having—.

b. The Cancellation-And-Substitution Method

The cancellation-and-substitution method can be used under any circumstances, but specifically if you are adding more than five words to the claim. To use this method, you cancel the claim in question and substitute an entirely different claim. The new claim should be given the next-highest unused claim number. Thus, if you originally submitted 12 claims and you want to cancel claim 1 and substitute a new claim, the new claim should be numbered 13. For example:

Claim 1, cancel and substitute new Claim 13 as follows:

13. An improved bicycle mechanism comprising [etc.]-(Make sure that you also amend all claims which were dependent on Claim 1 so that they're now dependent on Claim 13 and so that their terms conform with those in Claim 13.)

When the case is allowed, the clerk of your examining division will renumber all of your claims in order, starting with 1.

c. The Bracket-And-Underscore Method

The bracket-and-underscore method is used when you don't have too many amendments to make to your claim and you want to point out to the examiner exactly where you're making the amendments. Under this method you retype your entire claim with the notation "(amended)" after the number of the claim (use the same number), put brackets around words to be deleted, and underscore material to be added. If your typewriter doesn't have brackets (you're not permitted to use parentheses), you must make your brackets by hand or with virgules (slashes) and underscore lines, thus:

$$\underline{/\ \ /}$$

Some examiners don't like the bracket and underscore method of amending claims because it makes things look too confusing; I agree and therefore rarely use it. Note how the following claim reads originally (with bracketed words and without underscored words) and as amended (without bracketed words and underscored words):

Claim 1, rewrite as Claim 1 (amended) as follows:

1. (amended) A method [for] _of_ stimulating the growth _rate_ of swine by feeding them [aspirin in an amount effective to increase their rate of growth] _a daily dose of aspirin of 0.25 gram per kilogram of body weight._

Word Processing Note: With the availability of word processing, I now usually cancel all of the previous claims, even if I'm making one or two minor changes, and resubmit all of the claims in the amendment. This is easy to do by incorporating the claims from your disk file of the original application into your amendment and making the changes you desire. Examiners like this method since it presents all of the claims together and I strongly recommend it since it reduces errors in numbering, terms, etc. E.g., if your original claims were 1 (independent) and 2 to 10 (dependent) and you want to amend claim 1 only, you would cancel claims 1 to 10 and substitute new claims 11 to 20. Claim 11 would be a rewritten version of claim 1 and claims 12 to 20 would be identical to claims 2 to 10, respectively, except that they would be made dependent upon claim 11, rather than claim 1.

F. Drafting the Remarks

NEXT, ADD THE "REMARKS" PORTION of your amendment. Some general rules for drafting remarks which I'll state first may seem silly, but they're the customary practice and to deviate substantially may make the examiner feel uncomfortable and take a negative attitude toward your invention.

1. General Rules for Drafting Remarks

Rule 1: As stated before, when writing your remarks observe Inventor's Commandment #15 by never admitting that any prior art anticipates or renders any part of your invention obvious. Similarly, never derogate your invention or any part of it.

Rule 2: Never get personal with the examiner. If you must refer to the examiner, always use the third person. For example, never state "You rejected…"; instead, state "The Examiner [note the capitalization] has rejected…"

Better yet, state "The office action rejects…" or "Claim 1 was rejected…" Never, never address the examiner by name (except in the caption), and never make your amendment a "Dear Mr. [Examiner's Name]" letter. See the sample amendment of Fig. 13-G set out above for how it's done.

Rule 3: If there's an error in the OA, refer to the error in the OA, and don't state that the examiner made the error. Even if you find the examiner made a completely stupid error, just deal with it in a very formal way, keep emotions and personalities out of your response, and don't invalidate the examiner. Remember, you've probably made some stupid errors in your life also, and you wouldn't want your nose rubbed in them.

Rule 4: When referring to yourself, always refer to yourself in the third person as "Applicant" and never as "I."

Rule 5: Stick to the issues in your remarks. Be relevant and to the point and don't discuss personalities or irrelevant issues. Never antagonize the examiner, no matter how much you'd like to. It's improper and, if you turn the examiner against you, it can considerably narrow the scope of claims which are ultimately allowed.

Rule 6: Whenever you write any new claims or make any additions to a present claim, you must tell how they distinguish over the prior art the examiner has cited under Sections 102 and 103. Note Patent Rule 111(b) and (c):

(b) In order to be entitled to reexamination or reconsideration, the applicant must make request therefor in writing, and he must distinctly and specifically point out the supposed errors in the examiner's action; the applicant must respond to every ground of objection and rejection in the prior office action (except that request may be made that objections or requirements as to form not necessarily to further reconsideration of the claim be held in abeyance until allowable subject matter is indicated), and the applicant's action must appear throughout to be a bona fide attempt to advance the case to final action. A general allegation that the claims define a patentable invention without specifically pointing out how the language of the claims patentably distinguish them from the references does not comply with the requirements of this section.

"(c) In amending an application in response to a rejection, the applicant must clearly point out the patentable novelty which he thinks the claims present in view of the state of the art disclosed by the references cited or the objection made. He must also show how the amendments avoid such references or objections."

Rule 7: If you do disagree and think the OA was wrong, you must tell exactly why you disagree. If you agree that a claim is obvious over the prior art, don't admit this in your response; simply cancel the claim and don't give any reason for it, or if you must comment, state merely that it has been cancelled in view of the coverage afforded by the remaining claims.

Rule 8: Make a careful, complete, and convincing presentation, but don't agonize about words or minutiae. The reality is that many examiners don't read your remarks or else skim through them very rapidly. This is because they're generally working under a quota system, which means they have to dispose of (finally reject or allow) a certain number of cases in each fiscal quarter. Thus, the examiners are under time pressure and it takes a lot of time to read remarks. It's important to cover all the substantive points in the office action and to deal with every objection and rejection. If you do make an error, as stated, the PTO will almost always give you an opportunity to correct it, rather than forcing you to abandon your application.

Rule 9: If possible, thank or praise the examiner if you can find a reason to do so with sincerity. Examiners get criticized and told they're all wet so often that they'll welcome any genuine, deserved praised.

Rule 10: Don't emphasize your beliefs; they're considered irrelevant. E.g., don't say "Applicant *believes* this invention is patentable." Rather say, "Since the claims define novel structure which produces new and unexpected results as described above, Applicant *submits* that such claims are clearly patentable."

Note: You may wonder whether it makes sense to put much effort into your remarks even though the chances are great they won't be carefully read. My opinion is that it does, because you never know. Think of your effort as a kind of insurance against being the one in five (or whatever) whose remarks are in fact subjected to close scrutiny.

Although it's difficult , I recommend that you do the best job you possibly can in Amendment A since it will probably be the last chance you get to amend your claims in this application. After you draft your amendment, I suggest that you wait a few days and come back and review it again, pretending that you're the examiner. This will probably give you important insights and enable you to improve it further.

2. How to Draft Your Remarks

Your remarks should first provide a brief summary of what you've done to the specification and claims. For example, you can state: "The specification has been amended editorially and to correct those errors noted by the examiner. Claims 1 to 5 have been rewritten as new claims 13 to 18 to more particularly define the invention in a patentable manner over the cited prior art." Then briefly summarize what each claim recites, as done in Fig. 13-G above. If the drawing has been objected to, state that it will be corrected after allowance. If you want to make a voluntary amendment to the drawings, state that a request is attached (Form 13-2) and tell why you want to amend the drawing.

a. Restate First Rejection

Restate the first rejection of the OA. For example, state

"Claims 1 to 5 were rejected as unpatentable over references A and B." The examiner, thus oriented, saves the time it would take to reread the last OA.

b. Review Each Reference Relied on in the Rejection

One or two sentences for each is sufficient. For example: "Reference A (Clark patent 3,925,777) shows a clock having a sequential single-digit readout…[etc.]".

c. Specifically Describe Any Changes and Argue Sec 102 And Then Sec 103.

Discuss specifically how the claim in question has been amended and how it recites structure which physically distinguishes over the references under Section 102. For example, "Claim 1, now rewritten as new claim 5, recites '...' This language distinguishes over reference A under Section 102 because reference A does not show [etc.]. These distinctions are submitted to be of patentable merit under Section 103 because [discuss new results which flow from your novel structure, giving as many reasons as you can from Fig. 13-E, Part 1, and your completed Form 4-2.]." See Inventor's Commandment #5 in Chapter 5. I can't emphasize enough that you should discuss how your invention, *as claimed*, distinguishes over, i.e., has novel physical features not shown in , the reference, not how the reference differs form your invention, and not, at this stage, why your invention is better than the reference. Remember that under Sec 112, a means plus a function is considered a physical recitation.

Also be logical in your arguments. E.g., if you're claiming B and a reference shows A and B, don't argue that A is no good. Also don't argue that a reference should be taken lightly, i.e., it's a "paper patent," because its invention was never put into commercial use *unless* you're *absolutely sure* of your facts.

d.   Refute Any Improperly-Combination of References

If a combination of several references has been cited against your claim, first state why the combination cannot properly be made and then discuss your distinctions under Section 103. For example: *The combination of Smith's lever with Jones's pedal mechanism is submitted to be improper because neither Smith nor Jones suggests such a combination, and one skilled in the art would have no reason to make such a combination. Moreover, the combination could not be made physically because the lever of the Smith type would not fit in or work with Jones's pedal mechanism because...However, even if the combination could be made, Claim 1 distinguishes because the combination does not show [here quote language], and these distinctions are patentable under Section 103 because*

[discuss new results and give as many reasons as you can from Fig. 13-F and Form 4-2].

If the references themselves don't suggest that they should be combined (Reason 20 in Fig. 13-F), you can use the arguments and cases from pp. 7 and 8 of Fig. 13-G; this is a very common defect in many current rejections and the cases cited give powerful arguments.

e.   Note Secondary Factors of Unobviousness

If your invention has achieved any commercial success or has won any praise, this is relevant, and you should mention it here. If possible, submit copies of advertisements for your invention, copies of industry or trade praise, sales figures, a commercially-sold sample, etc. These things reify the invention (i.e., make it a "fait accompli") and impress most examiners.

f.   Draft Any Needed Declaration Under Rule 132 to Refute Technical Points Raised By Examiner

If you want to challenge any technical points raised by the examiner, such as proving that your invention works in a superior manner to a reference, that two references can't be combined, that a cited reference works in a far inferior way to yours, etc., you or an expert in the field should do the necessary research and make the necessary tests (including building and testing a model of the cited reference) and then submit a "Declaration Under Rule 132." The Declaration should have a caption as in Form 13-1 and an appropriate heading, such as "Rule 132 Declaration Regarding Inferior Performance Of Elias Patent." The body of the Declaration should start,

"Jane Inventor declares as follows:

1. I am the inventor [or I am a mechanical engineer (state education, experience, and awards).] in the above patent application.

Then, in numbered paragraphs, detail your technical reasons, including tests you made, etc. but state facts, not conclusions or arguments. Whenever you make any legal declaration or affidavit (as opposed to a brief or remarks), heed the words of the immortal Joe Friday: "Just the facts, m'am." You can attach and refer to "exhibits," i.e., documents in support of your arguments.

Then conclude with a "declaration paragraph," as in the last paragraph of Form 10-C and sign and date the declaration.

g. Request Reconsideration

Then request reconsideration of the rejection(s) and allowance of the claim: "Therefore Claim 5 is submitted to be allowable over the cited references and reconsideration and allowance are respectfully solicited."

If you have dependent claims which were rejected, treat these in the same manner. Since a dependent claim incorporates all the limitations of the parent claim, you can state that the dependent claim is patentable for the same reasons given with respect to the parent claim, and then state that is is even more patentable because it adds additional limitations, which you should discuss briefly.

Discuss each of the other rejections in a similar manner, that is, review the rejection, review the reference, review your new claims, discuss why they distinguish, and request reconsideration and allowance.

h. Responding to Rejections Under Section 112 for Lack of Clarity or Conciseness

If a technical rejection has been made (under Section 112), discuss how you've amended your claim and why your new claim is clear and understandable.

i. Requesting Claim Drafting Assistance from PTO

Once again, I emphasize that if you feel you have patentable subject matter in your application but have difficulty in writing new claims, you can request that the examiner write new claims for you pursuant to MPEP Section 707.07(j). Your remarks are the place to do this. For example, state, "Therefore it is submitted that patentable subject matter is clearly present. If the examiner agrees but does not feel that the present claims are technically adequate, the examiner is respectfully requested to write acceptable claims pursuant to MPEP 707.07(j)." If the examiner writes any claims for you, don't rest on them unless you're sure that the broadest one is as broad as the prior art permits, using the criteria above and in Chapter 5. Remember, if you are dissatisfied with the examiner's claims, you can once again submit your own claims, you can submit the examiner's claims with whatever amendments you choose, or you can interview the examiner to discuss the matter. You should request claim drafting assistance after the first OA, not after a final OA.

j. Repeat the Above for Any Other Rejections in the Office Action

After you've covered and hopefully decimated the first rejection in the manner discussed in subsections a. to j. above, then do the same for each additional rejection.

k. Distinguishing Irrelevant References

If a reference of interest has been cited but not applied against any claim, state that you've reviewed it but that it doesn't show your invention or render it obvious.

l. Conclusion

Last, provide a conclusion which should repeat and summarize, e.g., "For all the reasons given above, it is respectfully submitted that the errors in the specification are corrected, the claims comply with Section 112, the claims define over the prior art under Section 102 [briefly repeat why], and the claimed distinctions are of patentable merit under Section 103 because of the new results [repeat them briefly again] provided. Accordingly, this application is now submitted to be in full condition for allowance, which action is respectfully solicited." Then add the closing, "Very respectfully," followed by your signature, typewritten name, your address, and telephone number on the left-hand side. If you have a co-inventor(s), all of you must sign the amendment.

G. Drawing Amendments

IF YOUR OFFICE ACTION includes any objections to the drawings (or drawing), you must correct these before the case can issue and usually as soon as allowable subject matter is indicated. In addition, if you want to make any voluntary amendments to the drawings, you must get the examiner's approval in advance.

To deal with drawing objections, merely state in the beginning of your Remarks (see Section F above) that the drawing objections are noted and are corrected with new drawings submitted herewith or will be corrected after allowance. To make the corrections you must file new (corrected) drawings for substitution for your original drawings. This is

easy to do. Merely correct your bristol board or Mylar film originals and file new, good xerographic (or CAD output) copies. All lines must be crisp, black and sharp. Use Form 13-3 to submit your corrected replacement drawings.

If you have to correct your drawings, and you're doing so after allowance (the usual case), you should do so promptly after you receive the Notice of Allowance. This will give the PTO's drawing checkers time to review your corrected drawings and let you know if they're still informal within the statutory three-month period to pay the issue fee. If your corrected drawings aren't approved, the PTO will give you until the end of the three-month period, or an additional 15 days, to file proper drawings.

If you want to make voluntary amendments or corrections to your drawings, use the Drawing Amendment Approval Request Form 13-2, filling in all necessary blanks; mention the request and the reasons in the beginning of your Remarks. Indicate your desired changes in red ink on a photocopy of your drawing and attach it to Form 13-2.

Then, if the examiner approves your drawing amendments in the next OA, you can make the changes after (or before) allowance, by filing new, corrected drawings Remember that you can't add any new matter to the drawings. However, you can correct obvious errors, such as a reversed diode, a missing reference numeral, a missing line, etc.

H. Typing and Mailing the Amendment

THE AMENDMENT SHOULD BE TYPED with double- or 1.5 line-spacing on legal- or letter-size paper with 1.5-inch top and 1-inch left, right, and bottom margins. I number my paragraphs and include plenty of boldface or underlined "arguing" headings, e.g., **"The Elias Patent Fails To Show Any Schmitt**

Trigger." Don't forget to keep an identical copy of your amendment mounted in your file; the PTO won't return any paper you send them, although they will make a copy of any paper or record for the per-sheet photocopy charge in the Fee Schedule. Again, I recommend using a word processor or typing the amendment on easily-erasable paper on which you can readily make corrections, and then sending a photocopy of the original, since easily-erasable paper is not accepted by the PTO. Remember that the original signatures of all inventors must be on the copy you send to the PTO.

After your signature, add a "Certificate of Mailing" (don't use Express Mail) as follows:

CERTIFICATE OF MAILING

I hereby certify that this correspondence will be deposited with the United States Postal Service by First Class Mail, postage prepaid, in an envelope addressed to "Box Non-Fee Amendments, Commissioner of Patents and Trademarks, Washington, DC 20231" on the date below.

Date:_____

Inventor's Signature_____

_____

When you use this certificate, you can mail your amendment even at 23:59 on the last day of your response period (it doesn't have to go out on the day it's mailed). Even if you're mailing the amendment two months ahead of time you should use the Certificate anyway since if the amendment is lost in the mail, causing your application technically to go abandoned, you can get it revived easily by filing a declaration stating the full facts—see Rule 8(b). Don't forget to attach a postcard to your amendment reading as in Fig. 13-H.

Amendment A (__pp) in Application of John A. Novel, Ser. Nr. 999,999, filed 1985 Jan. 9, received today:

Fig. 13-H—Back of Receipt Postcard for Amendment

Make sure your amendment won't cause the total number of claims of your application to exceed twenty, or the number of independent claims to exceed three. Otherwise you'll have to pay an additional claims fee (strongly not recommended, since three independent and twenty total claims should be more than adequate).

Checklist for Sending In a Regular Amendment

Before you mail your amendment, please check the following list carefully to be sure that the amendment's complete and properly done.

[ ] A01. All points in the O.A. have been responded to.

[ ] A02. Any needed drawing objection has been responded to.

[ ] A03. The specification has been re-proofed and any needed corrections have been made.

[ ] A04. The prior-art portion of the specification has been amended to account for any significant new prior art (optional.)

[ ] A05. No new matter is included in any amendments to the specification

[ ] A06. All new claims have been checked against the checklist in Chapter 9.

[ ] A07. All new claims recite structure which is physically different from every cited reference (Sec. 102).

[ ] A08. The physically different structure in every claim is sufficiently different to produce new and unexpected results or otherwise be considered unobvious (Sec. 103).

[ ] A09. The case includes several very narrow dependent claims with a variety of phraseologies so that you won't have to present them for the first time if the next action is made final.

[ ] A10. The wording in the remarks is clear, grammatically correct, and understandable.

[ ] A11. The remarks are written in short paragraphs with ample "arguing" headings.

[ ] A12. The patentability of all new claims is argued with respect to the references using a two-part approach: (a) The claim has physical distinctions over the references under Sec. 102. (b) The claimed physical distinctions produce new and unexpected results or are otherwise unobvious under Sec. 103.

[ ] A13. All possible arguments for unobviousness (Fig. 13-f) have been presented.

[ ] A14. A request for claim-drafting assistance under MPEP 707.07(j) has been made, if desirable.

[ ] A15. The amendment is 1.5 or double spaced with an ample top margin.

[ ] A16. The last page of the amendment includes your name, address, and phone number.

[ ] A17. If the amendment will cause the case to have over 20 total or over three independent claims, the proper additional fee is included.

[ ] A18 The amendment is signed and dated in ink by all applicants.

[ ] A19. An identical file copy of the amendment has been made.

[ ] A20. The amendment is being mailed on time or a properly completed Petition to Extend with the proper fee is included.

[ ] A21. A Certificate of Mailing is typed in the amendment.

[ ] A22. All pages are complete and present..

[ ] A23. A receipt postcard is attached to the amendment.

[ ] A24. The envelope is properly addressed and stamped and addressed to "Commisioner of Patents and Trademarks, Washington, DC 20231." Add "Atn: Box Non-Fee Amendments" if you aren't sending any money with your amendment.

I. If Your Application Is Allowable

HOPEFULLY, your first amendment will do the trick and the examiner will decide to allow the case. If so, you'll be sent a Notice of Allowability and/or a formal Notice of Allowance (N/A), the latter accompanied by an Issue Fee transmittal form. You have a statutory period of three months to pay the issue fee; the forms are self-explanatory. Be sure to send in a receipt postcard with your issue fee transmittal. You can also place an advance order for printed copies of your patent (a space is provided on the issue fee transmittal form) at this time; and the minimum order is 10. However the printed copies aren't necessary as you can make photocopies from your patent deed. Also, be sure to attach a Certificate of Mailing to the Issue Fee Transmittal; put it on a separate sheet with the caption of Form 13-1, but headed "Certificate of Mailing".

When you receive your N/A, make any needed drawing corrections at once (see Section F above) and review the application and drawings once again very carefully to make sure everything is correct, logical, grammatical, etc. If you want to make any amendments at this time, you can still do so, provided they don't affect the substance of the application. Generally, only grammatical changes are permitted after N/A. The format of the amendment should be similar to that dictated in Form 13G, except that the first sentence should read, "Pursuant to Rule 312, it is respectfully requested that the above application be amended as follows":

Then make any amendments to your specification and claims in the previously used format. Under "Remarks," discuss the amendments, stating that they are not matters of substance and noting that they will require very little consideration by the examiner.

If you've amended your claims in any substantial way during prosecution, after the Notice of Allowance is received you should also file a Supplemental Declaration (Form 13-4) to indicate that you've invented the subject matter of the claims as amended and that you know of no prior art which would anticipate these claims.

Prior to sending in the issue fee, you should go through the following checklist:

Checklist for Paying an Issue Fee

[ ] A01. All needed drawing corrections have been made.

[ ] A02. Any needed specification or claim amendments have been made (PTO Rule 312).

[ ] A03. Issue Fee Transmittal Form is properly filled out and signed.

[ ] A04. A completed Supplemental Declaration is being filed if any significant claim changes have been made during prosecution.

[ ] A05. Check is attached for correct issue fee amount and signed; adequate funds are on deposit.

[ ] A06. Receipt postcard is attached, stamped, and addressed.

[ ] A07. Certificate of mailing is attached, completed, signed, and dated.

[ ] A08   Papers are mailed by due date (no extensions allowed).

[ ] A09. A file copy of all issue fee transmittal papers has been made.

Once your issue fee is renewed, your application goes to the Government Printing Office and no further changes are permitted.

Several months after the issue fee is paid, you'll receive an Issue Fee Receipt slip, which will indicate the amount of the issue fee you paid and will also indicate the number of your patent and the date it will issue, usually a month after you receive the receipt. A few days after your patent issues, you'll receive the deed, or letters patent, and, separately, any additional printed copies you've ordered. See Chapter 15, Section G for a discussion of maintenance fees.

J. If Your First Amendment Doesn't Result in Patent Allowance

IF YOUR FIRST AMENDMENT doesn't place the application in condition for allowance, the examiner will usually make the next OA final unless it cites any new references which weren't necessitated by new limitations in your amended claims. If your second OA isn't made final, you should respond to it in the same manner as you responded to the first OA. However, if the second OA is called final—and it usually will be—note the provisions of Rules 113 and 116(a) and (b), which govern what happens after a final action is sent:

Rule 113—Final Rejection or Action

(a) On the second or any subsequent examination or consideration, the rejection or other action may be made final, whereupon applicant's response is limited to appeal in the case of rejection of any claim (Rule 191), or to amendment as specified in Rule 116. Petition may be taken to the Commissioner in the case of objections or

requirements not involved in the rejection of any claim (Rule 181). Response to a final rejection or action must include cancellation or appeal from the rejection of, each claim so rejected, and, if any claim stands allowed, compliance with any requirement or objection as to form.

(b) In making such final rejection, the examiner shall repeat or state all grounds of rejection then considered applicable to the claims in the case, clearly stating the reasons therefor.

Rule 116—Amendments After Final Action

(a) After final rejection or action (Rule 113) amendments may be made cancelling claims or complying with any requirement of form which has been made, and amendments presenting rejected claims in better form for consideration on appeal may be admitted; but the admission of any such amendment or its refusal, and any proceedings relative thereto shall not operate to relieve the application from its condition as subject to appeal or to save it from abandonment under Rule 135.

(b) If amendments touching the merits of the application be presented after final rejection, or after appeal has been taken, or when such amendment might not otherwise be proper, they may be admitted upon a showing of good and sufficient reasons why they are necessary and were not earlier presented.

These rules mean, in effect, that "final" isn't final after all. It's just that the rules shift a bit. If you want to continue prosecuting your patent application after a final OA, you must take one of the following actions:

1. Narrow, cancel, or fix the claims *as specified by the examiner.*

2. Argue with and convince the examiner to change position.

3. Try a further amendment narrowing the claims.

4. Appeal to the Board of Appeals and Patent Interferences (BAPI).

5. File a continuation application (see Chapter 14).

6. Petition the PTO Commissioner.

7. Abandon the application.

Let's examine these options in more detail.

1. Comply With Examiner's Requirements

If the examiner indicates that the case will be allowed if you amend the claims in a certain way, for example, if you cancel certain claims or add certain limitations to the claim, and you agree with the examiner's position, you should submit a complying amendment similar to the previously-discussed amendment. However, instead of stating, "Please amend the above application as follows:" (Form 13-1), state "Applicant requests that the above application be amended as follows:" This is because the clerk won't enter any amendments after a final OA unless the examiner gives permission.

Generally no other amendments after a final OA are permitted unless you can show very good reasons why they weren't presented earlier. If your amendment changes the claims in the manner required by the examiner to get them allowed, this will clearly entitle it to entry. You should file your complying amendment as soon as possible, since you have to get the case in full condition for allowance within the three-month period, plus any extensions you've bought. If you file an after-final amount near the end of the 3-month period and the examiner agrees that it places the application in condition for allowance, but the period has expired, you'll have to buy an appropriate extension (Form 13-5): a case can't be allowed when it's technically abandoned. If you file an amendment or argument and it doesn't convince the examiner to allow your case, the three-month period will continue to run.

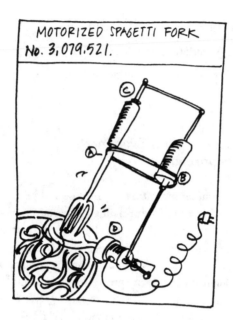

MOTORIZED SPAGETTI FORK
No. 3,079,521.

2. Convince the Examiner

You can try to convince the examiner to change position, either by written argument, by phone, or in person. Try to come to some agreement to get the case allowed. This is often an excellent, effective choice, especially if you have a friendly examiner and· you're willing to compromise. Do this as soon as possible so you'll have time to appeal or file a continuation application (Ch. 14, Sec B).

3. Amendment After Final Rejection

You can try a further amendment, narrowing your claims or submitting other claims, provided no new issues are raised. If the examiner agrees that the amendment narrows or changes the claims sufficiently to place the case in condition for allowance, the examiner will authorize its entry and allow the case. Otherwise the examiner will send you a further OA, called an "advisory action," reiterating

the examiner's former position, and you'll still have the opportunity to exercise the other choices. Even if the examiner disagrees and thus doesn't want to enter the amendment because it raises new issues and allow the case, the advisory action will state that whether you can or cannot still have the amendment entered for purposes of appeal. The examiner will allow it to be entered for appeal if it places the case in better condition for appeal and neither raises any new issues nor requires further search or consideration.

You should file any amendment as soon as possible. The PTO will try to reply to After-Final amendments within one month if you do the following with a *red marker*: (1) mark the upper right of p.1 of your amendment "RESPONSE UNDER 37 CFR 1.116— EXPEDITED PROCEDURE-EXAMINING GROUP NUMBER [insert #]," (2) address the envelope and the amendment "Box AF, Commr. of Pats... [etc.]," and (3) write "BOX AF" in the lower left of envelope.

If you do send in an amendment after final you should head it "Amendment Under Rule 116," request (not direct) that the case be amended as follows to place it in condition for allowance. Also follow the following checklist.

Checklist for Sending an After-Final Amendment

[ ] A01. All points on the checklist for "regular" amendments, except point A09, have been considered.

[ ] A02. The amendment requests (rather than directs) entry of the amendment.

[ ] A03. The claim changes or cancellations either comply with the examiner's requirements or otherwise narrow or revise the claims to obviate the outstanding rejections.

[ ] A04. The remarks state why the claim changes, if any, were not presented before.

[ ] A05. The claims don't contain any new limitations or radical changes which would raise new issues.

[ ] A06. The amendment is being sent in as soon as possible after final action.

[ ] A07. The first page of the amendment and the envelope are marked in red as indicated above.

4. Appeal

If you don't see any further way to improve the claims, and if you believe the examiner's position is wrong, you can appeal to the BAPI (Board of Appeals and Patent Interferences), a tribunal of examiners-in-chief in the PTO. To appeal, you must:

- File a notice stating that you appeal to the BAPI from the examiner's final action;

- Enclose an appeal fee (See Fee Schedule in Appendix);

- File an appeal brief in triplicate, describing your invention and claims in issue and arguing the patentability of your claims. This brief is due within 60 days after you file your notice of appeal;

- Enclose a brief fee;

- If you desire it, request an oral hearing and enclose a further hearing fee (see Fee Schedule); and

- As always, include a Certificate of Mailing.

For more information on appeal procedure, see PTO Rules of Practice 191 to 198.

After you file an appeal brief, the examiner must file a responsive brief (termed an "Examiner's Answer") to maintain the rejection. Prior to doing so, the examiner must have a conference with another examiner and take another good hard look at your case. Often this review will result in changing the examiner's mind. More commonly, the examiner will maintain the rejection and file an Examiner's Answer. You may then file a reply brief limited to any new arguments raised in the Examiner's Answer.

If you do have a hearing, you will be allowed 20 minutes for oral argument, and 15 minutes will be allowed for the examiner's presentation.

If the Board disagrees with the examiner, it will issue a written decision, generally sending the case back with instructions to allow the case. If it agrees with the examiner, its decision will state why it believes your invention to be unpatentable. The Board upholds the examiner in about 65% the appeals.

If the Board upholds the examiner and you still believe your invention is patentable, you can take a further appeal within 60 days of the date of the BAPI's decision to the Court of Appeals for the Federal Circuit (CAFC). As stated, the CAFC is located in Washington, but sits in local areas regularly. If the CAFC upholds the PTO, you can even request the United States Supreme Court to hear your case, although the Supreme Court rarely hears patent appeals. See Chapter 15, Section K for more on the CAFC.

Appeal briefs aren't easy to write, so I suggest you consult professional help if you want to appeal.

If the examiner has issued a ruling on a matter other than the patentability of your claims—for example, has refused to enter an amendment or has required the case to be restricted to one of several inventions—you have another option. Although you can't appeal from this type of decision. you can petition the Commissioner of Patents and Trademarks to overrule the examiner; see "Petitions to the Commissioner," below. You can also appeal this type of decision to the BAPI if you're also appealing a decision on patentability.

5. Continuation Application

If you want to have your claims reviewed further in another round with the examiner, you can file a "continuation application." Filing a continuation application is a relatively simple procedure involving writing new claims, paying a new filing fee, and sending in a special form requesting that a continuation application examination be prepared (see Chapter 14 for how to do this). You'll receive a new serial number and filing date for the purpose of your patent's duration, but you'll be entitled to the benefit of the filing date of your original application for the purpose of determining the relevancy of prior art. Your application will be examined all over again with the new claims. You must actually file the continuation application before the end of the three-month period or any extensions you buy (see Section P below). The Certificate of Mailing (CM) should not be used. According to the PTO's Rules (8 and 10), a CM isn't effective when an application is being filed; you must actually get it physically on file before the other case goes abandoned, unless you use the Express Mail certificate (Chapter 10, Section J, and Chapter 14, Section B).

6. Petitions to the Commissioner for Non-Substantive Matters

The Commissioner of Patents and Trademarks has power to overrule almost anyone in the PTO except the BAPI (Rules 181-183). Thus, if you think you've been treated unfairly or illegally, you can petition the Commissioner to overrule a subordinate. For example, if the PTO's application branch has made a ruling regarding your patent application, such as that it's not entitled to the filing date you think you're entitled to (but not a rejection of your claims), you can petition the Commissioner to overrule this ruling.

If you petition the Commissioner for any reason, you must do so promptly after the occurrence of the event forming the subject matter of the petition, and you must make your grounds as strong and as complete as possible. Generally, most petitions must be accompanied by a verified showing, which means a statement signed by you and either notarized or containing a declaration such as that in the last paragraph of Form 10-2 (Chapter 10, Section G) and a petition fee—(See Fee Schedule) under Rule 17(h)), except in certain specific areas, such as making the application special, where it's under Rule 17(i)).

7. Abandon Your Application

Any action you wish to take in response to a final OA must be made within the three-month period for response or any time extensions you buy (see Subsection 14); otherwise the application will go abandoned. That is, you must either appeal, file a continuation application, or get the examiner to allow your application within the period for response. However, if you're going to file an amendment or an argument, you should do it as soon as possible, preferably within one month, so the examiner's reply will reach you in time for you to take any further needed action within the three-month period.

If all claims of your application are rejected in the final OA, and you agree with the examiner and can't find anything else patentable in your application, you'll have to allow the application to become abandoned, but don't give up without a fight or without thoroughly considering all factors involved.

If you do decide to allow your application to go abandoned, it will go abandoned automatically if you don't file a timely reply to the final action, since the ball's in your court. You'll be sent a Notice of Abandonment advising you that the case has gone abandoned because you failed to reply to an outstanding office action.

K. Interferences

AS MENTIONED IN CHAPTER 5, Section G, an interference is a proceeding conducted by the PTO (a Patent Interference Examiner and the BAPI). An interference is instituted to determine priority of inventorship, i.e., who will get the patent when two or more inventors are claiming the same invention.

The PTO generally institutes an interference when they discover two patent applications claiming the same invention. However, since the PTO is such a large, complex, and populous organization, and since its employees do not always do perfect work, mistakes are sometimes made, and an application that should have been involved in an interference with another application may be allowed to issue as a patent without an interference being declared.

If this occurs and then an examiner or other patent applicant sees the patent and believes it claims the same invention as a pending application, an interference can be declared with the patent, provided the issued patent has not been in force for more than one year.

How is the interference instituted by you, the applicant, if you believe that you, rather than someone else, deserves the patent? Simple. You merely copy (present) the claims of the in-force patent in your application, informing the patent examiner about the patent from which you copied the claims, and showing the examiner how such claims are supported in your application.[4]

Remember, you must copy the claims of any patent within one year after it issues.

On the other hand, if you've been granted a patent, be aware that there may be other patent applicants whose applications contain the same invention as yours. All such applicants have one year from your patent's date of issuance to copy your claims in their applications to get their application into interference with your patent.

Procedurally, an interference is a very complex proceeding. Unless you have an exceptional grasp of patent law and formal advocacy techniques, definitely seek help from a patent attorney who's experienced in trial work. Unlike some of the other situations where I've recommended professional help, representation in an interference proceeding is usually very costly, usually running $10,000 to $25,000 or more.

Despite the need for professional help should an interference occur, there's much you can do on your own to help your case. The boy scout motto will do nicely here: Be prepared. If your application is one of the two percent that becomes involved in interference, sufficient advance preparation will go a long way toward helping your case. As I stressed in Chapter 3:

- Record all steps in your invention development (conception, building, etc.) carefully (Inventor's Commandment #1); and

- Be diligent in building, testing, and recording such for your invention (Inventor's Commandment #2); and

- File a patent application promptly.

Who wins an interference? As briefly stated in Chapter 5, the winner in an interference will not necessarily be the first to file a patent application on the invention. Rather, the first inventor to "reduce the invention to practice" (file a patent application or build and test the invention) will prevail, unless the other party conceives the invention first and has been diligent in effecting a reduction to practice. This means that the typical interference involves lots

---

[4]If you really want to do a bang-up job of patent prosecution, you should find the class and subclass of your patent application (you can find this by calling the clerk of the examining division to which your application is assigned) and then monitoring all patents which issue in the class/sub while your application's pending.

of testimony and introduction of documents by both sides, all for the purpose of proving priority. It's this aspect of the interference that virtually necessitates professional help.

Although there are certain advantages to the US's "first to invent" system, almost all other countries have a "first to file" system, which eliminates interferences and their attendant tremendous expense, complexity, time delays, etc. Some have called the interference laws a "patent attorney's relief act." If you agree, write your Congressperson or have your inventor's club launch an effort to simplify this area of the law.

L. Statutory Invention Registration (SIR)

IF YOU INTEND to abandon your application, but want to prevent anyone else from ever getting a valid patent on your invention, you can have an abstract and one drawing figure of your application published in the OG—Patents (see Chapter 6 and Bibliography) and your application printed like a patent. This is called "converting your application to a Statutory Invention Registration (SIR)." I strongly recommend against ever using a SIR for reasons stated in Chapter 14, Section F.

M. If Your Application Claims More Than One Invention

OFTEN PATENT APPLICATIONS CLAIM several embodiments of an invention, and the PTO will regard these embodiments as separate inventions. The PTO will thus require you to "restrict" the application to just one of the inventions. The theory is that your filing fee entitles you to have only one invention examined. If two of your claims are directed to the same invention, but the examiner

feels that the two claims are directed to subject matter which is classified in two separate subclasses (see Chapter 6), you can be required to restrict the application, that is, to eliminate one set of claims.

Another situation in which restriction may be required occurs when your application contains both method and apparatus claims. Even when both sets of claims are directed to the same invention, the examiner will usually consider them two separate inventions and require you to eliminate either the method or the apparatus claim.

Generally speaking, it's very difficult to fight a PTO-imposed restriction. Fortunately, it's possible to file a second application (called a divisional application—see Chapter 14) if you think pursuing the restricted claims is worth the cost (new filing fee plus new drawings) and if present indications are that your divisional application will comprise allowable subject matter.

Another, related situation occurs when you claim several embodiments or "species" of one invention. In the first OA, the examiner may require you to elect claims to one species for purpose of examination; this is to facilitate her search. If you don't get any generic claim allowed, i.e., a claim which covers all of your different species, you'll be allowed to claim only the elected species; you can file divisional applications on the non-elected species. (In this case, the PTO will consider each species to be a separate invention.) If you do get a generic claim allowed, you'll be allowed claim to up to five different species of the invention (Rule 146).

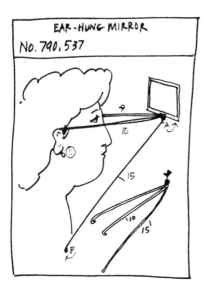

N. Protests Against Allowance of Your Patent Application

MOST OTHER COUNTRIES have a practice under which they permit the public to see pending and allowed applications before they issue in order to give the public a chance to cite prior art or otherwise object to the allowance of the application. However, there's no authorization for this practice in the U.S. Still, the PTO has occasionally instituted voluntary protest programs under which you're given the option of having your application published for protest. This means, among other things, that the confidentiality of your invention is given up since an abstract of your application will be published in the OG. Then copies of your application can be obtained by any member of the public who wants to order them; anyone can then protest against the allowance of your application by citing reasons to show why your invention isn't patentable.

If you do have an opportunity to have your application published for protest, I advise you to elect the procedure—the PTO will give you full instructions—since a patent application which

survives the protest procedure (most do) will become a stronger patent. The disadvantages are delay, that more examination may be required, that members of the public may cite fatally damaging prior art against your application, etc., but I believe the advantages outweigh these disadvantages. You won't lose any trade secret rights you wouldn't otherwise give up since your application was allowed and was going to issue (be published) anyway.

O. NASA Declarations

IF YOUR INVENTION RELATES to aerospace, the PTO will send you a form letter (PTOL-224) after your application is allowed. The letter will state that because your invention relates to aerospace, you'll have to file a declaration stating the "full facts" regarding the making of your invention. This is to be sure NASA has no rights in it. If you don't file the declaration, you wont get a Notice of Allowance. You should draft a brief declaration, along the lines of Fig. 10-K, simply stating that you made the invention on your own time, and with your own facilities, materials, and not in performance of any NASA contract. (One greedy patent agent I know of wanted to charge an inventor $300 to prep are such a declaration.)

P. Design Patent Application Prosecution

DESIGN PATENT APPLICATION prosecution is much simpler than regular patent application prosecution, and, armed with the instructions of this chapter, you'll find it to be duck soup. Design patent application prosecution will never require anything but the most elementary changes to the specification and claims; the examiner will tell you exactly what to do.

To be patentable, the *appearance* of your design, as a whole, must be unobvious to a designer of ordinary skill over the references (usually earlier design patents) which the examiner cites. If your design has significant differences over the cited prior art designs, it should be patentable; if not, you'll have to abandon your application, as there's no way to narrow or change the substance of the claim or drawings of a design patent application. If the examiner rejects your design as obvious in view of a reference, you should use the 102-then-103 attack as explained in Sections F and J and Inventor's Commandment #5 for utility patent applications, i.e., point out the differences in your design and then argue their importance and significance, from an aesthetic viewpoint.

If your design case is allowed, you must pay an issue fee (see Fee Schedule) for a term of 14 years. There are no maintenance fees for a design patent. You cannot convert a design application to a utility application or vice-versa.

Q. What to Do If You Miss or Want to Extend a PTO Deadline

BY STATUTE, you have six months after an OA is sent to file an amendment (35 USC 133) and three months (35 USC 151) to pay your issue fee after the Notice of Allowance is sent. However, the PTO can (and usually does) shorten the statutory period for replying to OA's to three months (30 days if it makes a restriction requirement— see Section L). If you miss any PTO deadline,e.g., the three-month period to reply to an O.A., your application technically becomes abandoned but the PTO won't send you a Notice of Abandonment until after the six-month statutory period expires. If your application goes abandoned, or if you want more time to reply to an OA, it can be "revived" or extended in any of three following ways:

1. Buying an extension;

2. Petition to Revive if delay was "unavoidable";

3. Petition to Revive if delay was avoidable but unintentional.

Let's look at these separately and in more detail.

1. Buy an Extension (Rules 136(a) and 17(a)-(d))

You can reply at any time up to the full six-month statutory period by buying an extension of up to three months[5] at the prices indicated in the Fee Schedule. To buy an extension in this manner (either to get more time to reply to an OA or to revive your case if you forgot to act within the three-month period), merely mail your reply (amendment) by the last day of the first, second, or third extended month with a "Petition For Extension of Time" (Form 13-5) completed as necessary. Make sure you include a Certificate of Mailing on your amendment. You should calculate your total number of months from the date of the OA; don't add your extension months to your original due date. E.g., assume your OA was mailed 1988 Feb 25 so that your three-month prior originally expired 1986 May 28 (May 25 fell on a Saturday and May 27 was a holiday). You want to buy a three-month extension. Your total period is then six months *from February 25*, i.e., to 1988 Aug 25, *not* three months from May 28.

---

[5]Up to four months if the shortened statutory period was less than three months; the fee is very high for the fourth month.

2. Petition to Revive If Delay Was "Unavoidable" (Rules 137(a), or 316(b) and 17(c))

If you failed to send in your amendment or issue fee within the three-month period and your delay was "unavoidable," e.g., you never received the OA, you had a death in the family, a severe illness, your home burned down, etc., you can petition to revive the application. The fee is indicated in the Fee Schedule and you should file three papers (a) your reply, (b) a petition to revive, and (c) a declaration. The petition (use the heading of Form 13-1) should petition to revive the above application, state that the delay was unavoidable because (give the reason), as explained in the attached declaration. The declaration (use heading of Form 13-1 and make last paragraph the same as that of Form 10-3) should state in detail the specific facts which caused the delay. Use numbered paragraphs and start it as follows:

"A.B. declares as follows:

1. I am the inventor in the above application."

Then, give your reasons in short, specific, numbered, factual paragraphs. Refer to and attach copies of any documents you feel are relevant. Your petition and paper must be promptly filed after you become aware of the abandonment. If the case has been abandoned over six months, you must also include a fourth paper disclaiming the terminal part of the term of any patent granted on the application for a period equal to the period of abandonment and include a terminal disclaimer fee (Rules 137(c), 321, and 20(d)). If your petition under this paragraph is denied, you can still petition under the next paragraph if you do so within three months.

3. Petition to Revive If Delay Was Avoidable but Unintentional (Rules 137(b) or 316(c) and 17(m))

If you failed to send in your amendment or issue fee within the three-month period and your delay was "avoidable but unintentional," e.g., you merely dropped the ball, misinterpreted the time to reply to the OA, etc., you can still petition to revive the application, albeit at a much higher cost. You should file three papers:

a. Your reply;

b. A petition to revive (same as petition in preceding paragraph, except state the delay was "unintentional"); and

c. A declaration similar to that of the preceding paragraph, except you need merely state that the abandonment was unintentional (no reason is needed—the stiff fee (see Fee Schedule) is ample). Your petition must be filed within one year of the date it went abandoned.

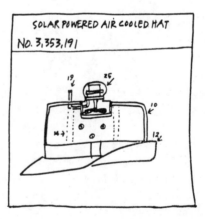

Chapter 14

YOUR APPLICATION CAN HAVE CHILDREN

A. Available Extension Cases

AS WE SAW IN CHAPTER 13 (application prosecution), the patent laws and PTO rules allow you to do much more than either getting a patent or abandoning the application. In this sense, perhaps, a patent application can best be understood by comparing it to a family tree, as shown in Fig. 14-A. The basic application is like a parent, and just like a parent has children, the parent application can be used to produce offshoots. Depending on the situation, the parent application is called by many names (e.g., "parent," "prior," "basic," or "original" application), while the new applications are referred to as "daughter," "continuation," "divisional," "reissue," "independent," "substitute" applications. If there are several successive extensions, the basic application is called the "grandparent" or "great-grandparent" application and the latest-filed application can be called a "granddaughter," "great-granddaughter," "continuation-of-a-continuation," etc., application.

Fig. 14-1 shows all of the different extensions which you may file. Note that some extensions come from the bottom point of the Basic Application (BA) or the basic patent, these are "sequential" extensions since they replace the BA or its patent. Other extensions come from the sides of the BA; these are "parallel" extensions since they can exist in addition to the BA or its patent..

- If your basic application was held to cover two or more inventions, and you've had to restrict it to one of these inventions, and you want to file a separate application on the other or "non-elected" invention, you should file a *divisional application* (left side of chart). As indicated, your divisional patent can be in addition to your original patent.

- If you want another round with the examiner, or a chance to try a new and different set of claims after a final Office Action (OA), you should file a *continuation application*.

- If you can't get or for some reason don't want a patent once you've filed your application, but want to be sure no one else will ever get a patent on the invention, you can have the PTO publish your patent application by converting it to a *Statutory Invention Registration* (SIR) or you can have your invention published as an Independent Defensive publication.

- If you've received an original patent (middle of Fig. 14-A), but you want to revise the claims of the patent or correct significant errors in the specification for some valid reason, you should file a *reissue application*. As indicated, your reissue patent takes the place of your original patent.

- If you've improved your basic invention in some material way during the pendency of your application, and you want to obtain specific claims to the improvement, you should file a *continuation-in-part* (CIP) application (right side of chart). As indicated, your CIP patent can exist with your original patent.

- If you abandon your application and later refile a new application on the same invention, the new application, which, as indicated by the broken line, has no copendency or continuity with the original application, is termed a *substitute application*. Of course no patent on your original application is possible.

- If you've made a major improvement in your basic invention which uses new concepts and can really stand by itself, you should file an *independent* application.

Now that I've identified the major types of patent applications, its time to examine each one in more detail. Before we do, however, a word of advice. As suggested in Chapter 13, the types of problems that will occasion your using the information in this chapter may make it appropriate for you to at least consult with an expert prior to making a decision. In other words, before you decide file a continuation, etc., you should seriously consider seeing a patent

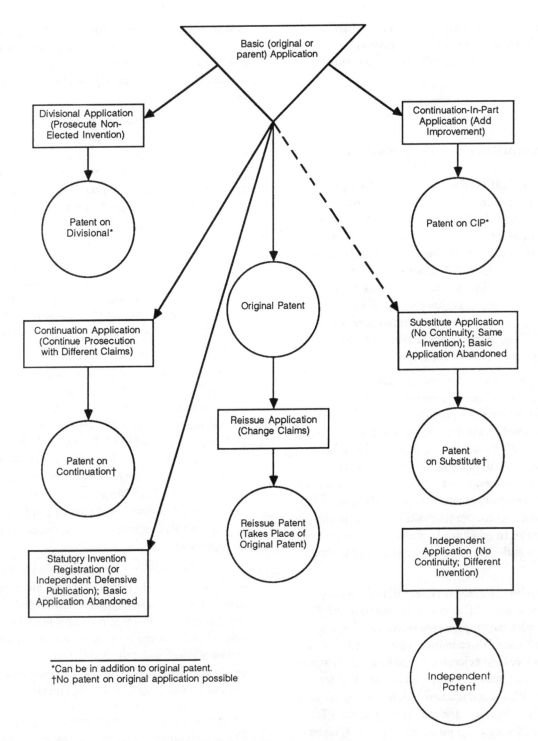

*Can be in addition to original patent.
†No patent on original application possible

Fig 14-A—Available Extension Cases

lawyer. Also note that, of necessity, the chart is abbreviated (it doesn't cover extensions of extensions), so rely primarily on the text, rather than the chart.

B. Continuation Applications

A CONTINUING OR CONTINUATION application is concisely defined in the *Manual of Patent Examining Procedure* (MPEP), Section 201.07, as "a second application for the same invention claimed in a prior application and filed before the [prior application] becomes abandoned." A continuation application is almost always filed in response to a final rejection when you want to have another round with the examiner. If you don't file it within the response period (three months unless extended for a fee), you give up your right to file it at all.

If you think that it's inconsistent for the PTO to allow you to continue prosecuting claims to an invention after it has supposedly declared an office action "final," a word of explanation is in order. The word "final" is a word of art, meaning that it has a special, unusual meaning. A "final" action doesn't mean that the examiner has given the final word on your *invention*, but merely has decided to cut off your right freely to change your claims in your *current application*. In other words, you've gotten as many go-arounds as they're going to give you for your filing fee.

An historical explanation will make it even clearer. In the "old" days when I worked in the PTO (early 60s), patent prosecution proceeded at a leisurely pace. We examiners were allowed to send four or five OAs before we had to issue a final action. We issued a final OA only after an issue had been clearly defined and reached, or if it was a fourth or fifth OA. However, since the late 1960s, the PTO instituted a "compact prosecution" practice; under this practice the examiner is almost always supposed to make the second OA final. The purpose of this

change was to obtain more income for the PTO (a continuation application gets the PTO an additional filing fee), to reduce the amount of work the PTO performed, and to shorten the backlog of pending applications.

However, two OAs are often not enough to define the invention adequately, reach an issue with the examiner, and complete the prosecution in a proper manner. Therefore, continuation applications are often filed nowadays, especially since the process has been made very simple by a new procedure, called the "File Wrapper Continuing" (FWC) procedure.[1]

When you file an FWC continuation, the PTO uses your same file jacket and papers and merely assigns a new serial number and filing date. The procedure is covered by the PTO's Rule 62 and an FWC Request Form which I've provided as Form 14-1.

A continuation application must cover the same invention as the parent or basic application, and the parent or basic application must be abandoned when the continuation is filed. The continuation application is entitled to the benefit of the filing date of the parent application for purposes of overcoming prior art, although, as stated, the continuation receives its own filing date and serial number. The continuation's separate serial number and filing date are solely for PTO administrative purposes and have no legal effect.

You can also file a continuation of a continuation application. In fact, it's theoretically possible to file an unlimited sequence of continuation applications. But note that if an issue has been reached in the parent application, the examiner can, and usually will, make the first OA in a continuation application final. In other words, each continuation application will be quickly rejected unless you truly

---

[1]There are other ways to file a continuation, division, CIP, etc., but they're not at all needed, except as covered below, in view of the flexibility of the FWC procedure.

come up with a different slant on or definition of your invention not previously considered by the PTO.

When a patent issues on a continuation application, the heading of the patent will indicate that it's a continuation application. Also, the filing date and serial number of the parent (original, basic) application will be given in order to apprise the public of the patent's effective filing date. To file an FWC continuation application, do the following:

1. Complete Form 14-1;

2. Attach a new filing fee (large or small entity—see Fee Schedule in Appendix); and

3. Attach a preliminary amendment containing the new claims you desire to prosecute or check the appropriate block on Form 14-1 if you want to have your Amendment under Rule 116 from the parent case entered.

To complete Form 14-1, check "continuation" in paragraph 1 and fill out all the other self-explanatory blanks in the form. As always, attach a receipt postcard (see Chapter 10). For the Preliminary Amendment, use Form 13-1. You must either get your FWC Request in before the period for response to the final rejection expires (Or any extensions you've bought—see Chapter 13.) or you can mail your FWC Request on the last day of the period for response if you use the Express Mail Certification at the bottom of Form 14-1, as explained in Chapter 10. (*Don't* use the Certificate of Mailing of Chapter 13.) Send the papers to the usual PTO address, "Attention Box FWC."

You should complete Form 13-1 exactly as you would do with a regular amendment (see Chapter 13), with the following three exceptions:

1. Leave the serial number and filing date lines blank;

2. Below the caption, insert the serial number and filing date of the parent as follows: "(A continuation of Ser. Nr. ____,____, Filed 198_ ___ ___)";

3. Amend the specification by adding the following sentence before line 1 on page 1: "This is a continuation of application Ser. Nr. ___,___, filed 198_ ___ ___." (Insert the serial number and filing date of the parent case.)

You should then proceed as usual: cancel the old claims and insert the new claims you desire in the normal-amendment manner, numbered in sequence after the highest numbered claim of the prior application. Under "Remarks," you should state, "The above new claims are being submitted as part of the continuation application; these claims are submitted to be patentable over the art of record in the parent cases for the following reasons." Then give your reasons and arguments in the same manner as you would for a regular amendment, as stated in Chapter 13.

Be sure to include all the claims you desire in the preliminary amendment since the first OA in the continuation may be made final if the examiner doesn't cite any new prior art.

Note that when you file a continuation application, the PTO will transfer the drawings from your parent application and abandon it as of the date of transfer.

As with a regular application, you'll receive your postcard back with the stamped serial number and filing date of your continuation application. Thereafter you'll receive a blue filing receipt and first office action in due course.

If the claims that are finally allowed in a continuation application, or divisional application (see Section B below) differ significantly from the claims originally presented in the parent application, you should file a Supplemental Declaration (Form 13-4) before or when you pay the issue fee.[2]

---

[2]In Chapter 13, I pointed out that whenever you change claims significantly during prosecution, you should file a new declaration to cover the changes.

Note on Changing Examiners: If you feel that the examiner in your parent case was unduly tough, it may be possible to get a different examiner in your continuation case by claiming your invention differently. The examining division to which a patent application is assigned is determined by the class and subclass to which the application is assigned, and this is in turn determined by the subject matter of the narrowest (longest or most specific) claim in the case.

Example: Suppose you've invented a new gear for a bicycle and the narrowest claim of your parent case recites the fine details of the gear per se. Your case will be assigned to an examining division in the "gear" arts. If your "gear" examiner is a hardnose, you'll probably be able to get it into bicycles, a different examining division, by adding the bicycle to your narrowest claim. You can do this by providing a "bicycle" preamble for the claim (see Chapter 9) or by actually reciting other parts of the bike in the body of the claim. This may sometimes run you into an "old combination" rejection (Chapter 10), but it's worth a try. If your narrowest claim is directed to a bike, your whole case will be classified in the bike division, and you'll have a different examiner.

Obviously, this maneuver can't be done in every instance, and you should do some research on the PTO's Examining Division art assignments (see the status page of any recent *Official Gazette*—Fig. 13-A) to make sure your end run around a particular examiner will work. Lastly, in an effort to get a new examiner, it also helps to change the title of your invention to one which is commensurate with your revised narrowest claim, e.g., "change "Gear with Anti-Backlash Pawl" to —Bicycle Pedal Drive Gear—."

Warning! This is one of several situations where I believe a consultation with a patent attorney or agent may be called for, due to the artsy nature of claims drafting (see Chapter 6, Section F).

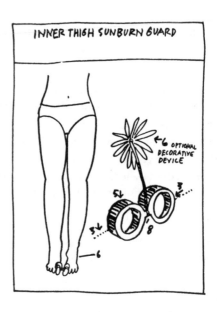

INNER THIGH SUNBURN GUARD

C. Divisional Applications

NOW LET'S TURN OUR ATTENTION to the divisional application.

A divisional application or "division" is "a later application for a distinct or independent invention, carved out of a pending application and disclosing and claiming only subject matter disclosed in the earlier or parent application." (MPEP 201.06). Divisional applications are filed when the PTO decides that two separate or distinct inventions have been claimed in the parent application (not permitted, since your filing fee entitles you to get only one invention examined), and you've agreed to restrict the parent application to the set of claims to one of the inventions. You have the option to file a separate, divisional application on the claims to the other invention. Divisional applications are so called because they cover subject matter which is "divided out" of the parent case.

A divisional application is entitled to the filing date of the parent case for purposes of overcoming prior art, although, like the continuation application, it receives its own serial number and filing date for

PTO administrative purposes. A patent issuing on a divisional application will show the serial number and filing date of the parent application; this will be the divisional's effective filing date. The parent application of a divisional application can either issue as a patent or become abandoned if you feel it's not patentable over the prior art. But remember that the divisional application must be filed while the parent is pending. Also note that you can file a division of a continuation application, and a continuation of a divisional application. (Definitely consult an expert if you get into these murky waters.)

1. Use File Wrapper Continuing Procedure If the Parent Case Is To Be Abandoned

If the parent application is to be abandoned, e.g., if you give up on the invention you elected, and unsuccessfully prosecuted in the original case, you can file a sequential divisional application very easily by using the FWC procedure, i.e., Form 14-1. Merely check the "Division" block in paragraph 1 of the form, and check and fill in the appropriate "drawing" blocks at the end of paragraph 1—most of the time you'll be using the same drawings of the parent case, but if you want to submit new drawings, e.g., to delete extraneous figures to an abandoned embodiment, you can do so at this time. Check the other blocks as appropriate.

As with the continuation, you should enclose a preliminary amendment to the daughter (here divisional) case. Use Form 13-1. Amend the specification by inserting, at the beginning, the following sentence: —This is a division of Ser. Nr. ___,___, Filed 198_ ___ ___, now abandoned.— Amend the claims by cancelling all of the claims to the elected invention of the parent case and submitting or leaving only claims to the divisional invention.

For example, suppose the parent case had claims 1 to 20, of which claims 1 to 10 were directed to the elected invention of the parent case (on which you couldn't get a patent) and claims 11 to 20 were to the non-elected invention which you now want to prosecute in the divisional case. Your amendment would simply read: "CLAIMS: Cancel claims 1 to 10." This will leave only claims 11 to 20 pending. Alternatively, you can cancel all of the claims of the patent case and submit entirely new claims in your divisional invention.

In the "Remarks" section, you can summarize what you've done and briefly point out why you think the divisional's claims are patentable. And don't forget the postcard and filing fee!

2. File a Complete Copy of the Divisional Application If Parent Case Won't Be Abandoned

If you're not abandoning the parent case, e.g., if a patent is going to issue on the parent, and you want to get another, parallel patent on the invention which is the subject matter of the divisional application, things get a bit more complex, but they're still bearable. Instead of using the FWC procedure (Form 14-1) you'll have to file a complete copy of the divisional application, including a transmittal letter (Form 10-1), drawings (see below), filing fee, specification, claims, and abstract, PAD (Form 10-2), and SED (Form 10-3), postcard, and optionally, a Preliminary Amendment. Everything should be the same as if you were filing a completely new application (use the checklist of Ch. 10), with the following exceptions:

a. Add the following sentence to the transmittal letter: "This is a division of Ser. Nr. ___,___, Filed 198_ ___ ___."

b. Add the same sentence to paragraph 1 of the specification, either by typing it in the actual specification or by way of a Preliminary Amendment if you're using a copy of the specification from the parent case. Later on, after your patent on the parent case issues, add the following to the end of this sentence: "—, now patent 4,___,___, granted 198_ __ __.—.

c. Delete any non-applicable figures from the drawing, i.e., any figures directed exclusively to the embodiment of the parent case.

d. Amend the specification, either directly on the copy you file, or by a Preliminary Amendment, to remove any matter directed exclusively to the embodiment or invention of the parent case, and to make any editorial amendments you desire or which you've made in the parent case.

e. Add the sentence of "1." above to the Patent Application Declaration Disclosure Form (Form 10-2).

To supply drawings for the parallel divisional case, you have two choices:

a. If you've made formal Mylar film or bristol board originals of your drawings, you can file very good xerographic copies of these for your divisional's formal drawings.

b. You can file rough xerographic copies as informal drawings and can file formal drawings later, as explained in Chapter 10.

Double Patent Warning: You're not permitted to obtain two patents on the same invention; if you do, it's called "double patenting," a situation in which both patents may be held invalid. However, if in your parent case the examiner required you to restrict the application to one of several inventions, due to a special statute (35 USC 121), you can file your divisional(s) on the non-elected invention(s) with total immunity from double patenting. However, if the examiner didn't require you to restrict, and you're filing your divisional "voluntarily," you must be sure that it's to a patentably different invention than that claimed in the parent case; otherwise, both patents can be held invalid for "double patenting."

Attorney Note: Once again, I recommend that you consult with a patent attorney in the event you

(or the PTO) decide that a divisional application is indicated.

D. Continuation-In-Part And Independent Applications

AS DEFINED IN MPEP 201.08, "a continuation-in-part" (CIP) is an application filed during the lifetime of an earlier application by the same applicant, repeating some substantial portion or all of the earlier application and adding matter not disclosed in the earlier application. CIP applications are not common; they're used whenever you wish to cover an improvement of your basic invention, for example, if you've discovered a new material, better design, etc. (Remember, you can't add these to a pending application because of the proscription on "new matter" discussed in Chapter 13.)

Generally, the parent application should be allowed to go abandoned when a CIP is filed, but if you do want the parent application to issue, you must be sure that the claims of the CIP application are patentably different—that is, they define subject matter which is unobvious over that of the parent application—or else the CIP and parent application patent can both be held invalid for double patenting.

The advantage of a CIP application over a separate application is that the CIP is entitled to the filing date of the parent application for all subject matter common to both applications. However, if any claims of the CIP cover subject matter unique to the CIP, such claims are entitled to the filing date of the CIP only.

If your "improvement" of your basic application is different enough to be unobvious over the basic invention, an entirely separate, independent, application, rather than a CIP, can be filed, but it's better to use a CIP application since the common subject matter gets the filing date of the parent application.

CIP Example 1: Suppose you've invented a bicycle gear with a new shape. You've claimed this shape in a patent application, which I'll call the parent application. After you file the parent application, your research shows you that the gear works much more quietly if it's made of a certain vanadium alloy (VA). The VA isn't patentable over the invention of the parent case and your parent case's claims cover the gear no matter what material it's made of, but since the VA works much better, you'd like to add a few dependent claims specifically to cover a gear made of the VA. In this way, if there's an infringer who copies your gear made of the VA, you can show the judge that the infringer is infringing your specific as well as your broad claims and you may have specific claims to VA to fall back on if your broad claims are held invalid. You can't add the VA to the specification or the claims of the parent case since it would be verboten "new matter." The solution: file a CIP, describing the VA in the specification, and add a few dependent claims which recite that the gear is made of VA. To avoid any possibility of double patenting you should abandon the parent case since the VA isn't patentable over the invention of the parent case. For purposes of clearing the prior art, your broad claims to the gear shape per se will get the benefit of the parent case's filing date and the claims to the gear made of the VA will be entitled to the later filing date of the CIP.

CIP Example 2: On the other hand, suppose your gear shape works well, but you've come up with a related, but unobviously-different shape which works better, i.e., the new shape is patentable over the invention of the parent case. You would file a CIP with claims to the new shape and continue to prosecute the parent case to a patent. The CIP's claims generally will be entitled to only the CIP's filing date, but their CIP status will entitle you to refer back to the parent's filing date to show when you came up with underlying concept common to the parent and CIP gears in case the CIP is ever involved in litigation or an interference.

CIP Example 3: Lastly, suppose your gear shape works well, but you come up with an unrelated, and unobviously-different shape which works better. You would file a new, independent, application, not related to the "parent," with claims to the new gear shape. The two applications would be entirely separate.

You can file a CIP of a continuation or divisional application or vice-versa in either case. It's also theoretically possible to file an unlimited number of successive CIP applications to cover successive improvements; there have been rare cases where inventors have filed chains of CIPs with as many as eight or more applications, each of which issued into a patent.

As with divisional applications, you can use the File Wrapper Continuing procedure for your CIP if you're abandoning the parent case, but if the parent will issue, you'll have to file a complete new set of papers for the CIP.

If you're allowing the parent case to go abandoned, i.e., you're filing a sequential CIP, you should:

- File the CIP using Form 14-1, in the same manner as you would file a FWC divisional, except check the "CIP" space in Form 14-1;

- Provide a Preliminary Amendment referring to the parent case, and with the new subject matter and claims you desire to add;

- Add any new drawings you desire;

- Attach a new declaration; use Form 13-4, which should refer to your Preliminary Amendment by its date; and

- Don't forget the postcard and filing fee.

If your parent case will issue, i.e., you're filing a parallel CIP, you can't use the FWC procedure or Form 14-1. Instead use the same procedure as outlined above for filing a divisional when the parent case will issue, except substitute "continuation-in-part" for "divisional" in the transmittal letter and specification and add the following sentence to your

PAD (Form 10-2): "This application in part discloses and claims subject matter disclosed in my earlier filed pending application, Ser. Nr. ___,___, Filed 198_ ___ " Don't forget the filing fee, postcard, and SED (Form 10-3). Also use the checklist of Ch. 10. The new subject matter of the CIP and any claims directed to it will be entitled to the CIP's filing date, not the filing date of the parent case.

If you're filing an independent application (rather than a CIP), do it in the usual manner (Chapters 7 to 10) except that you can add a heading and sentence to the specification as follows:

Cross-Reference To Related Application: This application is related to application Ser. Nr. ___, Filed ___, now patent Nr. ___, granted ___.

E.   Reissue Applications

AS STATED IN MPEP 201.05, "a reissue application is an application for a patent to take the place of an unexpired patent that's defective in some one or more particulars." Parts 1400 to 1401.12 of the MPEP discuss reissue applications extensively. If you've received a patent and believe that the claims are not broad enough, that they're too broad (you've discovered a new reference), or that there are some significant errors in the specification, you can file an application to get your original patent reissued at any time during its 17-year term. The reissue patent will take the place of your original patent and expire the same time as the original patent would have.

However, if you wish to broaden the claims of your patent through a reissue application, you must do so within two years from the date the original patent issued. Moreover, anyone who manufactures anything between the issue dates of the original patent and the reissue patent that infringes the broadened but not the original claims is entitled to "intervening rights." These preclude a valid suit against this person for infringement of the reissue patent's broadened claims.

Example: Suppose you invent a new gear shape and get a patent, but unfortunately you included an unnecessary limitation in your independent claims, namely they all recite that the gear is made of carbon steel. If you discover your error within a two-year period after your patent's issue date, you can file an application to reissue the patent with broader claims, i.e., claims which specify only the gear's shape and not their material. Your patent will be reissued with the broader claims. However, suppose that an infringer (Peg) made gears with your inventive shape, but out of aluminum, between the date of your original and reissue patents. Peg's aluminum gears would not infringe the claims of your original patent, but they would infringe the broader claims of the re-issue. Nevertheless, Peg can continue to make her aluminum gears with impunity since she has "intervening rights" by virtue of her manufacture of the aluminum gears in the interim.

To file a reissue application you must:

1. Reproduce the entire specification of the original application (a copy of the printed patent pasted one column per page is acceptable), putting brackets around matter to be cancelled and underlining matter to be added. When the reissue patent issues, it will include the brackets and underlining;

2. Supply a request for a title report on the original patent (see Fee Schedule for amount) and offer to surrender the original patent deed; and

3. Provide a detailed showing in your declaration as to why you believe the original patent to be wholly or partially inoperative or invalid; see Patent Rules 171-179.

Reissue patents are relatively rare and are identified by the letters "RE" followed by a five-digit number, for example, "Patent RE 26,420."

Due to the relatively complicated nature of reissue applications, I suggest that you consult a patent lawyer here also.

F. Statutory Invention Registration

IF YOU'VE FILED A U.S. APPLICATION and for some reason don't wish it to issue as a patent or can't obtain a patent on it, but want to be absolutely sure that no one else will ever be able to obtain a patent on it (for instance, you're manufacturing a product embodying the invention), you can elect to have an abstract of your application published in the *Official Gazette* and have your entire application published like a patent. Called a "Statutory Invention Registration" (SIR), this purely defensive procedure will cause your invention to become a prior-art reference, effective as of its filing date. The SIR will thus preclude anyone else from obtaining a patent on the invention, provided no application on the invention was filed earlier than yours. Your application will then be printed and published like a patent, but you won't have any monopoly rights. (You will retain the right to revive your application and get into interference if a patent or application is discovered which claims your invention.)

I don't recommend use of the SIR procedure because of the generally higher fee required—it's cheaper to publish your own book about your invention or to list it with an Invention Register, such as ITD, Inc., P.O. Box 371-0371, Tinley Park, IL 60477; Technotec, 8100 34th Ave. South, Minneapolis, MN. 55440 ($160), or Research Disclosure Magazine, Industrial Opportunities, Ltd., Homewell, Havant, Hampshire, PO9 1EF (about $100). If you have your invention published this way, the effective date of publication will be later than your filing date. However, the cost is generally much less and the later date won't make any difference

unless someone has filed on the same invention before publication.

If you do choose to convert your application to an SIR, follow PTO Rules 293-297 and 17(n) or (o).

G. Substitute Applications

THE TERM "SUBSTITUTE" is defined in MPEP 201.09 as "an application which is in essence a duplicate of an application by the same applicant which was abandoned before the filing date of the later case." A substitute can be filed for the same purpose as either a continuation, division, or CIP.

I hope you never have to file a substitute application, since it doesn't get the benefit of the filing date of the earlier case. This is because it wasn't filed while the earlier case was pending. Thus any prior art which issues after the filing date of the earlier case and before the filing date of the substitute case is good against the substitute case. If, however, you somehow allow your application to become abandoned and you can't successfully petition the Commissioner of Patents to revive the application (see Chapter 13), you still may be able to cover your invention by filing a substitute application, assuming significant prior art hasn't been published in the meantime.

There are no special forms or procedures for filing a substitute application; just file it like you would a regular patent application, except that you can add a reference in the specification to the "parent" case. As stated, you won't get the benefit of your "parent" case's filing date, but the date of the parent case may be useful if you ever have to swear behind a reference (see Chapter 13) or prove earlier conception and/or reduction to practice, e.g., in case of an interference (see Chapter 13, Section J).

Chapter 15

After Your Patent Issues: Use, Maintenance and Infringement

A. Always on Tuesdays

SEVERAL MONTHS AFTER YOU PAY the issue fee (Chapter 13), you'll receive an Issue Fee Receipt. This will indicate the amount you paid, and the number and issue date of your patent (about one or two months after you receive the receipt). On the issue date, which will almost always be a Tuesday the patent will be granted, published, and mailed to you so that several days later you'll receive your patent deed (also called "letters patent"). This consists of a copy of your patent on stiff paper, a fancy jacket, rivets, seal, and ribbon. You'll also receive (separately) the printed copies of your patent if you ordered them when you paid your issue fee. The highlights of your patent will also be listed in the *Official Gazette Patents* which is also almost always published on the Tuesday of grant.

B. Press Release

YOU MAY WISH, when you learn the number and date of your patent, to prepare press releases about it. See any book on advertising to learn how to prepare a press release; it should cover the five "W's" of reporting: Who, What, When, Where, and Why. Make it interesting and catchy, e.g., "Midgeville Gets Patent On Jam-Free Bike Mechanism." If you have an interesting or important invention, send a letter or copy of your PR (as soon as you get the receipt) to Mr. Edmund Andrews, N.Y. Times Patent Columnist, 229 West 43d St., New York, NY 10036; they may mention your patent in their regular Saturday column when your patent issues. Also send the PR to your local papers and trade magazines (each with a copy of your patent) on the day you get the patent.

C. Check Your Patent for Errors

FIRST, PROOFREAD YOUR PATENT carefully, preferably out loud with a friend or co-worker. Carefully examine the information in the heading of the patent—serial number, filing date, title, your name, etc.—to make sure all is correct. Then read the patent word for word and compare it with the the application in your file as amended during the prosecution phase.

If you find errors, you have several possible courses of action.

1. If the Errors Aren't Significant

If the errors aren't significant, that is, if the meaning you intended is obvious and clear, the PTO won't issue a Certificate of Correction, but you should make the error of record in the PTO's file of your patent. To do this, simply write a "make of record" letter to be put in the file of your patent, listing the errors you found. This letter should be captioned similarly to Form 15-1 (see below) with the patent number, issue date, and patentee(s) name(s) and should be headed, "Notation of Errors in Printed Patent." It should then list all the errors in the patent.

2. Certificate of Correction

If any of the errors you discover are significant, that is, if the meaning is unclear because of a wrong reference numeral, missing or transposed words, failure to include a significant amendment, etc., you may obtain a Certificate of Correction. If the errors are the fault of the printer, the Certificate of Correction will be issued free. If the errors are your fault, that is, they appear in your file as well as in the printed patent, you still can get a Certificate of

Correction, provided the error is of a clerical or minor nature and occurred in good faith. Examples are a wrong reference numeral, an omitted line or word, etc. The fee for a Certificate of Correction to fix your error is listed in the Fee Schedule in the Appendix. To obtain one (in either case), do the following:

Step 1: Fill out Forms 15-1 and 15-2. In Form 15-1 (the request letter), insert the patent number, issue date, patentee(s), Ser. Nr., filing date, and the date you mailed the form. Check paragraph 2 if the error is the fault of the PTO; check paragraph 3 and insert the amount from the Fee Schedule if the error is your fault.

In either case (whether you checked paragraph 2 or 3), in paragraph 4 below paragraph 3 list the places in the application file where the errors occurred and explain who was at fault, e.g.:

"4. Specifically, on p. 4, line 12 of the specification, applicant erroneously typed '42' instead of '24' and neither applicant nor the examiner detected this error during prosecution."

or

"4. Specifically, on p. 4, line 12 of the specification, the reference numeral '24' has been erroneously printed by the GPO in the patent as '42' instead of '24.'"

Step 2: Then complete the caption of Form 15-2 with the patent number, issue date, and inventor(s). (The PTO also furnishes carbon sets of the Certificate of Correction form gratis). In the body of Form 15-2, make the necessary corrections as if you were making an amendment (see Chapter 13) to the actual printed patent, e.g.:

"Col. 3, line 54, change "the diode" to —varistor 23—."

Put your return address and the patent number on the bottom of Form 15-2.

Step 3: Send one copy of completed Form 15-1 and *two* copies of completed Form 15-2 to the PTO

with a receipt postcard, and a check for the correct amount if the error was your fault. You'll get an approved copy of your Form 15-2 back in several months and the PTO will affix copies of it to the copies of your patent that it maintains in its storage facilities.

D. Patent Number Marking

IF YOU ALREADY HAVE SALES BLURBS promoting your invention, change them to indicate that your invention is "patented" rather than "patent pending." If you, or a licensee of yours, is manufacturing a product embodying the invention, you should consider marking your product with the patent number.

A section of the patent laws (35 USC 278) states that products embodying a patented invention may be marked with the legend "Pat." or "Patent," followed by the patent number. If you make or sell products embodying your invention which are properly marked, you can recover damages from any infringers you sue from the date you began marking. If you make or sell products but don't mark them with your patent number, or mark them "Patented" without the number, you can recover damages only from the date you notify the infringer of infringement, or from the date you file suit against the infringer, whichever is earlier.

The actual marking should be done on the product itself, on its package, or by means of a label affixed to the product. If you don't manufacture any product embodying the invention, or if the invention relates to a process that's not associated with a product and hence can't be marked, you can recover damages from an infringer for the entire period of infringement without marking.

The disadvantage of patent marking is that any sophisticated person who wants to copy your product can easily see the number of your patent, order the patent, read its claims, and attempt to design around your claims or some other aspect of your patent. If you don't mark your product, the potential infringer can still probably get this same information, but only through a lot more expense and effort. In other words, by not marking, you may depend in part on human inertia to protect your invention from being copied. Many companies, therefore, favor *not* marking their patented products, or simply marking them "Patented" without including the number, relying on their own familiarity with the field to enable them to quickly spot and promptly notify any infringer of the existence of the patent.

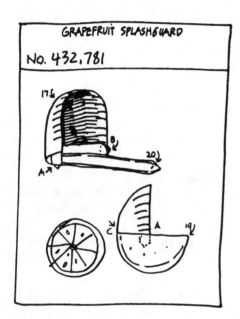

E.   Advertising Your Patent for Sale

IF YOU STILL HAVEN'T LICENSED or sold your invention by the time your patent issues, you can advertise the availability of your patent for license or sale in one or more of several publications, such as:

- *Patent Official Gazette*, Commissioner of Patents and Trademarks, Washington, DC 20231;

- The International Invention Register, P.O. Box 547, Fallbrook, CA 92028;

Write for listing information and fees.

F.   What Rights Does Your Patent Give You?

NOW THAT YOU'VE actually obtained a patent, you'll undoubtedly want to know exactly what rights you receive under it. While I've indicated earlier that a patent provides a 17-year monopoly on the manufacture, use, and sale of your invention, I'll now specifically discuss what this means in the real world.

1.   Enforceable 17-Year Monopoly on Manufacture, Use, and Sale

The grant of a patent gives you, or the person or corporation to whom you "assigned" (legally transferred) your patent or patent application, a 17-year monopoly on the invention *defined by the claims of the patent*, beginning with the date of issuance of the patent. This monopoly gives you the right to bring a valid suit against anyone who makes, uses, or sells your invention in the US.

You can use your ownership of the patent to make money in either of three ways:

1. Sell the patent outright

2. License others to make, use, and/or sell the patented invention in return for royalties under a variety of conditions, subject to the antitrust laws mentioned in the note below. See Chapter 16 for a more detailed discussion about the sale and licensing of patent rights

3. You can also use your patent to create a monopoly by preventing anyone else from making, using, or selling the invention. In this case you would manufacture the invention yourself and charge more than you'd have to in a competitive situation, as

Xerox did, and as Polaroid and Sony now do with their film and one-gun Trinitron CRT.[1]

Antitrust Note: Occasionally, companies or individuals who own a patent or manufacture a patented invention use their patent in ways that the antitrust laws prohibit (e.g., monopolistic behavior and behavior that imposes restraints on free trade). This is very rarely a problem for the independent inventor but can occasionally raise problems for large corporations. For a discussion of antitrust law as it affects the use of patents, go to any law library and look for any books on patent-antitrust law, such as *Intellectual Property and Antitrust Law* by William C. Holmes (Clark Boardman, 1985), or look under the heading "Patents," subhead, "antitrust" in any legal encyclopedia, such as *Corpus Juris Secundum.*

2. Property Rights

The law considers a patent to be personal property which can be sold, given away, willed, or even seized by your creditors, just like your car, a share of stock, or any other item of personal property. Even though it's personal property, the actual patent deed you receive from the PTO has no inherent value; thus you need not put it in your safe-deposit box or take any steps to preserve it against loss. Your ownership of the patent is recorded in the PTO (just like the deed to your house is recorded by your county's Recorder of Deeds). If you lose the original deed, the PTO will sell you copies of the printed patent or certified copies of a title report showing that you're the owner.

----

[1]If you want to continue to make money from your creativity after your patent expires, you should plow back some of your royalties or proceeds from the sale of the patent for research so that you can invent further developments and improvements, and thereby get more and later patents so as effectively to extend your monopoly beyond its relatively short 17-year term.

G. Be Wary of Offers to Provide Information About Your Patent

SOON AFTER BEING AWARDED A PATENT, a client of mine received an offer by mail, advising that an "article" about her patent was published and offering to send her a copy of the article for $3.95. After anxiously sending in her money, she received the "article," a photocopy of a page from the PTO's *Official Gazette*, showing the usual main drawing figure and claim of her patent! Fortunately she was able to obtain a refund by threatening to call in the FTC and postal inspectors, but you may not be so lucky; new rackets originate all the time.

Another offer frequently received by patentees, usually about a year or more after their patent issues, comes as a postcard which states something similar to the following: "In recent weeks the US Patent Office has issued a new patent which cites your patent as a reference, indicating a relationship between yours and the new patent. The card offers to send you the number or copy of the new patent for a fee, usually $20 to $80. It states that you should review the new patent for a possible infringement, interference, improvement of your invention, etc." This type of offer does provide a somewhat useful service, albeit at a very high cost. Personally I wouldn't accept the offer since almost all patents in which earlier patents are cited as references are very different and extremely unlikely to be of any value to the owner of the earlier patent. You can obtain an enhanced version of the same service at a much lower cost per patent by using a "forward search" service, such as Search Check, Inc., 2111 Jeff. Davis Hwy., P.O.B. 2327, Arlington, VA 22202, (703) 979-7230. They'll sell you a list of *all* patents which have issued after yours in which your patent was cited as a reference for about $50.

A third service is the "Patent Certificate." This offer is sent to many patentees in an official-looking letter from Washington marked "U.S. Patent Certificate," "For Official Use Only" (next to the postage stamp), and "Important Patent Information."

In reality, it's from a private company which wants to sell you a nicely-framed version of your patent. Needless to say, this product is of no official value.

A fourth service, definitely of questionable value, also comes on a post card which states something similar to the following: "Our search of your patent has located X companies that manufacture, market, or sell products in a field allied to your invention." It offers to sell you the names of the X companies for a stiff fee, usually about $80. If you want to find the names of the companies which are in a similar field, I strongly advise that you save your money and instead take a trip to a store or library where you'll find plenty of suitable companies for free; use the techniques of Ch. 11.

H. Maintenance Fees

INVENTOR'S COMMANDMENT #17:
You must pay maintenance fees if you want to keep your patent in force. These fees are due as follows: Maintenance Fee (MF) I: due 3.0 to 3.5 years after issuance, MF II: due 7.0 to 7.5 years, and MF III: due 11.0 to 11.5 years.

IN 1983 CONGRESS allowed the PTO to institute a maintenance fee system. While maintenance fees are totally new for the U.S., their use has been commonplace all over the world (except Canada) for decades. Under the U.S. maintenance fee system, your patent, when granted, will subsist in force for 17 years from its date of issuance, provided three maintenance fees are paid.

If no maintenance fees are paid, it will expire four years from grant. If a first maintenance fee is paid between years 3.0 and 3.5 from grant, the patent will be extended to expire eight years from grant. If a second maintenance fee is paid between years 7.0 and 7.5, the patent will expire 12 years from grant. And if a third maintenance fee is paid between years 11.0 and 11.5, the patent will expire at the end of the full 17 years from grant. This information is succinctly presented in Fig. 15-A, a maintenance-fee timing chart.

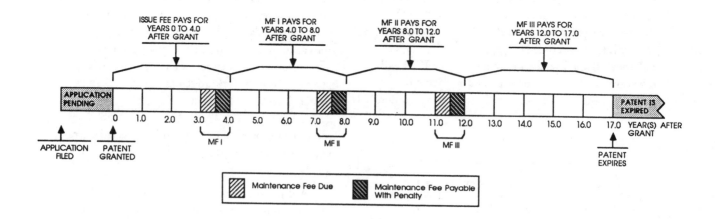

Fig. 15-A—Fee Maintenance Timing Chart

If you forget to pay any fee during its normal six-month payment period, you can pay it in the six-month period (grace period) following its normal six-month payment period, provided you pay a surcharge, see Fee Schedule.

If you forgot to pay a maintenance fee in the normal and grace periods, the patent will have expired at the end of the grace period. However, an expired patent can be revived on petition if you can show, by declaration, that the delay was "unavoidable." See Chapter 13, Section P for how to prepare such a petition and declaration. The petition must be accompanied by a very stiff surcharge; see Fee Schedule and Rule 378.

Use Form 15-3 to pay maintenance fees; the fees (large and small entity) are listed in the Fee Schedule. Be sure to complete every blank in the form, including the serial number of the application; otherwise the PTO won't accept your fee and the delay may carry you into the grace period, costing you a surcharge. If the PTO accepts you maintenance fee, they'll send you a Maintenance Fee Statement to this effect. Anyone can sign Form 15-3.

If you use the Certificate of Mailing at the end of the form, you can send in the fee on the last day of the period. If the last day of the period falls on a Saturday, Sunday or holiday, it's extended to the next business day. And don't forget a postcard! Of course, if you feel, at any time a maintenance fee is due, that your invention's prospects have become nil, you shouldn't pay the fee. In this case your patent will expire as indicated above.

The PTO won't accept a maintenance fee before its due period and may send you a reminder only after the due period expires, when you're in the six-month grace (penalty) period. They also may send you a Notice of Expiration if you don't pay the fee in either the regular or grace periods.

I. Legal Options If You Discover an Infringement of Your Patent

IF SOMEONE INFRINGES YOUR PATENT, you may wish that the earth would shake, the skies thunder, and the infringer be swept away in a storm of fire and brimstone. In fact, of course, nothing will happen and the infringement will continue unless you affirmatively do something about it. Although some think that the PTO plays a role in infringement situations, it doesn't. Rather, the patent owner must assume the full burden for stopping the infringer and obtaining damages. Here, viewed broadly, are the possible steps you can take:

- Ask the infringer to stop and pay you compensation for the past infringement; or

- Ask the infringer to pay you compensation for past infringement and royalties for future activity; or

- Ask the infringer to buy your patent for a sum which will cover past infringement and the present value of future activity; or

- If you're a manufacturer and the infringer has a patent of interest to you, exchange licenses with the infringer; or

- Sue the infringer in federal court in the district where the infringer resides or has committed infringement in the event your request is unsuccessful. If your suit is successful, you'll be awarded damages for the period covered by the patent and also get an injunction, precluding the infringer from using your invention in the future, during the remaining term of the patent.

The injunction is an order signed by a federal court, which, if violated, can subject the violator to contempt-of-court sanctions, including imprisonment and fines. Damages will generally be equivalent to a reasonable royalty you could have gotten had you licensed the patent. In exceptional cases—if the infringer's conduct was flagrant or in bad faith—you may also be able to recover attorney fees and/or triple damages.

J.   What to Do About Patent Infringement

LET'S NOW TAKE A CLOSER LOOK at what to do if your patent is infringed.

Step 1:   Obtain Details of the Infringement

If you discover what you believe to be an infringement of your patent, you should obtain as many details and particulars about the infringing device or process and infringer as possible. To do this, procure service manuals, photographs, actual samples of the infringing device, advertisements, product-catalog sheets, etc., plus details of the individual or company which is committing the infringement.

Step 2:   Compare Your Broadest Claim with Infringing Device

You should now compare your patent's claims with the physical nature of the infringing device. Remember that to infringe your patent, the device in question must physically have or perform all of the elements contained in your patent's main or broadest claims. Even if the infringing device has additional elements, it will still infringe. For example, if your claim recites three elements, A, B, and C, and the infringing device has four elements, A, B, C, and D, it will infringe. But if the infringing device has only two of the three elements, A and B, it won't infringe. Similarly, if the supposed infringing device has three elements A, B, and C', it won't infringe, provided element C of your claim doesn't read on element C' of the supposed infringing device.

Step 3: Apply the Doctrine of Equivalents

Even if your claims don't literally read on the infringing device, but the infringing device is the equivalent of your invention in structure, function, and result, it still may be held to infringe under the "doctrine of equivalents."

Example: Minerva Murgatroid of San Francisco has a patent on a mechanism for bunching broccoli. Its main claim recites the mechanism, including a recitation that the broccoli is banded with a wire-reinforced paper band. She didn't claim the band more broadly because she didn't think to do so, this being the only type of band which would work at the time she got the patent.

A few years later, LeRoy Phillips of Philadelphia discovers a plastic broccoli band which will work just as well as Minerva's wire-reinforced band. He makes broccoli-banding machines and sells them, with his plastic bands, to Fred Farmer, who uses them to band broccoli on his farm in Fresno. Minerva can sue either LeRoy in Philadelphia or Fred in Fresno. Even though her main claim doesn't literally read on LeRoy's machine (i.e., describe all of its physical elements), she can win the infringement suit since his machine is equivalent in structure, function, and result to the patented version, the band material being a relatively minor change that won't get LeRoy or Fred off the hook.[2]

Step 4: Consider Whether a Contributory Infringement Has Occurred

If your claims don't read on the infringing device, but the infringing device is a specially made component that's only useful in a machine covered by your patent, the infringer may be liable under the doctrine of "contributory infringement."

Example: In the example above, LeRoy makes an entire broccoli-banding machine like Minerva's, except that he doesn't sell or supply any bands. Minerva's claims don't literally read on LeRoy's machine since her claims recite the band. Nevertheless, Minerva can bag Fred under the doctrine of contributory infringement since his broccoli-banding machine is useful only in the machine of Minerva's patent claim and since it has no other non-infringing use.

Step 5: Find a Patent Attorney

When you first reasonably suspect that an infringement is occurring, you should promptly consult with a patent attorney (see Chapter 6, Section F). This is because you'll need to embark on a course of

---

[2]There's a rarely-used converse of the doctrine of equivalents, the so-called negative doctrine of equivalents. Under this, even if your claims literally read on the infringing device, but the infringing device has a different structure, function, or result than your invention, the device may be held not to infringe.

action which is very difficult for the non-lawyer to perform in its entirety. Unfortunately, the cost is high, and few patent attorneys will take this type of case on a contingent fee (you pay them only if you win). This means you'll have to pay the attorney up front (or at least partially up front). It depends on the complexity of your case, but an initial retainer of $10,000 would be typical. However, if you've got a very strong case and if the infringer is solvent (would be good for the damages if you win), it's possible that you may find an attorney who will take your case on a contingent fee basis. If you do get an attorney to do this, you still may have to pay the out-of-pocket costs through trial; these can run as high as $50,000, so be sure you can afford them.

If, as will usually happen, you can't get a contingent-fee arrangement, you should be prepared for a shock: patent trial attorneys generally charge about $150 to $250 per hour, and a full-blown infringement suit can run to hundreds or even thousands of hours' work, most of it before trial! You should be sure that your damages, if you win, will make this worthwhile. Also, be sure the defendant can pay any judgment you obtain. And don't depend on getting attorney fees or triple damages; these are awarded only in "exceptional cases," i.e., those where the defendant's conduct was flagrant.

Of course, what's sauce for the goose is sauce for the gander: since the infringer will usually have the same fee burden, it may be inclined to settle if your attorney writes a few letters and it thinks you're serious about suing. A substantial number of cases are in fact settled before suit is even brought and most are settled before trial.

Suits for patent infringement must be brought in federal court in the district where the infringer (who may be a corporation) resides, or is headquartered, or where the infringer has any place of business and has made, used, or sold the patented invention. Thus, if possible, you should select an attorney whose office is in one of these locations. If you do decide to do the substantial part of this yourself, especially where filing and prosecuting your case in court is involved, you'll need expert guidance that's beyond the scope of this book. I recommend *Litigation: Procedure and Tactics* by R. S. White, as a primary resource if you plan to conduct your own litigation.

Note: The material in the following steps is not intended to help you do your own patent infringement litigation (it would take a big book just to get you started), but to give you an overview of what's involved so that you can play an active role in deciding on your course of action and, if a lawsuit is brought, helping your attorney bring its prosecution to a successful conclusion.

Step 6: Write a Letter

The first step to follow in the event an infringement has occurred is to write a letter. This letter can:

- Ask the infringer to stop infringing your patent and to pay you royalties for past activity; or

- Offer the infringer a license under your patent for future activity and again ask for a settlement for the past. Remember, any infringer is a potential licensee so don't make war right away.

As is often the case, the letter may go unanswered, or your demands may not be acceded to. If so, you'll have to sue for patent infringement if you want to recover damages or an injunction. Also, if you actually charge anyone with infringement, rather than just offer a licence, be prepared to follow through with suit, since the infringer can sue you to have your patent declared invalid under what is known as a "declaratory judgment action."

Step 7: Act Promptly

The statute of limitations for patent infringements is six years, which means that you cannot recover damages that occurred more than six years back from the date you filed suit. However, despite this rather lengthy limitations period, it's important that you not wait six years, but act rapidly once you become aware of an infringement. Otherwise, the infringer may reasonably argue that it continued to infringe because you appeared not to be concerned, and you may be prevented from collecting the bulk of the damages you would otherwise be entitled to. This would occur under the legal doctrines known as *estoppel* and *laches*, which generally mean that a court won't award you damages if your action (or lack of action in this case) in some way brought them about.

If you're selling a product embodying the invention and you failed to mark it with the patent number (see Section D above), the six-year term of damages can be considerably shortened as a practical matter by application of the patent-marking statute (35 USC 287). On the other hand, you can bring suit even after your patent has expired and still go back six years during the time the patent was in force (again, provided that you had some valid reason for delaying your action).

Step 8: Who Should Be Sued?

Obviously, you can sue any manufacturer who makes any device or practices any process covered by the clause of your patent. What you may not know is that you can also sue the retailer or ultimate purchaser of the invention (including a private individual) as well as the manufacturer. Suits against the retailer or customer are sometimes brought in order to find a court that's favorable, or at least geographically close, to the patent owner. If a suit is brought against the retailer or customer of a patented invention, the manufacturer of the patented invention will usually step in and defend or reimburse the customer's suit. If your infringer is an out-of-state manufacturer and you can sue its local retailer, it puts a tremendous burden on the manufacturer to defend at a distance.

If the infringer of your patent is a company or individual who's making products embodying your invention under a government contract, you can sue only the government in the Court of Claims in Washington. You can't sue the company and you can't sue in your local jurisdiction. Moreover, you can't get an injunction prohibiting the company from manufacturing your invention since the infringing device may be useful for national defense. You can, however, recover damages and interest.

Lastly don't be afraid to take on a big company simply because they have more resources to defend a patent infringement suit than you have to prosecute it. You have the right to a jury trial—see Sec. N below, which more than equalizes the odds.

Step 9: Stop Importation of the Infringing Device

In addition to suing, if a device covered by your invention is being imported into the U.S. and the effect of such importation is to harm or prevent the establishment of a U.S. industry, or restrain or monopolize trade in the U.S., you can bring a proceeding before the International Trade Commission to have the device stopped at the port of entry. While such a proceeding is complex and expensive, it provides a remedy which is extremely powerful. The pertinent statute is 19 USC 1337(a), and two articles about ITC actions can be found in the *Journal of the Patent Office Society* for 1979 Mar., p. 115, and 1984 Dec., p. 660.

K. Product Clearance (Can I Legally Copy That or Make This?)

THIS IS THE OTHER SIDE of the coin: Here I'll assume that, instead of having your own invention,

you're interested in copying the invention or product of someone else or making a new product which you feel may be covered by someone else's patent. What can you legally do and how do you find out?

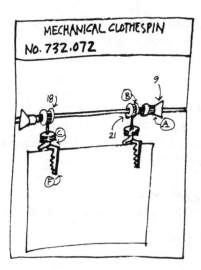

1. Common Misconceptions

Before giving you the applicable rules and information, first I want to dispel some widespread misconceptions so you'll start from neutral territory.

▲

Common Misconception: *If you make an identical copy of a device or circuit, you can be validly sued for infringement, even if the copied device is not patented.*

Fact: *You are free to copy any device or circuit, even to the minutest details, so long as you do not infringe any applicable patent, trademark, or copyright, and so long as you don't copy any features which have a "secondary meaning" (see Chapter 1, Section Q).*

▲

Common Misconception: If a product is not marked "patented" and it does not have a patent number, you are free to copy the product since the law requires patented products to be marked with the legend "Patented" and the patent number.

Fact: Patented products don't have to be marked as such: See "Patent Number Marking" (Section D).

▲

Common Misconception: If a product which you intend to make is shown in the drawing of another's patent, you would be an infringer of that patent if you made the product.

Fact: Only the claims of a patent determine infringement. See Chapters 9, 13, and Section J above..

▲

Common Misconception: That which you do in your own home or for your own personal use will not infringe a patent which is otherwise applicable.

Fact: While "home infringement" may be difficult to detect, nevertheless it is a form of infringement which is legally actionable and can subject the infringer to paying damages and/or an injunction prohibiting further infringement.

▲

Common Misconception: If you change a patented product a fixed percentage, say 20%, you won't be an infringer.

Fact: The amount you'll have to change a patented product to avoid infringement is not subject to quantitative analysis, but rather is determined by the breadth of the patent's claims (see Section J above).

2. Find Out if There's an Applicable Patent and Whether You Will Infringe It

If you do want to manufacture a specific product or perform a specific process commercially, and you have some reason to believe it may be covered by an in-force patent, how can you find out whether you can proceed without infringing the patent in the process

Unfortunately, there is no way to be 100-percent sure, because no search can cover pending patent applications. However I can give you some pretty specific instructions and guidelines.

If the process or product you wish to duplicate is already manufactured or used, look at the product, the literature accompanying it, and the packing material, to see if any patent number is given. If you can get the patent number, order the patent from the PTO. If the patent issued over 17 years ago, it's expired and the invention is in the public domain; you can manufacture or use the product or process with impunity.(Provided that there's no improvement which is covered by an later patent whose number isn't marked on the product. Here's a rough guide which will help you make a rough determination as to when any patent issued: Patent #1 issued in 1836, #100,000 in 1870; #500,000 in 1893; #1,000,000 in 1911; #1,500,000 in 1924; #2,000,000 in 1935; #2,500,000 in 1950; #3,000,000 in 1961; #3,500,000 in 1970; #4,000,000 in 1977; #4,500,000 in 1985. Patent 3,700,000 (1972) expires in 1989.)

If the patent is in force, things usually aren't as bad as they seem. Often a patent which supposedly covers a product in reality may cover only one minor aspect of the product (such as the power plug) which is easy to design around. Sometimes the patent doesn't even cover the product at all: How can you be sure? The only way is to read its claims carefully, diagramming them if necessary, to know exactly what they cover. If what you want to manufacture is not covered by the claims, and if you feel there is probably no other patent on the thing you wish to manufacture, you are free to do so.

If the product or process you wish to manufacture is simply marketed "Patented" and carries no number, your task is more difficult. You can write to the company, asking for the number and date of their patent, or whether their patent is in force, but they're not bound to answer, and you'll have tipped your hand by communicating with them.

You can have a (relatively cheap) search made in the PTO of all of the patents issued to the company in question. Computer searches are good for this; see Chapter 6, Section L, but there is no guarantee that this will uncover the manufacturer, since the patent may not be owned by the company in question; the manufacturer may simply be a licensee. The best way to determine whether an in-force patent is applicable is to make a search in the relevant classes and subclasses of the PTO (see Chapter 6), or have someone make the search for you. The search should seek to find any patent on the invention in question. This will involve a greater expenditure, but at least you will be fairly certain of your position. (If, however, there is a patent pending on the product or process, there is no way to obtain any details, even if the manufacturer makes the product "patent pending"; thus, not all risks can be eliminated.)

If the product or process you wish to manufacture has been known or used in the marketplace for over 17 years, you can be pretty sure that no in-force patent will be applicable, or that even if one is applicable, it is just about to expire anyway.

If you can't find any US patents and the product or process is relatively new, you shouldn't feel free to copy it because it may be subject of a pending patent application. Although you can't search pending patent applications since they're kept secret, you can often find out about pending US applications by searching for a published corresponding foreign application, which are published 18 months after filing. To search published foreign applications, use one of the database searching services listed in Ch. 6, Sec. L.

3. What to Do If an In-Force Patent Is Applicable

If there is an in-force patent applicable, and you still wish to manufacture the product, you have several alternatives:

- You can manufacture or use the product or process and hope that the patentee won't catch you. If you do this, it is wise, as well as good accounting practice, to keep reasonable royalty reserves (see Chapter 16 for how to determine a reasonable royalty) in case you're ever caught. Also, you should analyze the patent, or have a patent attorney do so, to set up defenses to show that you were not a "willful" infringer, since willful infringers may be subject to triple damages or attorney fees in a lawsuit. If you do manufacture without a license, be aware that the patent owner may discover and sue you and get an injunction against you prohibiting you from further manufacturing. Although the idea of manufacturing without a license may seem deceitful, risky, and inadvisable, you should be aware that it is done all the time in the industry; the infringer simply takes the full-speed-ahead-and-damn-the-torpedos attitude and hopes to be able to negotiate a favorable settlement or break the patent if caught.

- You can also ask the patent owner for a license to manufacture under the in-force patent. However, here you take the risk, if you aren't familiar with the patent owner's practices, of being refused a license. Moreover, you'll have shown your hand, so that if you do manufacture, the patent owner will be looking out for you and will certainly sue or accuse you of infringement in short order.

- You can also make an extended validity search to try to "break" the patent. You should use a professional, experienced searcher to do this and should expect to spend a thousand or more dollars in order to make the widest and most complete search possible. Also, you should order a copy of the PTO's file of the patent (see Fee Schedule for cost) to see if there are any weaknesses or flaws in the patent that are not apparent from the printed patent itself. Again, the services of an experienced attorney should be employed here, because breaking patents requires a highly-skilled practitioner.

• If you find highly relevant prior art, you can bring it to the attention of the patent owner and ask it to disclaim or dedicate the patent to the public. Or, you can send the art to the PTO to be put in the file of the patent (35 USC 301) or apply to have the patent re-examined (35 USC 302; see Sec. M below). You can even sue the patent owner, if it has asserted the patent against you, for a judgment declaring the patent invalid.

• Your last alternative is to review the claims of the patent and then try to design around them. Often, the claims of a patent, upon analysis, will be found to have one or more limitations that can be eliminated in your product or process so that you can make the patented invention even cheaper than the patentee. Alternatively, you can design around one of the elements of the patent, make an improved device, and get your own patent on it. Remember that if you don't infringe the independent claims, you won't have to worry about the dependent claims (see Chapters 9, 13 and Section J above).

4. If No In-Force Patent Is Applicable

Unless there is an in-force patent covering an item, anyone is free to make and manufacture identical copies of it, provided:

• One doesn't copy the trademark of the product;

• The shape of the product itself is not considered a trademark (such as the shape of the *Fotomat* huts); and

• You don't copy "secondary meaning" features (see Chapter 1).

If you buy a product from another in the course of business, you don't have to worry about patent infringement. Under the Uniform Commercial Code, your vendor is obligated to indemnify you for any such infringement, although, technically, you can be validly sued for infringement.

I am reminded of the story of one manufacturer's effort to copy a small hardware item by having it manufactured cheaply in the Orient. He sent the item overseas with instructions to make several thousand identical copies of the item. Since he didn't give any further instructions, the Oriental manufacturer did as instructed, manufacturing and shipping back several thousand copies of the item, including a faithful copy of the embossed trademark of the manufacturer's competitor. The manufacturer then had to spend significant money obliterating the trademark, thereby losing his entire profit in the process.

L. The Court of Appeals for the Federal Circuit (CAFC)

FORMERLY, FULL-BLOWN patent infringement suits were very expensive and could cost each side $150,000 or more in attorney fees, travel and deposition expenses, witness fees, and telephone and secretarial expenses. Thus the law worked very much in favor of wealthy or large corporations which are far better equipped to defend and maintain patent infringement suits than a single individual. Therefore, if you discovered an infringement, it was usually to your advantage not to sue and to accept a settlement that was less than you thought you were owed. In short, "gold ruled the law."

The pendulum recently has swung back a good deal in favor of the patent owner, however, primarily because of several important statutory changes, one of which is the new Court of Appeals for the Federal Circuit (CAFC). All patent appeals, both from the PTO's refusal to grant a patent, and from judgments in infringement suits brought in the U.S. District Courts around the country[3] are now heard by the CAFC, which is headquartered in Washington, DC, but which travels around the U.S. to hear appeals in major cities. This means that one court decides all appeals and thus creates a body of legal interpretation that's uniform. Previously, appeals were decided by the various Circuit Courts of Appeal covering the area of the country where the U.S. District court was located, with a resulting patchwork quilt of inconsistent decisions.

One happy result of the uniformity brought by CAFC has been the upholding more of patents and higher damage awards for infringement. While as mentioned, patent infringement lawsuits and trials must still be brought and conducted in local U.S. District Courts, all of these courts are bound to follow the more pro-patentee decisions of the CAFC (under principles of common law precedent).

M. Using the Re-Examination Process to Reduce the Expense of Patent Infringement Suits

ANOTHER VALUABLE STATUTORY CHANGE is the re-examination process (35 USC 302) in which the PTO can be asked to re-examine any in-force patent to determine whether prior art newly called to its attention knocks out one or more of the patent's claims. How does this help the patent holder? Suppose the patent holder decides to go after an infringer. Very soon after the first demand letter is sent, or suit is filed, the infringer will make a search and tell the patent holder of prior art which the infringer feels invalidates one or more claims in the patent. Formerly, if it still thought its patent was valid, the patent holder had to push ahead with an expensive patent infringement lawsuit and hope that the U.S. District Court judge (who is often unfamiliar with patent principles) would decide its way instead of for the infringer.

Now, instead of leaving the matter up to the judge, the patent holder can request a re-examination by the PTO—the request can be made before or during an infringement lawsuit. The PTO will re-examine the claims in light of the prior art and either issue a certificate of patentability, or unpatentability. In the case of the former, this opinion will weigh almost conclusively in favor of the patent holder in the ensuing litigation and quite often will lead to a favorable settlement beforehand. In the event the latter occurs (certificate of unpatentability), the unpatentable claims would be cancelled automatically by the PTO. This would not result in victory for the patent holder, but would save time and money that otherwise would have been spent haggling in court.

---

[3]U.S. District Courts are the federal trial courts where patent infringement actions are first brought and decided.

The re-examination process can also be used to your advantage if you're accused of infringement. By obtaining a PTO certification that the patent holder's claims are unpatentable over the prior art, you may have saved yourself an expensive defense in court.

To institute a re-examination of any patent, anyone can file a request, together with the patent number, prior art, and the fee (see Fee Schedule). The fee appears huge, but is small compared with the expense of litigation. If the PTO feels that the prior art is relevant, it will conduct the re-examination; both sides can file a brief setting forth their arguments. If the PTO feels the newly-cited art isn't relevant, it will terminate the proceeding and refund a large part of the fee.

N. Jury Trials

AS A RECENT ARTICLE IN FORBES (1985 Jun 17) noted, juries love the patent holders and have been awarding very large damages in patent infringement actions, especially where an individual patent holder has sued a large corporation. Thus, if you sue on a patent, always demand a jury trial: juries love injured individual inventors.

O. Arbitration

INSTEAD OF A LAWSUIT, if both parties agree, the entire infringement dispute can now be submitted to arbitration. In this case, an arbitrator, usually a patent attorney, will hear from both sides in a relatively informal proceeding. The arbitrator will adjudicate the patent's validity, infringement, etc. The arbitrator's fee (about $5,000 to $20,000 and up) is far cheaper than the cost of a regular lawsuit complete with depositions, formal interrogatories, and other formal proceedings, etc. In addition, it's

much faster. Any arbitrator's award must be filed with the PTO. The American Arbitration Association is frequently used and has rules for arbitration of patent disputes, but the parties can use any arbitrator(s) they choose.

EYE PROTECTOR FOR FOWL NO. 620,832

Fig 2.

P. How Patent Rights Can Be Forfeited

PATENTS AND THEIR CLAIMS can and often are retroactively declared invalid or unenforceable by the PTO or the courts for various reasons, such as:

1. Relevant prior art which wasn't previously uncovered (Chapter 6);

2. Public use or sale of the invention prior to the filing date of the patent application (Chapter 5, Section E);

3. Misuse of the patent by its owner, e.g., by committing antitrust violations; and

4. Fraud on the PTO committed by the inventor, e.g., by failing to reveal relevant facts about the invention and the prior art (Chapter 10, Section L).

In addition to losing your patent rights, you may discover that what you thought was a broad claim is so narrow as to be virtually useless. Generally you'll discover this when your airtight infringement action goes down the drain when the PTO or the judge declares your patent non-infringed.

Most people are surprised to learn that patents, even though duly and legally issued, can be declared invalid, nonenforceable, or non-infringed. In other words, a patent isn't the invincible weapon many believe it to be. Rather a patent can be defeated if it has weaknesses or if it isn't used in the proper manner. Formerly over 50 percent of the patents that got to court were held invalid, but this figure is declining greatly, so that now a substantial majority are upheld. In addition, the percentage of patents which are held invalid is lower than the court statistics would indicate since they don't count the many valid patents that don't get to court because the infringer saw the impossibility of invalidating them or didn't want to spend the $150,000 or more necessary to fight the patent.

An accused infringer of a patent can avoid liability (i.e., defend against the infringement action) in three different ways:

1. By showing that the claims of the patent aren't infringed;

2. By showing that the patent isn't enforceable;

3. By showing that the patent is invalid.

Non-infringement (1) has already been covered above in Section J. A patent can be declared non-enforceable (2) if the owner of the patent has misused the patent in some way or has engaged in some illegal conduct that makes it inequitable for the owner to enforce the patent. Some examples of conduct that will preclude enforceability of a patent are:

- False marking (marking products with patent numbers that don't cover the patent marked);

- Illegal licensing practices, such as false threats of infringement;

- Various antitrust violations;

- Extended delay in bringing suit, which works to the prejudice of the accused infringer; and

- Fraud on the PTO, such as withholding a valuable reference from your Information Disclosure Statement, or failing to disclose the full and truthful information about your invention in your patent application.

The validity of the patent being sued on can be challenged by:

- Prior-art references that the PTO didn't discover or use properly;

- Proof that the specific machine covered by the patent is inoperable;

- A showing that disclosure of the patent is incomplete, that is, it doesn't teach one skilled in the art to make and use the patented invention;

- A showing that the claims are too vague and indefinite under Section 112;

- A showing that the patent was issued to the wrong inventor, etc.

As you can imagine, the subjects of patent non-enforceability and invalidity are also complex and difficult. In fact, it has been said that if enough money is spent, almost any patent can be "broken." However, patents are respected in many quarters and, as stated, billions of dollars change hands in the United States each year for the licensing and sale of patent rights.

Q. Adverse Interference Decision

FOR ONE YEAR following the issuance of your patent, you are potentially subject to losing it if another inventor who has a pending application on the same invention can get the PTO to declare an interference. If the other inventor wins "priority" in the interference, i.e., the PTO finds that the other inventor was entitled to a patent instead of you, you'll effectively lose your patent. While conflicts between patent applications and in-force patents are relatively uncommon, they do occur. See Chapter 13, Section K for more on this.

R. Taxes

I INCLUDE THIS BRIEF SECTION because, unfortunately, most inventors give no thought whatever to taxes, either with regard to the money they spend to get their patents, or to the money they make when they sell or license their patents. I say "unfortunately," because the government will effectively subsidize your patent expenses by allowing you to deduct them. Because of space limitations, I can't provide a full guide to all of the patent-tax rules here, but here's the basics. You should consult a tax professional or the IRS for the final word:

1. Your patent and invention expenses (search, drafting, models, testing, attorney fees, PTO filing fee, etc.) are deductible. Small disbursements (up to a few hundred dollars) will not generally be questioned if they are expensed (deducted 100 percent for the year incurred), but larger disbursements must be capitalized and depreciated over the life of the patent, or its estimated life (about 19 years), if it hasn't issued yet. Prosecution expenses can usually be expensed, however. Use Schedule C; the IRS considers that you're in the business of inventing. You can even deduct the cost of this book!

2. If you license your invention on a non-exclusive basis (see Chapter 16), you haven't given away all of your rights, so your royalties are considered ordinary income. Report on Schedule C.

3. If you sell all of your patent rights, or grant a full exclusive license, the IRS considers that you've sold it all; your receipts or royalties, even though received over a long number of years, are considered capital gains; report on Schedule D.

S. Patent Litigation Reimbursement

BECAUSE OF THE HIGH COST of litigation, two cost assistance services are now available: (1) HLPM Insurance Services, Inc., 10509 Timberwood Circle, Louisville, KY 40223, 1-800 537-7863 will, in return for annual premiums, reimburse the cost of patent enforcement litigation, up to the policy limit. (2) Patent Protection Institute, 1368 Lincoln Ave., San Rafael, CA 94901, 415 459-3855 will finance patent enforcement litigation in return for a percentage of any recovery.

Chapter 16

Ownership, Assignment, and Licensing of Inventions

IN THE SIMPLEST POSSIBLE SITUATION, a single inventor invents something, obtains a patent on it, manufactures it, and markets it directly to the public for the full period that the patent remains in force. In most instances, things are not that simple. Two or more people may be involved in the conception of the invention, and many more in its development and marketing. Established businesses may want to use the invention and be willing to pay large sums for the privilege. Employees and employers may disagree over who owns a particular invention which was developed at least partially on company time or with company materials or facilities. Thus, the entire question of invention ownership and utilization can at times become quite complex.

In this chapter, I outline some of the ways to deal with these various ownership questions and how to deal with the more common sorts of agreements that will be necessary to accomplish this. However, the subject of invention ownership, licensing, and transfer is complicated, so if you are new to it, you'll probably want to retain a lawyer, if only to review your plans and paperwork.

A. The Property Nature of Patents

BEFORE I BEGIN EXPLAINING who owns an invention, it might be helpful to review exactly what patent ownership means. A patent should be thought of as a valuable right. The right, as I've stressed elsewhere in this book, gives you a monopoly over the manufacture, use, and distribution of your invention. This means that you have the right to exclude others from making, using, or distributing the invention without your permission. If you do grant a company permission to use your invention, the law terms this permission a "license." As with most other intangible economic rights—e.g., the right to operate a business, the right to withdraw money from a bank account, the right to vote stock in a corporation—patent rights, or a portion of them,

can be sold to others, or licensed for a particular use over a particular period of time.

Without the rights granted by a patent, an invention has virtually no economic value to its inventor unless it is preserved and worked or sold as a trade secret (Chapter 1, Section F). Because most inventions cannot be thus preserved (due to reverse engineering considerations), patent ownership and invention ownership therefore often amount to the same thing.

B. Who Can Apply for a Patent?

ONLY THE TRUE INVENTOR can apply for a patent. As mentioned in Chapter 1, when it comes to eligibility to apply for a patent, the status of the applicant-inventor makes no difference, so long as he or she is the true inventor. I.e., the applicant can be of any nationality, sex, age, or even incarcerated, insane, or deceased. (Insane and dead people can apply for patents through their personal representative.)

What happens to patent ownership if more than one person is involved in a particular invention? If other people are involved in the inventing stage, they're considered co-inventors. Most often, the trick is to determine what type of activity constitutes invention. For instance, if one person came up with the conception of the invention, while the other merely built and tested it, not contributing any inventive concepts but merely doing what any skilled artisan or model maker could do, the second person is not a co-inventor. Similarly, financiers, or others who provided financial, but not technical, input should not be listed as co-inventors.

On the other hand, if one person came up with the idea for an invention and the model maker then came up with valuable suggestions and contributions that went beyond the skill of an ordinary model maker or machinist and made the invention work far better, both people should be named as co-inventors on the patent application (see Chapter 10, Section F).

The PTO and the courts don't recognize degrees of inventorship. The order in which the inventors are named on a patent application is legally irrelevant, although the first-named inventor's name will be more prominent in the printed patent.

If the joint inventors invented different parts of the claimed invention, they should keep good records as to what part each invented so that the named inventors can be changed if one inventor's part is dropped later in the prosecution. Joint inventors need not have worked together either physically or at the same time, and each need not have made the same type or amount of contribution. To qualify as a co-inventor, an inventor need merely have contributed something to at least one claim of the application, even if it's a dependent claim.

When two or more persons work on an invention, disputes regarding inventorship sometimes arise later. E.g., a model builder may later come back, after an application is filed, and claim to have been wrongfully excluded as a joint applicant. As I stated in Chapter 3, the best way to avoid such problems is for all inventors to keep a lab notebook, i.e., a technical diary, which faithfully records all developments and is frequently signed by the inventor(s) and witnessed. In that way the complaining model builder can be answered by positive proof from the real inventor(s). Also using the Consultant's Agreement (Ch. 4, Sec. F) will eliminate many potential disputes. Absent such documentation, or agreement and expensive disputes can arise, with only vague memories to deal with.

It's important to include in your application all the inventors who are true inventors and to exclude those who aren't inventors. If it is discovered later that your inventorship is incorrect, and that the mistake resulted from bad faith, your patent could conceivably be held invalid, although this rarely happens. (If you do discover that the wrong

inventor(s) is(are) named on a patent or patent application, this can be corrected under Patent Rules 48 or 324.)

▲

Common Misconception: *If you want to make your financier a 50% owner of your invention, it is okay to do this by filing the patent application in both of your names.*

Fact: *A U.S. patent application must be filed in the name(s) of the true inventor(s) only. There are several legitimate ways to convey an interest in your invention to a non-inventor (see Sections E and F below).*

▲

Common Misconception: *If you came up with a bare idea for a valuable invention and your associate "took your ball and ran with it," i.e., built and tested the invention after hundreds of hours of work leading to final success, then your associate must be named as a co-inventor with you.*

Fact: *As stated above, only the true inventor(s) i.e., the one(s) who came up with the inventive concepts created in the claims, should be named as applicants. An associate who did only what any model maker would have done should not legally be named as co-inventor, no matter how much work was involved. On the other hand, if your associate contributed inventive concepts which made the invention workable, and which are recited in one or more claims, then the associate should be named as a joint inventor with you.*

C. Joint Owners' Agreement

PROBLEMS COMMONLY ARISE in situations where there are two or more inventors or owners of a patent application or patent. These include questions as to who is entitled to commercially exploit the invention, financial shares, what type of accounting must be performed on partnership books, etc. Fortunately, most of these predictable problems can be ameliorated, if not completely prevented from arising, by the use of a Joint Owners' Agreement (JOA).

A JOA is also desirable since a federal statute (35 USC Section 262) provides that either of the joint owners of a patent may make, use, or sell the patented invention without the consent of and without accounting to (paying) the other joint owner(s). This statute seems unfair since it can work a severe hardship on one joint owner in either of two ways:

1. If one joint owner exploits and derives income from the patent while pushing the other aside, the passive joint owner will not be rewarded for any inventive contribution (if an inventor) or any capital

contribution (if an investor, i.e., someone who has bought part of the patent).

2. If one joint owner works hard to engineer and develop a market for the patented product, the other joint owner can step in as a competitor without compensating the engineer or marketing pathmaker for the efforts accomplished.

The JOA that I provide as Form 16-1 prevents these results from occurring and also accomplishes the following:

- Prohibits any joint owner from exploiting the patent without everyone's consent, except that if there is a dissenter, a majority can act if consultation is unsuccessful;

- Provides that in case of an equally-divided vote, the parties will select an arbiter, whose decision shall control;

- Provides that disputes are to be resolved by mediation or binding arbitration if mediation fails.

- Provides that the parties shall share profits proportionately, according to their interests in expenditures and income, except that if one party does not agree to an expenditure, the other(s) can advance the amount in question, subject to double reimbursement from any income;

- Provides that if an owner desires to manufacture or sell the patented invention, that owner must pay a reasonable royalty to all owners, including himself; and

You should not regard this agreement as cast in stone, but merely as one solution to an unfair statute. You may modify, add to, or replace this agreement with any understanding you desire, so long as you're aware of the problems of Section 262, as paraphrased above.

The manner of completing the JOA of Form 16-1 is straightforward. It should be completed after or concurrently with an assignment (Section E below) or a joint patent application (Chapter 10). Fill in the names and respective percentages owned by each at the top of the form, identify the patent application (or patent) next, and have each joint owner sign and date the end of the form. As with all agreements, each joint owner should get and preserve an original signal copy. The JOA should not be filed in the PTO.

Caution: Time and space do not allow me to freely explain the possible ramifications of each paragraph in the JOA, or the many possible variations that might be more appropriate to your situation. If you want to be sure that your joint owners agreement accurately reflects your needs, consult a patent attorney.

D. Special Issues Faced by the Employed Inventor

MANY INVENTORS ARE EMPLOYED in industries which are at least somewhat related to the inventing they do on their own time. Such inventors naturally have a strong desire to learn what rights, if any, they have on inventions which they make during their employment, both on their own time and when they are on the job. This complex subject is covered in detail in Spanner, *Who Owns Innovation? The Rights and Obligations of Employers and Employees* (Dow Jones Irwin 1984). I'll just cover the high points here.

Generally, the rights and obligations of employed inventors are covered by the Employment Agreement (EA) they sign with their employer, i.e., the EA prevails, unless it conflicts with state law (see below). Below is an example (Fig. 16-B) of a typical EA.

If you have no EA, you'll own all your inventions, subject to the employer's extensive "shop rights" (i.e., a right to use the invention solely for the employer's business, without paying the employee) on inventions made using company time, facilities, or materials.

If you have an EA, it will almost certainly require that you assign (legally transfer) to your employer all inventions either:

1. Made during the term of employment;

2. Which relate to the employer's existing or contemplated business;

3. Which use the employer's time (i.e., the time for which the employee is paid), facilities, or materials; or

4. Are within the scope of the employee's duties.

Note that under items 1, 2, and 4 even if an employee makes an invention at home, on the employee's own time, the employer still can be entitled to ownership.

Also, you'll usually be bound to disclose *all* inventions to the employer (so the employer can determine if they're assignable). Lastly, most EAs will require you to keep your employer's trade secrets confidential during and after your employment. Some states, such as California, have enacted statutes (Calif. Labor Code, Section 2870 et seq.) prohibiting the employer from requiring the employee to sign any EA which is broader than the foregoing, i.e., the employee can't be made to turn over all inventions, no matter where and when made, to the employer.

Suppose an employee makes an invention outside of the scope of her EA; say Roberta, an employee of Silicon Valley Chips (SVC) invents a new toilet valve, and her EA requires her to disclose to SVC all inventions made during the term of her employment at SVC. Roberta should disclose the valve to her employer (regardless of whose time it was invented on) and then go ahead and do whatever she wants with it, since she owns it totally.

If the invention is clearly within the scope of the EA, or is in a grey area, I recommend first disclosing it to the employer. If the employer isn't interested in the invention after reviewing it, the employee can apply for a release, a document under which the

CORP 624 8/66	**EMPLOYMENT AGREEMENT**	**PHILCO CORPORATION** A SUBSIDIARY OF *Ford Motor Company*

NO ADDITIONS, DELETIONS OR CHANGES ARE PERMITTED

In consideration of my employment and the compensation paid to me as an employe by Philco Corporation (Philco), I hereby recognize as the exclusive property of, and assign, transfer, and convey to Philco without further consideration each invention, discovery or improvement (collectively called inventions) made, conceived or developed by me (whether alone or jointly with others) during the period of my employment which relates in any way to my work at Philco or to the business, work, interests or investigations of Philco or its subsidiaries. I will communicate to Philco's Patent Department promptly and fully, and preserve as confidential information of Philco, all such inventions. I will maintain adequate and current written records of all such inventions, which records shall be and remain the property of Philco. Except as Philco may otherwise consent in writing, I will not disclose, or make any use of except for Philco, at any time during the period of, or subsequent to, my employment at Philco, any confidential information, knowledge, trade secrets, or data of Philco I may produce or obtain during the period of my employment unless and until that information shall have become public knowledge. Upon request by Philco, I will at any time execute documents assigning to it, or its designee, any such invention or any patent application or patent granted therefor, and will execute any papers relating thereto. I also will give all reasonable assistance to Philco, or its designee, regarding any litigation or controversy in connection with my inventions, patent applications, or patents, all expenses incident thereto to be assumed by Philco. I agree that the provisions of this paragraph and the rights and liabilities of the parties under these provisions or with respect to the subject matter of this paragraph shall be interpreted according to and governed by the law of the Commonwealth of Pennsylvania.

I understand that my employment is not for any definite term, and may be terminated at any time, without advance notice, by either myself or Philco; that my employment is subject to such rules, regulations and personnel practices and policies, and changes therein, as Philco may from time to time adopt; and that my employment shall be subject to such layoffs and my compensation subject to such adjustments, as Philco may from time to time determine.

I agree to pay Philco, and hereby authorize Philco to deduct from any pay or moneys due me, the cost of equipment received by me while in its employ, which, as a result of my failure to exercise reasonable care under the circumstances, is lost, damaged, or not returned in good condition (except for ordinary wear and tear in the course of business) upon demand. I waive responsibility on the part of Philco for loss or damage to personal equipment.

I understand that medical information disclosed to Philco's examining Physician is not for treatment as a patient and is not privileged.

I elect to become subject to the Workman's Compensation Laws of the State of ....

I acknowledge that the terms contained herein are the entire terms of my employment agreement, that there are no other arrangements, agreements or understandings, verbal, or in writing, regarding my present or future employment with Philco, and that any purported arrangements, agreements or understandings made in the future shall not be valid unless evidenced by a writing signed by a properly authorized official of Philco.

I intend to be legally bound by this agreement and agree that it shall be binding upon my heirs, executors, administrators, other legal representatives, and assigns.

EMPLOYE'S SIGNATURE		SIGNATURE OF COMPANY REP. WITNESSING EMPLOYE'S SIGNATURE
	(SEAL)	
DATE OF SIGNATURES	PLACE OF SIGNATURES (CITY AND STATE)	

Fig. 16-B—Typical Employment Agreement

employer reassigns or returns the invention to the employee.[1] If the invention is in the grey area and the employer wants to exploit the invention, the employee can then try to negotiate some rights, such as a small royalty, or offer to have the matter decided by arbitration. Failing this, a lawsuit may be necessary, but I favor employees disclosing "grey-area" inventions so that their invention will not be engulfed in a cloud of ownership uncertainty.

Most EA's also require the invention-assigning employee to keep good records of inventions made and to cooperate in signing patent applications, giving testimony when needed, even after termination of employment, etc. Most companies give the employee a small cash bonus, usually from one hundred to several hundred dollars, when the employee signs a company patent application. This bonus is not in payment for the signing (the employee's wages are supposed to cover that) but to encourage employees to invent and turn in invention disclosures on their inventions. Some employers, such as Lockheed, give their inventor-employees a generous cut of the royalties from their invention, and some will even set up a subsidiary entity (partly-owned by the employee-inventor) to exploit the invention. Most, however, prefer to reward highly-creative employees via the salary route.

There is currently legislation as well as voluntary proposals within various engineering organizations to expand the rights of the employed inventor. One of these is to change the U.S. to the German system, where employees own their inventions but usually assign them to their employers in return for a generous cut, e.g., 20 percent of the profits or royalties.

E. Assignment of Invention and Patent Rights

SUPPOSE YOU'RE AN EMPLOYED INVENTOR and you make an invention on your employer's time and your employer wants to file a patent application on it in your name. This raises a problem. If it's filed in your name, how will the employer get ownership? Since inventions, patent applications, and patents aren't tangible things like a car, money, or goods, you can't transfer ownership by mere delivery, or even by mere delivery with a bill of sale or receipt. To make a transfer of ownership in the arcane patent world, you must sign an "assignment"—a legal document which the law will recognize as effective to make the transfer of ownership.

An assignment for transferring ownership of an invention and its patent application is provided in Form 16-2 (Appendix) and a completed assignment is shown in Fig. 16-C.

---

[1]Sometimes the employer will retain a "shop right" under the release, i.e., a non-transferable right to use the invention for its own purposes and business only.

Assignment of Invention and Patent Application

For value received, _____Minerva Murgatroid_____,

of _Merion Station, PA_____

(hereinafter **Assignor**), hereby sells, assigns, transfers, and sets over unto

_Leroy Phillips_____

of _Philadelphia, PA_____

and her or his successors or assigns (hereinafter **Assignee**) _100_ % of the following: (A) **Assignor's** right, title and interest in and to the invention entitled " _Fusion System Using_ _Psychic Field_____ "

invented by **Assignor**; (B) the application for Untied States patent therefor, signed by **Assignor** on ___1988 Dec. 16_____ ,U.S. Patent and Trademark Office Serial Number _06/771,432___;

Filed _1988 Dec. 21_____; (C) any patent or reissues of any patent that may be granted thereon; and (D) any applications which are continuations, continuations-in-part, substitutes, or divisions of said application. **Assignor** authorizes **Assignee** to enter the date of signature and/or Serial Number and Filing Date in the spaces above. **Assignor** also authorizes and requests the Commissioner of Patents andTrademarks to issue any resulting patent(s) as follows: ___O___ % to **Assignor** and ___100___ % to **Assignee**. (The singular shall include the plural and vice-versa herein.)

Assignor hereby further sells, assigns, transfers, and sets over unto **Assignee**, the above percentage of **assignor's** entire right, title and interest in and to said invention in each and every country foreign to the United States; and **Assignor** further conveys to **Assignee** the above percentage of all priority rights resulting from the above-identified application for United States patent. **Assignor** agrees to execute all papers, give any required testimony and perform other lawful acts, at **Assignees** expense, as **Assignee** may require to enable **Assignee** to perfect **Assignee's** interest in any resulting patent of the United States and countries foreign thereto, and to acquire, hold, enforce, convey, and uphold the validity of said patent and reissues and extensions thereof, and **Assignee's** interest therein.

In testimony whereof **Assignor** has hereunto set its hand and seal on the date below.

_____Minerva Murgatroid_____

_____

State: _Pennsylvania_____ :

:ss

County: _Montgomery_____ :

Subscribed and sworn to before me_ 1988 Dec. 28_____

(date)

_____Henrietta Bessomy_____.
Notary Public

SEAL

(My commission expires 1990 Aug. 15)

Fig. 16-C—Completed Assignment Form

As indicated, full assignments (transfer of 100% of the invention and its patent application) are usually made by employed inventors who have agreed, in their EA, to assign all inventions they make within the scope of their employment to their employer; in these cases the assignee is usually a corporation.

A partial assignment (transfer of less than 100%) is usually made where the "assignee" (the person getting the transferred interest) has financed all or part of the patent application.

The assignment document presented here, like the Joint Owners' Agreement, is but one of many possible alternatives. If you use it, you may want to change a number of provisions to fit your situation. Also, keep in mind my cautionary note regarding the joint owner's agreement, that is, a consultation with a patent attorney is advisable if you wish to fully understand how this agreement will affect your rights. For example, where there will be many owners of the patent application, the percentage interest of each should be specifically listed in the last sentence of paragraph 1.

To complete the assignment do the following:

Lines 1 and 2: Insert the names of the assignors (the inventor-patent applicants) on lines 1 and 2 of the first paragraph after "received," and insert their cities and states of residence after "of" on line 2.

Line 3 and 4: Do the same for the assignee on line 4.

Line 5: Put the percentage of the patent rights being assigned (normally 100%).

Line 6 and 7: Put the title of the invention on line 7.

Lines 8 thru 10: Put the date the patent application was signed (sometimes termed "executed" in the law) on line 9. If the application has already been filed, also put the serial number of the patent application on line 9 and put the filing date on line 10. Put the percentages owned by the assignor and assignee in the penultimate line of this paragraph.

Then, take the assignment to a notary, sign it before the notary, and have the notary write in the city, county, date, sign the assignment, and affix a notarial seal. While the PTO doesn't require the assignment to be notarized for it to be recorded (see below), it is wise to do so, as assignments that have been notarized are more readily admissible in court, and can be used in some foreign countries, should this ever be necessary.

F. Record Your Assignment with the PTO

TO BE FULLY EFFECTIVE, the assignment must be recorded in the PTO, just as the deed to your house must be recorded with your county clerk. If the assignment is not recorded, and the assignors make another assignment to a different assignee who is unaware of the first assignment, the second assignee's rights will prevail over those of the first assignee if the second assignee records the assignment first. This means the assignee should record the assignment as soon as possible after it's signed. To do this, merely send the assignment to the PTO with a the recordal fee (see Fee Schedule in Appendix) and request that it be recorded. The PTO will stamp the assignment with "recording indicia" (a record of the date and place it has been recorded, e.g., "Recorded 1986 Mar 14, Reel 2012, Frame 547"), make a copy of the assignment, and return the recorded assignment to the person requesting recording, just as your county's recorder did with the deed to the building in which you're living or working.

If the assignment has been made before the patent application is filed, it is permissible to send the assignment in with the application and have it recorded at that time (see Chapter 10, Section J), provided you fill in the date the application was signed. However even if the assignment is signed before filing, I prefer to wait until I can add the filing date and application (serial) number to it before sending it in for recordal, since this will connect it to the applicatin in an unequivocal way.

G. Licensing of Inventions— An Overview

USUALLY, THE OWNER OF A PATENT application or patent needs to allow others to make and sell the patented invention. Inventors, after all, are rarely also manufacturers. When an inventor gives another permission to manufacture and market an invention in exchange for compensation (e.g., a royalty, flat payment, etc.), it is, as stated, done with a document termed a "license." It is essential that a license agreement be written and signed by the inventor (the licensor) and the manufacturer (the licensee). Here are just a few major considerations and terms which can be written into a license agreement:

1. The proposed licensee can buy an option from you (the licensor) under which you give it the exclusive (or non-exclusive) right to obtain a license under your patent application or patent within a fixed time, say two years. The payment for this option can be merely the company's agreement to research and develop your invention (this is a typical arrangement), or it can involve a cash payment. The general rule is that the more you receive up front, the more seriously the licensee will look at your invention.

2. As noted, if you grant the company a license, the license can be exclusive, under which you agree to license only the company and no one else, or it can be non-exclusive, under which you license them but also have the right to license others. Exclusive licenses are more common, since manufacturers want to have a monopoly. Non-exclusive licenses are usually used where a very valuable invention exists and several manufacturers want licenses to get into the business, e.g., Pilkington Brothers, the great British glass company, granted many non-exclusive licenses under its float glass patents.

3. The license, if granted, can be for the life of the patent, or just for a limited term, say five years, with an option to renew for succeeding five-year terms.

4. The license can require the payment of an advance which may be recoverable against royalties, or may it be in addition to royalties. You, of course, want to get as much money at the beginning as possible under the old "bird-in-the-hand" theory.

5. The license can require the payment of minimum annual royalty payments during each year of its existence. This is usually done when an exclusive license is granted.

6. The license rights can be transferred ("assigned") by your licensee to another manufacturer, or any such assignment can be prohibited. From your point of view, it's a good idea to try to get a provision included in the agreement prohibiting assignment without your approval.

There are hundreds of other, less important, considerations in licensing, which I won't discuss here. Licensing, as you may have gathered by now, is a difficult, complex subject, and one that requires negotiation as well as skill. Unfortunately, most invention licensing agreements tend to be tailor-made by large corporations to protect their interests. To date, no good self-help law book deals with the ins and outs of doing this. However, I refer to several standard patent law treatises in the bibliography.

It's important to realize that even though you can make a great invention, prepare a patent application on it, and sell it to a manufacturer, you may not be able to represent yourself adequately in negotiating a license agreement unless you're familiar with licensing and adept at business. It's therefore often wise to hire a patent lawyer to review any contract which is offered to you. You'll find this will probably cost several hundred dollars but the money will be well spent, especially if you have a potentially good deal in the offing.

In fact, most reputable companies would prefer that you be represented by an attorney when you negotiate a license agreement and often give you money to pay an attorney. The reason for this is that an agreement between an unrepresented inventor and a much larger company will be interpreted against the company by the courts much more often than when the inventor is represented by an experienced lawyer.

H. Universal License Agreement

IF YOU DO FEEL CONFIDENT enough to represent yourself, and you're the type of person who can go through a long license agreement with nitpicking skill and then competently negotiate with corporate pros, more power to you. Start your quest by referring to the Universal License Agreement in the appendix (Form 16-3 in Appendix). This agreement can be used to exclusively or non-exclusively license your invention as well as to license know-how. It can also be used to grant a potential licensee an option to evaluate your invention for a given period in return for a payment. As I've said, most companies will either prefer their own license agreement or to make one up from scratch, but you can use the Universal License Agreement for purposes of comparison.

Do you find the agreement long and complex? So do I. To deal with it easily, it's best to consider each of its parts separately. The sample shown (Fig. 16-D) is for the first page of an exclusive license with an option grant and a know-how license.

Part 1: The licensor is the party, usually the inventor, who does the licensing, while the licensee is the party who is licensed, i.e., given permission to use the invention, patent, know-how, etc.

The Patent Royalty rate is the rate, in %, which is paid for use of the patent. The rate is made purposely low (2%) since a know-how license has been granted at a rate of 3% for an overall (total) royalty of 5%. It's usually to an inventor's advantage to license know-how, as well as patent rights, and to

make the know-how rate as high a proportion of the total rate as possible. This is because patents can be held invalid and can only be licensed for a limited term (the duration of the patent application plus the 17-year term of the patent), usually a total of about 19 years; whereas a know-how license can extend indefinitely.

A licensing fee (advance) is customarily paid to the licensor upon signing the agreement as a reward for past work. In the agreement, the licensing fee is computed as an estimate of the first year's sales by multiplying (a) the Patent Royalty Rate by (b) the Estimated First Year's Sales in Units by (c) the Estimated Unit Price in dollars. Again, it's usually in the inventor's interest to get as large a signing bonus as possible, and not to have this money be set off against later royalty payments.

The "Exclusive" box is checked, indicating that only the licensee will be licensed. If the "Non-exclusive" box is checked, the licensor would be able to license others and make the invention him, in addition to the licensee. The title, serial number, and filing date of the patent application are identified next.

The "Minimum Number of Units to Be Sold to Compute Minimum Annual Royalty" (whether or not they are actually sold) is provided to insure that the licensor receives an adequate income from the licensee in as much as he can't, under an exclusive license, license others to derive more income. This minimum annual royalty has been computed on the basis of a minimum annual number of units to be sold (rather than a fixed dollar amount) to give the licensor the benefit of inflation in unit price. While the manufacturer can cut the price of the licensed product and thereby reduce its royalty payments to you, it's generally not in its interest to do this, since it will be reducing its profits as well. However, if you want protection against this possibility, you can substitute a fixed dollar amount for the minimum annual royalty.

Universal License Agreement

1. Parties, Terms, and Parameters:

This agreement is between:

Licensor: _____HARRY BERESOFSKY_____, of
_____CHERNEGOV, UKRAINE_____.

Licensee: _____CHERNOBYL REACTOR WORKS, INC._____, of
_____RUSSIAN HILL, CA_____.

Patent Royalty Rate (%): ___2.00____ % x Est. 1st Yr's Sales (units): ___200____ x

Estimated Unit Price $_1,000.00____ = Resultant Licensing Fee $_4,000.00____

Type of License: [x] Exclusive   [ ] Nonexclusive

Invention Title: ___PERPETUAL ENERGY MACHINE_____.

Patent Application Ser. Nr.: _07/123,456_____, Filing Date: __1988 Aug. 9_____

Minimum Nr. of Units to be Sold to Compute Min. Annual Royalty: _300_____

Minimum Annual Royalties Start Year Commencing 19 89_____.

    [x] Option Granted: Premium $_5,000____ For Term Of (months): _18____

    [x] Know-How Licensed: Know-How Royalty Rate (%): _3.00____

Running Royalty (Patent Royalty and Know-How Royalty, if applicable) (%): _5.00_

2. Effective Date: This agreement shall be effective as of the latter of the signature dates below written and shall be referred to as the Agreement of such date.

3. Recitals:

A. LICENSOR has developed an invention having the above title and warrants that LICENSOR has filed a patent application on such invention in the U.S. Patent and Trademark Office, which patent application is identified by the above title, Serial Number, and Filing Date. LICENSOR warrants that licensor has full and exclusive right to grant this license on this invention and LICENSOR'S patent application. If the "Know-How" block above is checked, LICENSOR has also developed know-how in connection with said invention and warrants that LICENSOR owns and has the right to license said know-how.

B. LICENSEE desires, if the "Option Granted" block above is checked, to exclusively investigate LICENSOR'S above invention for the term indicated. If said "Option Granted" box is not checked, or if said box is checked and LICENSEE investigates LICENSOR'S invention for the term indicated and such investigation is favorable, LICENSEE desires to make, use and sell the products embodying such invention and covered by the claims of LICENSOR'S patent application and any patent(s) issuing thereon (hereinafter "Licensed Product").

4. If Option Granted: If the "Option Granted" box above is checked, then (A) the patent license grant of Part 5 below shall not take effect except as defined in this part, and (B) LICENSOR hereby grants LICENSEE, for the option premium stated above, an exclusive option to investigate LICENSOR'S invention for the term indicated above, such term to commence from the date of this Agreement. LICENSOR will furnish LICENSEE with all information and know-how (if any) concerning LICENSOR'S invention in LICENSOR'S possession. LICENSEE will investigate LICENSOR'S invention for operability, costing, marketing, etc. LICENSEE shall report the results of its investigation to LICENSOR at any time before the end of the option term. If LICENSEE'S determination is favorable, it may thereupon exercise this option and the patent license grant of Part 5 below shall become effective. If LICENSEE'S determination is unfavorable, then said option shall not be exercised and no patent license grant shall take effect and all rights hereunder shall revert to LICENSOR and LICENSEE shall deliver to LICENSOR all results of its investigations for LICENSOR'S benefit.

5. Patent License if Option Exercised or if Option Not Granted: If the "Option Granted" box above is checked and LICENSEE has investigated LICENSOR'S invention and such investigation is favorable and LICENSEE has exercised its option, or if said box is not checked, then LICENSOR hereby grants to LICENSEE, subject to the terms and conditions herein, a patent

Form 16-3

Fig. 16-D—Completed First Page of Universal License Agreement

For the privilege of obtaining an option to exclusively evaluate the invention for the Option Term, an Option Premium (a one-time cash payment) has been paid to the licensor.

The Know-How Royalty Rate is stated and is added to the Patent Royalty Rate to get the total, or Running Royalty Rate.

Part 2: The effective date of the agreement is the date when the last signature is made.

Part 3: Here the Recitals provide the reasons or premises for the agreement. The recitals simply state that the licensor has an invention, a patent application, and possibly know-how, and the licensee desires to evaluate licensor's invention (if an option has been granted) and to make, use, and sell the licensed invention.

Part 4: This covers the parties' rights if an option has been granted. In this case, the regular license grant doesn't take effect yet, but the licensee has the exclusive right to investigate the invention for the option term indicated in Part 1. If the invention is favorable, the licensee will exercise its option and the patent license grant of Part 5 will take effect. If not, the option will not be exercised and all rights will revert to the licensor and the licensor will get the results of the licensee's investigation of the invention.

Part 5: This contains the actual license grant. This comes into play immediately if the invention is licensed or if an option is granted or if the option is granted and exercised. Remember, if an option is granted, the actual license isn't granted until the option is exercised. The license granted (exclusive or non-exclusive) gives the licensee the right to make, use and sell the Licensed Product in the U.S., and it includes any derivative applications and patents (see Chapter 14). If the "know-how" box of Part 1 has been checked, then know-how is also licensed.

Part 6: Know-how is covered in this part. If know-how is licensed, then the licensor is obligated to communicate all of its know-how to the licensee within one month, plus provide up to 80 hours of consultation to the licensee, with travel and other expenses paid by licensee. The licensor disclaims any guarantee that the know-how is workable. The know-how royalty is to be paid for three years and thereafter for so long as the licensee enjoys a U.S. competitive market share of at least 15%. This means that the licensor can enjoy know-how royalty payments indefinitely, provided its know-how was valuable enough to give the licensee a market share of over 15% after three years has passed.

Part 7: This concerns royalties and is the heart of the agreement.

Subpart A: If a Licensing Fee is paid; it's an advance against future royalties. If the estimated Licensing Fee has been computed inaccurately (Part 1) then an adjustment is made when royalties are paid. (**Note:** It is possible to draft an agreement whereby the licensing fee is a one-time payment and not an advance against royalties.)

Subpart B: The running royalty is covered and is paid quarterly, within one month after the end of each quarter, together with a report of the sales made in the quarter.

Subpart C: The minimum annual royalty is to be paid if an exclusive license has been granted. The MAR payment is computed using the royalty rate times the minimum number of units of Part 1. Minimum annual royalties start as also stated in Part 1. If the minimum number of units in not sold in any year, the licensor must pay the appropriate makeup difference to the licensee with its payment for the fourth quarter.

Subpart D: If the minimum is not paid by licensee, either due to lack of sufficient sales or licensee's choice, then the license grant will be converted to a non-exclusive one, and the licensor can immediately license others.

Subpart E: If the license is or becomes non-exclusive, then the licensor may not grant more favorable terms to any other licensee.

Subpart F: Patent royalties are not due after the patent expires, or is declared invalid, or no patent is granted.

Subpart G: Late payments earn interest at 10%.

Subpart H: The "Net Factory Sales Price," on which royalties are based, is the factory selling price, less shipping, insurance, taxes, etc., if billed separately. If the units are imported, then the importer's gross selling price is the basis for royalties. The royalty paid on returns is deductible against future royalties.

Part 8: This requires the licensee to keep full records for at least two years so that the licensee can verify the royalty payments.

Part 9: Here the licensee's sublicensees are bound by all of the terms of the agreement and the licensee must notify the licensor if it grants sublicenses. A licensee will usually grant a sublicense when it has the licensed product made for it by a contracting company.

Part 10: This simply states the parties' responsibilities for patent prosecution.

Subpart A: Requires the licensor to pay for prosecution of the U.S. patent application, together with the patent maintenance fees which are payable after the patent issues. If the licensor intends to abandon the patent application, it must notify the licensee at least two months in advance to give it the opportunity to take over.

Subpart B: The licensor may file for patent coverage abroad, but if it doesn't do so, then the licensee may do so. If licensor wants to license any foreign licensees, it has to give the licensee the opportunity of first refusal.

Subpart C: If the licensee takes over the U.S. patent prosecution, and is successful, then it can reduce its royalties by 25%, and can deduct its patent prosecution expenses. If the licensee elects to file abroad, then the royalty rate on foreign sales is 50% of the U.S. rate, less foreign prosecution expenses.

Part 11: This requires the licensee to mark products sold with the legend "patent pending" while the patent application is pending and with the patent number (see Chapter 15) after the patent issues.

Part 12: This states that if the patent is infringed, the licensor can sue to enforce its patent rights. If it doesn't choose to do so, the licensee may do so. If the licensee sues, it can keep 75% of this recovery, less costs of the suit.

Part 13: This clause states that licensor doesn't guarantee that its patent is valid or that it has any particular scope (breadth).

Part 14: This clause states that the term or maximum duration of the agreement shall be until the last patent of licensor expires, unless know-how is licensed, in which case Part 6 governs the term.

Part 15: This clause covers the situations when the parties may terminate the agreement before the term expires. Under Subparts A and B, the licensor may terminate the agreement if the licensee defaults in making royalty payments, or if it ever declares bankruptcy, etc. Subpart C, the antishelving clause, is very important. This protects the licensor in case the licensee stops production for 1.5 years, or doesn't start production within 1.5 years from the date the license agreement is signed. In these cases, the licensor can terminate the agreement.

Part 16: This clause states how and where notices under the agreement are to be sent.

Part 17: This clause provides that if the parties have any dispute, they shall submit the matter to mediation. If mediation can't resolve the dispute, the parties must submit the dispute to binding and final arbitration. In no case will the dispute go to a court for resolution since litigation is extremely expensive and thus works to the detriment of the independent inventor.

Part 18: This clause allows the licensor to assign (legally transfer) its rights to anyone without permission, but the licensee needs advance permission of licensor, unless it makes an assignment to its successor in business.

Part 19: This clause specifies that the laws of licensee's state shall govern interpretation of the agreement. Normally, state law on the interpretation of contracts doesn't vary much, but since a licensor is usually at an economic disadvantage, I've given it the benefit here.

Part 20: This states that neither party shall take any action which hampers the rights of the other and that both parties shall engage in good faith and fair dealing. This clause is supposed to be read into any agreement, but I've expressly stated it in order to increase cooperation and reduce disputes.

Part 21: This one states that the parties have carefully read the agreement and have consulted, or have been given an opportunity to consult, counsel and that each has received a signed original. This makes a challenge to the agreement more difficult.

All that remains is to sign the agreement. Each party should get an original, ink-signed copy.

Note: Again, let me remind you that while the Universal Agreement incorporates most of the customary terms and covers many common licensing situations, it probably won't be appropriate for your situation without some modification. Obviously, if your arrangement won't fit within the terms of this agreement, or if you don't like any of the "fixed" terms, such as the 80 hours consultation (Clause 6), the 15% market share (Clause 6), compulsory arbitration (Clause 17), etc., you should propose changes, or hire an expert to help you.

I. How Much Should You Get for Your Invention?

Many inventors seem to believe that patents are almost always licensed at a royalty rate of 5%. The 5% royalty generally means that you would get five percent of the money received by the factory for its sales of the item embodying your invention. This is sometimes termed five percent of the "ex-factory" price. This assumption is simply not true. While 5% is often used as a starting point in many license negotiations, very few licenses are granted at this rate. I've seen them run from 0.1% to 15% of the factory price of the licensed item (as high as 30% of the retail price for software).

As you've guessed, many factors affect the royalty rate. Obviously, the more desirable your invention is to the licensee, the better royalty you'll get, subject to industry norms. Here's a list of some factors which militate in favor of increasing the royalty rate; you should use as many of these as possible in your negotiations: Sales volume, selling price, low competition, your bargaining skill, profit margin, ingeniousness of product, amount of development work inventor has done, degree to which invention pervades product, size of licensed territory, amount of services or material/ parts you furnish, absence of competition between licensee and licensor, degree of respect in field for patent, difficulty of avoiding patent to licensee, ease of making agreement, cost savings to licensee, and low start-up costs to implement invention. Also the custom of the industry will dominate, e.g., toys usually get a royalty rate of 2.5 to 4%; medical products 6 to 7%. An exclusive license will entitle you to about 50% more than a non-exclusive license.

Instead of a negotiated percentage, some experts advocate getting a royalty equivalent to "one-third of the manufacturer's profit rate." This means that the company will take its selling price for your invention, say $10, subtract its cost of manufacture, including overhead, say $7, and give you one-third of the difference, i.e., $1 = 1/3 of its $3 profit. This type of royalty is often enticing to a manufacturer since the company only contemplates parting with a portion of its profit, not paying a fixed sum per item, whether the particular product turns out to be profitable or not. If your licensee is willing to accept this type of royalty, you can substitute this language in the Universal Agreement. But, if you do so, be sure you include an auditing right (such as Clause 8) to insure that you can verify its cost of manufacture.

If you're offered a single lump-sum payment for all your rights (this is rare), should you take it, and if so, how much should you get? To answer the first question, only you can decide if a relatively large bird in the hand is worth more than a potential (but by no means assured) stream of smaller, but aggregatively heavier, smaller birds in the bush over the years. To grapple with the second question, estimate the potential sales of your invention for the life of your patent application (one to three years), plus the term of the patent (17 years), then apply your royalty to this figure, and be willing to take half of this as application single payment lump sum for application fully paid-up license.

For example, suppose you expect your widget to be sold for the next 19 years (two years during patent pendency, and 17 years during life of patent), for an average factory price of 50 cents and an average yearly quantity of 150,000 units, and that application five percent royalty is fair. Applying the formula, the substitute lump-sum payment for your royalty would be 0.5 x 19 x 150,000 x $0.05, or $35,625. If you are offered much less than this, it could very well be unwise to sell.

Don't make Mary Jacobs' mistake. She invented the bra (out of two hankies and a ribbon) and was able to sell her patent for $15K in 1914. Although this was a princely sum then, she practically gave it away since (as you know) her invention soon took hold and her patent eventually was worth $15 *million*!

The disadvantage with the alternative lump-sum calculation is that it's very hard to estimate anything about what will happen in the next 19 years. Will sales go up or down? Will the product become obsolete or even more popular? Will competition affect its price, etc.? These are just some of the imponderables and unknowables, so, as stated, be extremely careful before selling your rights for a single lump-sum payment.

Precautions: If you do have an opportunity to sell your invention, you should use the assignment form (16-2), changing "For value received" at the beginning of the form to — In exchange for $_____ —. For obvious reasons, make sure you actually receive the money by certified check or money order before you sign. Do not, under any circumstances, assign your patent in return for a series of payments: if your assignee defaults in the payments, you'll be left without your patent or your money, but with a big legal headache — getting your patent back. If someone wants to buy your patent for a series of payments, see a lawyer or legal forms book and make a suitable license with an agreement to assign only after all payments have been made.

It has been said that knowing where to look is half the battle of knowing the law. With this in mind, this section is provided to help you avoid having to hire a patent lawyer in case you encounter any situations or problems which this book does not cover. I've also provided a number of resources and publications I feel will be of interest to inventors and other creative people. I provide comment generally where the title of the book or source isn't self-explanatory. Most books which can't be found in a general or business library may be found in a law library. Prices aren't indicated since they change frequently. This list isn't exclusive by any means: if you browse in your bookstore or a patent depository library, you'll find many other valuable books of interest.

Government Publications

Annual Index of Patents. Issued yearly in two volumes: *Patentees* and *Titles of Inventions*. U.S. Government Printing Office (GPO), Washington, DC 20402. Comes out long after the end of year to which it pertains--for instance, in September. Available in search and public libraries.

Attorneys and Agents Registered to Practice before the U.S. Patent and Trademark Office. Annual. GPO. Contains alphabetical and geographical listings of all attorneys and agents.

Classification Definitions. Many looseleaf volumes. Contains definitions for each of 66,000 subclasses. Available in search libraries.

Rules of Practice in Patent Cases. (Title 37, Code of Federal Regulations) GPO. Revised annually. The PTO's Rules of Practice. A must for all who prosecute their own patent applications. Almost

always incomplete due to frequent rule changes. Look in *Official Gazettes* for later rules.

Index to Classification. Loose-leaf. Contains 66,000 subclasses and cross-references arranged alphabetically. Search libraries.

Manual of Classification. Loose-leaf. Contains 300 search classes for patents arranged numerically, together with subclasses in each class. Search libraries.

Manual of Patent Examining Procedure. Revisions issued several times per year. GPO. Called "the patent examiner's bible," the MPEP provides answers to most questions about patent prosecution.

Law Books Relating to Patents

Corpus Juris Secundum, vol. 69, Patents. A legal encyclopedia which will answer almost any question on patent law. West Pub. Co., St. Paul, 1958 (supplemented annually). Any law library.

Elias, S., *Intellectual Property Law Dictionary*, 1st. ed., Nolo Press 950 Parker Street, Berkeley, CA 94710.

Greene, A.M. *Patents Throughout the World*. Clark Boardman, 1980. Revised annually.

Greer Jr., T.J. *Writing and Understanding U.S. Patent Claims; A Programmed Workbook*. Michie, 1979.

Grissom, G.& Pressman, D., *The Inventor's Notebook*, 1st ed., Nolo Press, 1987.

Journal of the Patent Office Society. Monthly. Box 2600, Arlington, VA 22202. Contains articles on patent law and advertisements by patent services, for instance, draftspersons, drawing reproducers, searchers.

Landis, J.L. *The Mechanics of Patent Claim Drafting*, 2d ed., 1974, Practicing Law Instititute, 810 Seventh Avenue, NY 10019.

Patent and Trademark Laws. BNA. Revised annually.

Patent Official Gazette. Issued each Tuesday. GPO. Contains drawing and main claim of every patent issued each week, miscellaneous notices, new PTO rules; lists inventors, assignees, etc. Available in search and public libraries.

Questions and Answers about Plant Patents. PTO. Free.

Questions and Answers about Trademarks. PTO. Free.

Remer, D., & Elias, S., *Legal Care For Your Software*, 3d. ed., Nolo Press, 1987.

Salone, M.J., How to Copyright Software, 2d. ed., Nolo Press, 1987.

Small Business Administration. The SBA's list of free publications has three sections: "Management Aids," "Small Marketer's Aids," and "Small Business Bibiliographies." Listed are dozens of excellent, concise business pamphlets, such as no. 82, *Reducing the Risks in Product Development*, and 6.004, *Selecting the Legal Structure for Your Firm*. Order from your local SBA office or SBA, Washington DC 20416.

The Story of the U.S. PTO. GPO. 1985.

Drafting Patent License Agreements, 2d. ed., Mayers & Brunsvold. BNA

Trademark Official Gazette. GPO. Lists trademarks published for opposition and registered each week.

Martindale-Hubbell Law Directory. Martindale-Hubbell, New York. Annual. Lists patent attorneys by geographical area (including some foreign) and give ages, colleges, and sometimes other information about attorneys. Any law library.

Milgrim, R.M. *Trade Secrets*. Matthew Bender, New York, 1967.

Nordhaus, R.C. *Patent License Agreements*. Jural, Chicago, 60626, 1976.

Walker On Patents (2d. ed. by Deller; 3d. ed by Lipscome) Bancroft - Whitney. Acomprehensive legal treatise. Law Libraries.

White, R.S. *Patent Litigation Procedure & Tactics*. Matthew Bender, New York, 1974.

General Interest Books Relating to Patents and Inventions

Clark, R.W. *Edison--The Man Who Made the Future*. Putnam, New York, 1977.

Florman, S.C., *The Existential Pleasures of Engineering*. St. Martens, 1976. A brilliant, eloquent panegyric of technology; a crushing blow to Reich, Mumford, Rozak, et al.

Harness, Charles R., Esq. *The Catalyst*. Pocket Books, New York, 1980. Science fiction story involving a patent attorney, an invention, and an interference.

Inventing: How the Masters Did It. Moore Pub., Durham, N.C., 1974.

Lessing, L. *Man of High Fidelity: Edwin Howard Armstrong*. Lippincott, Philadelphia, 1956. Biography of the inventor of frequency modulation; he committed suicide because of the delays and difficulties of patent litigation against the large radio companies, but his widow eventually collected millions in settlements.

The National Inventors Hall of Fame. *Biographies of Inductees*. NIHF Foundation, Room 1D01, Crystal Plaza 3, 2001 Jeff Davis Hwy., Arlington, VA 22202. Free.

Ord-Hume, A., *Perpetual Motion: The History of An Obsession*, St. Martens, 1981. A must if you're filing on a perpetual motion machine.

Paige, R.E. *Complete Guide to Making Money with Your Ideas and Inventions*. Barnes & Noble, New York, 1976. Excellent guide to invention marketing.

Walsh, J.E. *One Day at Kitty Hawk*. Crowell, New York, 1975. The story of the development and sale of rights to the airplane.

Publications Relating to Business

Adams, A.B. *Apollo Handbook of Practical Public Relations*. Apollo Editions, New York, 1970. How to get publicity.

Applied Sciences and Technology Index. H.W. Wilson Co., Bronx, NY 10452. Lists engineering, scientific, and industrial periodicals by subject.

Ayer Directory of Newspapers and Periodicals. Annual. Ayer Press, Philadelphia, PA 19106. Lists United States newspapers and magazines geographically.

Bacon's Publicity Checker--Magazines, Bacon's Publicity Checker--Newspapers. Annual. Bacon Pub. Co., Chicago. Classifies all sources of publicity.

Bragonier Jr., R. and Fisher, J., *What's What; A Visual Glossary of Everyday Objects*. Ballantine, 1981.

California Manufacturers Register. Annual. 1115 S. Boyle Ave., Los Angeles, CA 90023.

Clifford, D. & Warner, J., *The Partnership Book*, 3d. ed., Nolo Press, 1988.

Conover Mast Purchasing Directory. Conover Mast, Denver, CO 80206. Annual. Three volumes. Manufacturers listed alphabetically and by products. Also lists trademarks.

Dible, D.M. *Up Your Own Organization*. Entrepreneur Press, c/o Hawthorn Books, New York. How to start and finance a business.

Drucker, P., *Innovation and Entrepreneurship*, Harper & Row, 1985. How any organization can become entrepreneurial.

Dun & Bradstreet Reference Book. Six issues per year. Lists three million businesses in the United States and Canada. D&B also publishes specialized reference books and directories, e.g., *Apparel Trades Book* and *Metalworking Marketing Directory*.

R. Fisher & W. Ury, *Getting to Yes; Negotiating Agreements Without Giving In*. Penguin, 1981.

Guide to American Directories, 9th ed. B. Klein Pubs., New York, 1975. Lists directories by industry, profession, and function.

International Yellow Pages. R.H. Donnelley Corp., NY 10017. Similar to local Yellow Pages, but provides foreign business listings.

McKeever, M.,, *How to Write a Business Plan*, 3d. ed., Nolo Press, 1988.

MacRae's Blue Book. MacRae's Blue Book Co., Hinsdale, IL 60521. Sources of industrial equipment, products, and materials. Also lists trademarks.

Petillon, L.R. & Hull, R.J., *R & D Partnerships*. Clark Boardman, 1985.

Phillips, M., & Rasberry, S., *Marketing Without Advertising*, 1st ed., Nolo Press, 1988.

Rich, S.R. & Gumpert, D., *Business Plans That Win Money: Lessons From the MIT Enterprises Forum*. Harper & Row, 1985.

Thomas Register of American Manufacturers. Thomas Pub., NY 1000. Eleven volumes. Similar to *Conover Mast Directory* above.

Trademark Register of the United States. Annual. Trademark Register, Washington Bldg., Washington, DC 20005. Lists all registered trademarks by subject matter classes.

Ulrich's International Periodicals Directory. R.R. Bowker Co., NY 10036. Lists periodicals by subject.

Venture Capital Monthly. S.M. Rubel Co., Chicago, IL.

Brown, D. *The Entrepreneur's Manual.* Ballantine, 1981.

Books Relating to Self-Improvement

I believe that the real key to success and happiness, in inventing as well as life, lies principally within each individual's own mind. A positive, optimistic attitude, hard work and perseverance, the willingness to take full responsibility for one's own destiny, and living and thinking mainly in the present time--rather than luck, inherited abilities, and circumstances--are principally responsible for success and happiness. I have therefore provided a list of books whose main purpose is to prime you with the attitude to secure such success and happiness so that you'll be able to use *Patent It Yourself* as effectively as possible.

Bois, S., *Explorations in Awareness.* Harper & Row, 1957.

Break through mental blocks and preconceptions.

Branden, Nathaniel. *The Psychology of Self-Esteem.* Nash, Los Angeles, 1971.

Dyer, W.W. *Your Erroneous Zones.* Funk & Wagnalls, New York, 1976.

Ellis, A., and R.A. Harper. *A New guide to Rational Living.* Wilshire Book Co., Los Angeles, 1975.

Harman, W. & Rheingold, H., *Higher Creativity--Liberating the Unconscious For Breakthrough Insights,* J.P. Tarcher, 1984.

Hayakawa, S.I., *Language in Thought and Action,* HBJ 1978.

Johnson, W., *People In Quandries.* Harper & Row, 1946. Classic book on emotional problem solving.

Weinberg, H. *Levels of Knowing & Existence.* Harper & Row, 1961. A new approach that answers many questions.

GLOSSARY

This Glossary[1] provides a list of useful words to describe the hardware, parts, and functions of your invention in the specification and claims. The most esoteric of these words are briefly defined. While some definitions are similar, this is due to space limitations; all words have nuances in meanings.

If you're looking for a word to describe a certain part, look through the list for a likely prospect and then check a dictionary for its precise meaning. If you can't find the right word here, look in your search patents, in "What's What" or another visual dictonary, or in a thesaurus. If you can't find an appropriate word, you'll probably be able to get away with "member" or "means-plus-a-function" language. Also, for new fields, you may invent words, preferably using Latin or Greek roots, as Farnsworth did with "television", or by extending the meaning of words from analogous devices (e.g., "base" for a part of a transistor.)Very technical or specialized fields have their own vocabulary (e.g., "catamenial" in medicine, "syzygy" in astronomy); look in appropriate tutorial texts for these. The words are grouped loosely by the following functions:

1. Structure
2. Mounting And Fastening
3. Springs
4. Numbers
5. Placement
6. Voids
7. Shape
8. Material Properties
9. Optics
10. Fluid Flow
11. Electronics
12. Movement
13. Rotation

---

[1]Expanded and used with kind permission and thanks from a list originally prepared by Louis B. Applebaum, Esq. of Newport, RI.

1. Structure

annulus (ring)
apron
arm
arbor (shaft)
bail (arch wire)
band
barrel
bascale (seesaw)
base
beam
—cantilever
—simple
belt
bib
blade
blower
board
body
boom
boss (projection)
boule (pear-shaped)
bougie (body-insertion member)
breech (back part)
branch
canard (front wing)
carriage
case
cincture (encircling band)
clew (sail part)
chord
column
configuration
container
conveyor
cornice (horiz. top of structure)
cover
cylinder
dasher (plunger,
churn)
detent
device
die
doctor blade (scraper)
disparate (dissimilar)
diversion
dog (holder)
drum
echelon (staggered line)
element
enclosure
fence (stop on tool)
flute (groove on shaft)
fillet (narrow strip)
fin
finger
finial
fircate (branch)
flange
frame
fluke (triangular part)
frit (vitreous substance)
frustrum (cut-off area)
frame generatrix (path traced)
futtock (curved ship timber)
gnomon (sundial upright)
gudgeon (pivot)
gauge
grommet
gusset (triangular insert)
handle
head
header (base, support)
homologous
housing
hub
jacket
jaw
lagging (support)
leg
lip
list (margin strip)
magazine
mandrel (tapered axle)
manifold
marginate (w/margin)
medium
member
mullion (dividing strip)
neck
nacelle (pod)
object
panel
particle
partition
piece
piston
platform
plug
pontoon
portion
post
projection
purlin (horiz. rafter support)
pylon (support)
rib
ring
rod
screed (guide strip)
scroll
sear (catch)
shell
shoe
shoulder
skeleton
sleeve
snorkel
spar (pole support)
spline (projection on shaft)
spoke
sprag (spoke stop)
spur
stanchion
station
stay
stent (stretcher)
step
stepped
stile (dividing strip)
stop
strake (ship plank)
strip
strut
tang (shank, tool)
tine
tip
tongue
track
trace (pivoted rod)
tracery (scrolling)
trave (crossbar)
truss
tuft
turret
tuyere (air pipe)
upright
wall
warp
woof (weft)

2. Mounting & Fastening

attach
bolt
billet (tip of belt)
busing
cable
clamp
cleat (reinforcer)
clevis (U-shaped pin)
connection
couple
coupling
demountably
docking
dowel
engage
fay (join)
ferrule (barrel)
ferruminate (attache,
solder)
fix
guy wire
harp (lamp shade support)
hold

holder
hook
imbricate (regular overlap)
joint
—universal
keeper
key

latch
lock
lug
matrix
mount
nail
nut
pin

ribband (holds ribs)
rivet
scarf (notched joint)
screw
seam
seat
secure
set

sliding
solder
springably
support
thrust
weld

3. Springs

air
bias
—element
coil
compressed

elastic
expanded
helical
—compression
—tension

leaf
press
relaxed
resitient
springably

torsional
urge

4. Numbers

argument
compound
difference
divisor
dividend
equation

formula
index
lemma
minuend
modulo
multiplicand

multiplicity
multiplier
plurality
power
product
quotient

remainder
subtrahend
variable

5. Placement (Relation)

adjacent
aft
aligned
angle
aposition (facing)
array
attached
axial
bottom
close
complementary
concentric
contracted
course
contiguous
crest
disposed
distal
divided
edge

engaged
evert (inside out)
extended
external
face
fiducial (reference)
film
fore
horizontal
integral
intermediate
internal
interposed
juxtaposed
layer
located
lower
mating
meshing
mesial (between)

normal
oblique
obtuse
offset
open
opposed
overlapping
parallel
perpendicular
positioned
projecting
prolapsed (out of place)
proximal
proximate
reference
removable
resting
rim
row

sandwich
section
slant
spacer
staggered
superimposed
supported
surface
surrounding
symmetrical
tilt
top
vernier (9:10 gauge)
vertical

6. Voids

aperture
bore
cavity
chamber
concavity
cutout
dimple
duct
embrasure (slant opening)
engraved

filister (groove)
foramen (opening)
fossa (depression)
gain (notch)
gap
groove
hole
hollow
intaglinated (engraved)
lumen (bore of tube)

mortise (cutout)
nock (notch on arrow)notch
opening
orifice
passage
placket (garment slit)
raceway
rabbet (groove)
recess
separation

slit
slot
sulcus (groove)
ullage (lost liquid)
via (path)

7. Shape

acclivity (slope)
acicular (needle shaped)
agonic (no angle)
annular
anticline (peak)
arch
arcuate
barrel
bevel
bifurcated (2 branches)
bight (bend)
buckled
bucket
chamfer (beveled)
channel
circular
coin
concave
convex
conical
convoluted (curled in)
corner (inside, outside)
corrugated
crest
crimp

crispate (curled)
cup
cusp (projection)
cylinder
depression
dihedral (two-faced)
direction
disc
dome
drawing (pulling out)
elliptical
fairing (streamlined)
fin
flange
fold
fork
fossa (groove)
fundus (base)
furcate (branched)
helical
hook
incurvate (curved in)
line
lozenge (diamond shaped)
mammilated (nipple-shaped)
notch
oblate (flattened)

ogive (pointed arch)
oblong
orb (globe)
oval
parabolic
parallelogram
plane
prolate (cigar-shaped)
rectangular
reticulated (gridlike)
rhombus (not lozenge)
rhomboid (non-parallel sides)
round
salient (standing out)
serrated
sheet
shelf
sinusoidal
slab
spherical
square
stamped
striated (grooved or ridged)
syncline (V-shaped)
swaged (flattened)
swale (depression)

taper terminus (end)
tesselated (tiled)
thill (horse joinder stake)
topology (unchangeable geometry)
tram (on wheels)
trefoil (three-leaved)
triangular
trihedral (3-sided)
trough
tumesence (deturmesence)
tubular
turbinate (top/spiral shaped)
twist
upset (distorted)
vermiculate (worm-eaten)
volute (spiral)
wafer
web
wedge
xyresic (razor sharp)

8. Materials & Properties

adhesive
concrete
cork
dappled (spotted)
denier (gauge)
dense
elastic
enlarged
fabric
fiber
flexible

foraminous
haptic (sense of
touch)
humectant
(moistener)
insulation
liquid
material
metal
nappy
opaque

pied (splotched)
plastic
porous
prill (pure metal)
refractory
resilient
rigid
rubber
sand
screen
shirred (gathered)

stratified (layered)
strong
sturdy
translucent
transparent
xerotic (dry)
wood

9. Optics

astigmatic
bezel
bulb
—fluorescent
—incandescent
fresnel

lamp
light
—beam
—ray
opaque
pellicle

pellucid (clear)
reflection
refraction
schlieren (streaks)
transparent
translucent

transmission
window

10. Fluid Flow

accumulator
afferent (to center)
aspirator
bellows
bibb (valve)
cock (valve)
conduit
connector
convection
cylinder
—piston
—rod
dashpot
diaphragm

discharge
dispenser
efferent (away from
center)
filter
fitting
flue
gasket
hose
hydraulic
medium
nozzle
obturator (blocker)
outlet

pipe
plunger
port
—inlet
—outlet
pump
—centrifugal
—gear
—piston
—reservoir
—seal
—siphon
—tank
—vane

sprue (vent tube)
tube
valve
—ball
—check
—control
—gate
—shutoff
wattle (intertwined
wall)
weir (dam)

11. Electronics

adder
amplifier
astable
capacitance
clipping
conductor

contact
control element
demodulator
diode
electrode
electromagnet

filament
flip flop
gate (AND, OR, etc.)
impedance
inductance
insulator

integrated circuit
laser
lead
light emitting diode
line cord
liquid crystal

maser
memory
motor
multiplier
multivibrator
pixel (CRT spot)
power supply
oscillator

read-only memory
read-and-write
memory
resistance
sampling
Schmitt trigger
Shottky diode
shift register

socket
solenoid
switch
terminal
thermistor
transformer
transistor
triode

valve
varistor
wire
Zener diode

12. Movement

alternate
articulate (jointed)
avulsion (tear away)
cam
compression
cyclic
detent (click)
downward
drag
drift pin
drill
eccentric
emergent
epicyclic (on circle)
extensible

extrude
grinding
impact
inclined plane
inertia
interval
lag
lead
lever
linkage
—parallel
longitudinal
machine
meeting
nutate (to and fro)

pressing
propelling
pulverize
sagging
sequacious (regular)
severing
skive (peel)
slidable
straight line
—motion
snub (stop)
terminating
toggle
torque
traction

transverse
traversing
triturate (grind to
powder)
trochoid (roll on
circle)
urging
vibrating
wedge

13. Rotation

antifriction
—ball
—needle
—roller
—tapered
bell crank
brake
—band
—disc
—shoe
bushing
cam
chain
clevis (circular
holder)
clutch

—centrifugal
—sprag (stop)
—toothed
—one-way
cog (tooth)
connecting rod
crank arm
drive
—belt
—pulley
—sheave
—toothed
flexible coupling
friction
fulcrum
gear

—bevel
—crown
—internal
—non-circular
—pinion
—right angle
—spur
—worm
—wheel
guide
intermittent
—escapement
—geneva
—pawl
—pendulum
—ratchet

jack
journal
orbit
pivot
pulley
screw
sheave (pulley)
sprocket
radial
radius bar
seal
tappet (valve cam)
variable speed
winch
yoke

Appendix

FEE SCHEDULE, TIMING CHART, AND FORMS

Note: Form numbers indicate the chapters in which the forms are discussed; e.g., Form 10-7 is discussed in Chapter 10.

Fee Schedule

These fees were good as of this printing, but you should check with the PTO, (703) 557-4636, or its PCT Office, (703) 557-2003, if it's much later. Two fees separated by a slash refer to large entity/ small entity; single fees apply to both entities. PTO fees are listed in order for the patenting process.

Service or Item	Fee ($)	Form/Chapter
PTO Fees		
Disclosure Document, filing	6	3-3
Printed Copy Of Patent or Patent Order Coupon		
Utility/Design; Also For Copy of SIR	1.50	Ch. 6
Application Filing Fees:		
Utility Patent (incl. reissue)	370/185	10-1, 14-1
Design Patent	150/75	10-8
Plant Patent	250/125	Ch. 10
Fee For Each Independent Claim Over Three	36/18	10-1, 14-1
Fee For Each Claim Over Twenty (Independent or Dependent)	12/6	Ch. 10
Surcharge — Multiple Dependent Claims In Any		
Application	120/60	Ch. 10
Surcharge If Filing Fee Or Declaration Late	120/60	Ch. 10
Recording Assignment Per Application or Patent involved	8	10-1
Surcharge If Any Check Bounces	50	Ch. 10
Petitions To Commissioner:		
Regarding inventorship, Misc.,		
Maint. Fees, Interferences, Foreign		
Filing Licenses, Access to		Ch. 13
Records, Foreign Priority Papers,		
Amendments After Issue Fee, Defer/		
Withdraw A Case From Issue	120	10-7
To Make Application Special	80	Ch. 10
Extensions To Reply To Office Action:		
1st Month	62/31	13-5
2nd Month	180/90	13-5
3rd Month	430/215	13-5
4th Month (if avail.)	680/340	13-5
Petition To Revive Abandoned Appn.:		
Unavoidable Delay	62/31	Ch. 13
Unintentional Delay	620/310	Ch. 13
Certified Copy Patent Application As Filed	10	Ch. 12
Appeal To Board Of Appeals & Pat. Intrfs.:		
Filing Notice of Appeal	140/70	Ch. 13
Filing Brief	140/70	Ch. 13
Oral Hearing	120/60	Ch. 13
Application Issue Fees:		
Utility Patent	620/310	Ch. 13
Design Patent	220/110	Ch. 13
Plant Patent	310/155	Ch. 13
Certificate To Correct Patent (Applicant's Mistake)	60	15-1
Reexamination Fee	2000	Ch. 15
Partial Refund (If Reexamination Failed)	1500	Ch. 15

Fee Schedule (continued)

Service or Item	Fee ($)	Form/Chapter
Utility Patent Maintenance Fees:		
I (3.5 years — pays for yrs 4 thru 8)	490/245	15-3
II (7.5 years — pays for yrs 9 thru 12)	990/495	15-3
III (11.5 years — pays for yrs 13 thru 17)	1480/740	15-3
Late Charge (in 6-month grace period)	120/60	15-3
Late Charge (after patent expires)	550	15
Copy Of PTO Records, per page	0.50	
Certified Copy Of File & Contents— Issued Patent	170	Ch. 15
Certified Copy of Patent Assignment Record	5	Ch. 14
Disclaimer Of Claims Or Terminal Part Of Term Of Patent	62/31	
Dedication Of Entire Term Or Terminal Part Of Term Of Patent	NC	

Other Fees

Service or Item	Fee ($)	Form/Chapter
Trademark Application Filing (In PTO)	175	Ch. 1
Trademark Application Filing (In California)	50	Ch. 1
Copyright Application Filing (In Copyright Office)	10	Ch. 1
Filing a European Pat. Appn., incl. agent's fee, approx.	3500-4000	Ch. 12

PCT Fees

Service or Item	Fee ($)	Form/Chapter
Transmittal Fee	170	Ch. 12
Search Fees:		
In US PTO		
—no corres. prior US appn. filed	550	Ch. 12
—corres. prior US appn. filed	380	Ch. 12
In European Patent Office	1140	Ch. 12
International Fees:		
Basic (First 30 Sheets)	436	Ch. 12
Each Additional Sheet Over 30	9	Ch. 12
Designation Fee, each country or office up to 10	106	Ch. 12
11th and additional countries or offices	NC	Ch. 12
Chapter II Fees:		
Handling Fee	134	Ch. 12
Examination Fee		
In US PTO	400	Ch. 12
In EPO	600	Ch. 12

Timing Chart

The following is a summary of some of the more important timing intervals that apply in intangible property law. This list in not intended to be comprehensive, and certain exceptions may be applicable, so check the pertinent parts of this book, or with a patent attorney if you have a special situation or need more precise advice.

From the date of first publication, offer of sale, sale, or public or commerial use (excluding experimental use) of anything embodying an invention, one must file a US utility, design, or plant patent application within ....1 year.

To preserve all foreign filing rights, one must not sell or publicly disclose details of an invention until .... after US filing date.

From the PTO's mailing date, one must file a response to most official actions within .... 3 months.

The maximum statutory time to reply to an Office Action, provided extensions are bought, is .... 6 months.

The full term of a utility or plant patent is ....17 years.

The full term of a design patent is .... 14 years.

From the date of issue (grant) the issue fee will keep a utility patent in force for the first.. 4 years.

Timely payment of Maintenance Fee I will keep a utility patent in force for another..4 years.

Timely payment of Maintenance Fee II will keep a utility patent in force for another.. 4 years.

Timely payment of Maintenance Fee III will keep a utility patent in force for the final...5 years.

For works not made for hire, the copyright term is .... author's life + 50 years.

For works made for hire, the copyright term is the shorter of .... 75 years from publication or 100 years from creation.

To get statutory damages and attorney fees, one must apply to register a copyright before infringement begins or within ....3 months of publication.

A California state trademark registration lasts for .... 10 years.

A US (federal) trademark registration lasts for ....20 years.

State and US trademark registrations can be renewed .... in perpetuity.

If kept secret, and provided it's not discovered independently, a trade secret will be enforceable against those who discover it illegally .... in perpetuity.

Unless a foreign filing license has been granted, after filing a US patent application, before foreign filing a patent application, you must wait .... 6 months.

From the US filing date, one must file a foreign Convention application (PCT, EPO, or industrial countries) within ....1 year.

Timing Chart (continued)

One must file a foreign Non-Convention
application (non-industrial countries) .... before invention becomes publicly known.

From the US filing date, after filing a PCT
application, if examination in the foreign jurisdiction
is desired, one must file abroad within ....20 months.

From the US filing date, after filing a PCT application,
if examination in the US PTO or the European Patent
Office is desired (Chapter II), one must file a request within ....19 months.

From the US filing date, after filing a PCT application
and electing Chapter II, one must file abroad within ....30 months.

Proprietary Materials Agreement

(Keep-Confidential/ Non-Disclosure Agreement)

Proprietary Materials (items, documents, or
models loaned—describe or identify fully):

_____

_____

Proprietary Materials loaned by (name and address):_____

_____ ("Lender")

Proprietary Materials loaned to (name and address): _____

_____ ("Borrower")

Borrower acknowledges and agrees as follows:

(1) **Borrower** has received the above **Proprietary Materials** from **Lender**.

(2) These **Proprietary Materials** contain valuable proprietary information of **Lender**. This proprietary information constitutes a trade secret of **Lender** and loss or outside disclosure of these materials or the information contained within these materials will harm lender economically.

(3) **Borrower** acknowledges that these **Proprietary Materials** are furnished to **Borrower** under the following conditions:

    (a) These **Proprietary Materials** and the information they contain shall be used by **Borrower** solely to review or evaluate a proposal or information from, supply a quotation to, or provide a component or item for **Lender**.

    (b) **Borrower** agrees not to disclose these **Proprietary Materials** or the information they contain to any other person, except persons within **Borrower's** organization having a good faith "need to know" same for the purpose of this Agreement. If these **Proprietary Materials** are disclosed under this part to any other person within **Borrower's** organization, **Borrower** shall have each such person also complete and sign a copy of this Agreement and furnish such copy to **Lender**.

    (c) **Borrower** shall exercise a high degree of care to safeguard these **Proprietary Materials** and the information they contain from access or disclosure to all unauthorized persons.

    (d) **Borrower** shall not make any copies of these **Proprietary Materials** except upon written permission of **Lender** and **Borrower** shall return all **Proprietary Materials** (including any copies made) to **Lender** at any time upon request by **Lender**.

(4) These terms shall not apply to any information which **Borrower** can document becomes part of the general public knowledge without fault of **Borrower** or comes into **Borrower's** possession in good faith without restriction.

Borrower _____

Date_____

Invention Disclosure

Inventor(s):_____

Address(es):_____

Title of Invention:_____

    To record **Conception**, describe: 1. Circumstances of conception, 2. Purposes and advantages of invention, 3. Description, 4. Sketches, 5. Ramifications, 6. Possible novel features, and 7. Closest known prior art. To record **Building and Testing**, describe: 1. Any previous disclosure of conception, 2. Construction, 3. Ramifications, 4. Tests, and 5. Test results. Include sketches and photos, where possible. Continue on additional identical copies of this sheet if necessary; inventors and witnesses should sign all sheets.

Inventor(s):_____

Dates of Signatures:_____

The above confidential information is

Witnessed and Understood: _____ __/__/__

_____ __/__/__

Date _____

Commissioner of Patents and Trademarks
Washington, District of Columbia 20231

Request for Participation in Disclosure Document Program:

Disclosure of _____
                              Your Name(s)

Entitled _____
                              Title of Disclosure

Sir:

Attached is a disclosure of the above-entitled invention (consisting of ___ sheets of written description and ___ separate drawings or photos), a check for $_____ a stamped, addressed return envelope, and a duplicate copy of this letter.

It is respectfully requested that this disclosure be accepted and retained for two years (or longer if it is later referred to in a paper filed in a patent application) under the Disclosure Document Program and that the enclosed duplicate of this letter be date stamped, numbered, and returned in the envelope also enclosed.

The undersigned understands that (1) this disclosure document is neither a patent application nor a substitute for one, (2) its receipt date will not become the effective filing date of a later-filed patent application, (3) it will be retained for two years and then destroyed unless it is referred to in a patent application, (4) this two-year retention period is not a "grace period" during which a patent application can be filed without loss of benefits, (5) in addition to this document, proof of diligence in building and testing the invention, and/or filing a patent application on the invention, may be vital in case of an interference, and in other situations, and (6) if such building and testing is done, signed and dated records of such should additionally be made and these should be witnessed and dated by disinterested individuals (not the PTO).

Very respectfully,

_____   _____
Signature of Inventor   Signature of Joint Inventor

_____   _____
c/o(Print name)   Print Name

_____   _____
Address   Address

_____   _____

Enclosures:

As stated above

Form 3-3 — Request For Participation In Disclosure Document Program

In the United States Patent and Trademark Office

Ser. No.: _____

Filed: _____

Applicant(s):_____

Title: _____

Group Art Unit: _____

Examiner: _____

Disclosure Document Reference Letter

Commissioner of Patents and Trademarks   Date _____
Washington, District of Columbia 20231

Sir:

    A disclosure document as identified below was previously filed in the Patent and Trademark Office. As this disclosure relates to the above patent application, it is requested that this Disclosure Document be retained and referenced to the above application.

Disclosure Document Title: _____

Disclosure Document Number:_____

Disclosure Document Filing Date: _____

Very respectfully,

_____   _____
Signed Name   Signed Name

_____   _____
Printed Name, First Applicant   Printed Name Joint Applicant

_____   _____
Address of First Applicant   Address of Joint Applicant

_____   _____

Inventor(s): _____   Invention: _____ .

Factor	Weight if Positive	Weight if Negative
1. Cost...	_____	_____
2. Weight..	_____	_____
3. Size..	_____	_____
4. Safety/Health..	_____	_____
5. Speed...	_____	_____
6. Ease of Use..	_____	_____
7. Ease of Production.................................	_____	_____
8. Durability..	_____	_____
9. Repairability...	_____	_____
10. Novelty...	_____	_____
11. Convenience/Social Benefit....................	_____	_____
12. Reliability...	_____	_____
13. Ecology..	_____	_____
14. Salability..	_____	_____
15. Appearance...	_____	_____
16. Viewability...	_____	_____
17. Precision..	_____	_____
18. Noise..	_____	_____
19. Odor...	_____	_____
20. Taste...	_____	_____
21. Market Size...	_____	_____
22. Trend of Demand...................................	_____	_____
23. Seasonal Demand...................................	_____	_____
24. Difficulty of Market Penetration.............	_____	_____
25. Potential Competition............................	_____	_____
26. Quality...	_____	_____
27. Excitement..	_____	_____
28. Markup...	_____	_____
29. Inferior Performance..............................	_____	_____
30. "Sexy" Packaging...................................	_____	_____
31. Miscellaneous.......................................	_____	_____
32. Long Life Cycle.....................................	_____	_____
33. Related Product Addability.....................	_____	_____
34. Satisfies Existing Need...........................	_____	_____
35. Legality..	_____	_____
36. Operability...	_____	_____
37. Development...	_____	_____
38. Profitability..	_____	_____
39. Obsolescence..	_____	_____
40. Incompatability.....................................	_____	_____
41. Product Liability Risk............................	_____	_____
42. Market Dependence...............................	_____	_____
43. Difficulty of Distribution.......................	_____	_____
44. Service Requirements.............................	_____	_____
45. New Tooling Required............................	_____	_____
46. Inertia Must Be Overcome......................	_____	_____
47. Too Advanced Technically.......................	_____	_____
48. Substantial Learning Required.................	_____	_____
49. Difficult To Promote..............................	_____	_____
Totals:..	_____	_____

Positive Total - Negative Total = Net: _____

Signed: _____   Date: _____

Inventor(s)

Form 4-1 — Positive And Negative Factors Evaluation

Inventor(s): _____   Invention: _____

_____

List Positive Factors	Weight		List Negative Factors	Weight
_____	____		_____	____
_____	____		_____	____
_____	____		_____	____
_____	____		_____	____
_____	____		_____	____
_____	____		_____	____
_____	____		_____	____
_____	____		_____	____
_____	____		_____	____
_____	____		_____	____
_____	____		_____	____
_____	____		_____	____
_____	____		_____	____
_____	____		_____	____
_____	____		_____	____
_____	____		_____	____
_____	____		_____	____
_____	____		_____	____
_____	____		_____	____
_____	____		_____	____
_____	____		_____	____
_____	____		_____	____
_____	____		_____	____
_____	____		_____	____

Totals: _____   _____

(Positive Total - Negative Total) = Net   _____

Signed:_____   Date: _____
     Inventor(s)

Consultant's Work Agreement

1. **Parties**: This Work Agreement is made between the following parties:
   Name(s) _____
   Address(es): _____
   _____
   (hereinafter Contractor), and

   Name(s): _____
   Address(es): _____
   _____

   (hereinafter Consultant).

2. **Project:**_____
   _____
   _____

3. **Work:**_____
   _____
   _____

4. **Work/Payment Schedule:**_____
   _____
   _____

5. **Date:** This Agreement shall be effective as of the latter date below written.

6. **Recitals:** Contractor has one or more ideas relating to the above project and desires to have such project developed more completely, as specified in the above statement of Work. Consultant has certain skills desired by Contractor relating to performance of the above Work.

7. **Performance:** Consultant will perform the above work for Contractor, in accordance with the above Work/Payment Schedule and Contractor will make the above payments to Consultant.

8. **Intellectual Property:** All intellectual property, including trademarks, writings, information, trade secrets, inventions, discoveries, or improvements, whether or not registerable or patentable, which are conceived, constructed, or written by Consultant and arise out of or are related to work and services performed under this agreement, are, or shall become and remain the sole and exclusive property of Contractor, whether or not such intellectual property is conceived during the time such work and services are performed or billed.

9A. **Protection Of Intellectual Property:** Contractor and Consultant recognize that under US patent laws, all patent applications must be filed in the name of the true and actual inventor(s) of the subject matter sought to be patented. Thus if Consultant makes any patentable inventions relating to the above project, Consultant agrees to be named as an applicant in any US patent application(s) filed on such invention(s). Actual ownership of such patent applications shall be governed by clause 9.

9B. Consultant shall promptly disclose to Contractor in writing all information pertaining to any intellectual property generated or conceived by Consultant under this Agreement. Consultant hereby assigns and agrees to assign all of Consultant's rights to such intellectual property, including patent rights and foreign priority rights. Consultant hereby expressly agrees, without further charge for time, to do all things and sign all documents deemed by Contractor to be necessary or appropriate to invest in intellectual property, including obtaining for and vesting in Contractor all U.S. and foreign patents and patent applications which Contractor desires to obtain to cover such intellectual property, provided that Contractor shall bear all expenses relating thereto. All reasonable local travel time and expenses shall be borne by Consultant.

10. **Trade Secrets:** Consultant recognizes that all information relating to the above Project disclosed to Consultant by Contractor, and all information generated by Consultant in the performance of the above Work, is a valuable trade secret of Contractor and Consultant shall treat all such information as strictly confidential, during and after the performance of Work under this Agreement. Specifically Consultant shall not reveal, publish, or communicate any such information to anyone other than Contractor, and shall safeguard all such information from access to anyone other than Contractor, except upon the express written authorization of Contractor. This clause shall not apply to any information which Consultant can document in writing is presently in or enters the public domain from a bona fide source other than Consultant.

11. **Return Of Property:** Consultant agree to return all written materials and objects received from Contractor, to deliver to Contractor all objects and a copy (and all copies and originals if requested by Contractor) of all written materials resulting from or relating to work performed under this Agreement, and not to deliver to any person, organization, or publisher, or cause to be published, any such written material without prior written authorization.

12. **Conflicts Of Interest:** Consultant recognizes a fiduciary obligation to Contractor arising out of the work and services performed under this agreement and accordingly will not offer Consultant's service to or perform services for any competitor, potential or actual, of Contractor for the above Project, or perform any other acts which may result in any conflict of interest by Consultant, during and after the term of this Agreement.

13. **Mediation And Arbitration:** If any dispute arises under this Agreement, the parties shall negotiate in good faith to settle such dispute. If the parties cannot resolve such dispute themselves, then either party may submit the dispute to mediation by a mediator approved by both parties. If the parties cannot agree to any mediator, or if either party does not wish to abide by any decision of the mediator, they shall submit the dispute to arbitration by any mutually-acceptable arbitrator, or the American Arbitration Association (AAA). If the AAA is selected, the arbitration shall take place under the auspices of the nearest branch of such to both parties. The costs of the arbitration proceeding shall be borne according to the decision of the arbitrator, who may apportion costs equally, or in accordance with any finding or fault or lack of good faith of either party. The arbitrator's award shall be non-appealable and enforceable in any court of competent jurisdiction.

14. **Governing Law:** This Agreement shall be governed by and interpreted under and according to the laws of the State of _____.

15. **Signatures:** The parties have indicated their agreement to all of the above terms by signing this Agreement on the respective dates below indicated.

Contractor:

Date: _____

_____

Consultant:

Date: _____

_____

Searcher's Worksheet

Inventor(s): _____

Invention Description (use key words and variations): _____

_____

_____

Selected Search Classifications

Class/Sub	Description	Checked	Comments

Patents (and Other References) Thought Relevant

Patent #	Name or Country	Date	Class/Sub	Comment

Searcher: _____   Date: _____

PART NAME   **PART NAME**

10_____   84_____
12_____   86_____
14_____   88_____
16_____   90_____
18_____   92_____
20_____   94_____
22_____   96_____
24_____   98_____
26_____   100_____
28_____   102_____
30_____   104_____
32_____   106_____
34_____   108_____
36_____   110_____
38_____   112_____
40_____   114_____
42_____   116_____
44_____   118_____
46_____   120_____
48_____   122_____
50_____   124_____
52_____   126_____
54_____   128_____
56_____   130_____
58_____   132_____
60_____   134_____
62_____   136_____
64_____   138_____
66_____   140_____
68_____   142_____
70_____   144_____
72_____   146_____
74_____   148_____
76_____   150_____
78_____   152_____
80_____   154_____
82_____   156_____

Form 8-1 — Drawing Reference Numerals Worksheet

In The United States Patent And Trademark Office

Mailed 198 _____

Commissioner of Patents and Trademarks
Washington, District of Columbia 20231

Sir:

Please file the following enclosed patent application papers:

Applicant #1, Name:_____

Applicant #2, Name:_____

Title: _____

( ) Specification, Claims, and Abstract: Nr. of Sheets _____

( ) Declaration: Date Signed:_____

( ) Drawing(s): Nr. of Sheets Enc. (In Triplicate): Formal:_____   Informal:_____

( ) Small Entity Declaration Of Inventor(s) ( ) SED of Non-Inventor / Assignee/Licensee

( ) Assignment; please record and return; recordal fee enclosed.

( ) Check for $ _____   for:

    ( ) $_____ for filing fee (not more than three independent claims and twenty total claims are presented).

    ( ) $_____ Additional if Assignment is enclosed for recordal.

( ) Return Receipt Postcard Addressed to Applicant #1.

Very respectfully,

_____   _____
Applicant #1 Signature   Applicant #2 Signature

_____   _____
Address (Send Correspondence Here)   Address
_____   _____

Express Mail Label #_____; **Date of Deposit 198___**

    I hereby certify that this paper or fee is being deposited with the United States Postal Service using "Express Mail Post Office To Addressee" service under 37 CFR 1.10 on the date indicated above and is addressed to "Commissioner of Patents and Trademarks, Washington, DC 20231."

Signed:_____
      Inventor

Form 10-1 — Utility Patent Application Transmittal

PART NAME   **PART NAME**

10_____   84_____
12_____   86_____
14_____   88_____
16_____   90_____
18_____   92_____
20_____   94_____
22_____   96_____
24_____   98_____
26_____   100_____
28_____   102_____
30_____   104_____
32_____   106_____
34_____   108_____
36_____   110_____
38_____   112_____
40_____   114_____
42_____   116_____
44_____   118_____
46_____   120_____
48_____   122_____
50_____   124_____
52_____   126_____
54_____   128_____
56_____   130_____
58_____   132_____
60_____   134_____
62_____   136_____
64_____   138_____
66_____   140_____
68_____   142_____
70_____   144_____
72_____   146_____
74_____   148_____
76_____   150_____
78_____   152_____
80_____   154_____
82_____   156_____

Form 8-1 — Drawing Reference Numerals Worksheet

In The United States Patent And Trademark Office

Mailed 198 _____

Commissioner of Patents and Trademarks
Washington, District of Columbia 20231

Sir:

Please file the following enclosed patent application papers:

Applicant #1, Name:_____

Applicant #2, Name:_____

Title: _____

( ) Specification, Claims, and Abstract: Nr. of Sheets _____

( ) Declaration: Date Signed:_____

( ) Drawing(s): Nr. of Sheets Enc. (In Triplicate): Formal:_____ Informal:_____

( ) Small Entity Declaration Of Inventor(s) ( ) SED of Non-Inventor / Assignee/Licensee

( ) Assignment; please record and return; recordal fee enclosed.

( ) Check for $ _____ for:

    ( ) $_____ for filing fee (not more than three independent claims and twenty total claims are presented).

    ( ) $_____ Additional if Assignment is enclosed for recordal.

( ) Return Receipt Postcard Addressed to Applicant #1.

Very respectfully,

_____   _____
Applicant #1 Signature   Applicant #2 Signature

_____   _____
Address (Send Correspondence Here)   Address

_____   _____

Express Mail Label #_____**; Date of Deposit 198___**

    I hereby certify that this paper or fee is being deposited with the United States Postal Service using "Express Mail Post Office To Addressee" service under 37 CFR 1.10 on the date indicated above and is addressed to "Commissioner of Patents and Trademarks, Washington, DC 20231."

Signed:_____
      Inventor

Declaration For Utility or Design Patent Application

As a below-named inventor, I hereby declare that my residence, post office address, and citizenship are as stated below next to my name and that I believe that I am the original, first, and sole inventor [if only one name is listed below] or an original, first, and joint inventor [if plural names are listed below] of the subject matter which is claimed and for which a patent is sought on the invention, the specification of which is attached hereto and which has the following title:

" _____ "

I have reviewed and understand the contents of the above-identified specification, including the claims, as amended by any amendment specifically referred to in the oath or declaration. I acknowledge a duty to disclose information which is material to the examination of this application in accordance with Title 37, Code of Federal Regulations, Section 1.56(a).

I hereby declare that all statements made herein of my own knowledge are true and that all statements made on information and belief are believed to be true; and further that these statements were made with the knowledge that willful false statements and the like so made are punishable by fine or imprisonment, or both, under Title 18, United States Code, Section 1001, and that such willful false statements may jeopardize the validity of the application or any patent issued thereon.

Please send correspondence and make telephone calls to the First Inventor below.

Signature: Sole/First Inventor: _____

Print Name:_____   Date: _____

Residence: _____   Citizen Of:_____

Post Office Address: _____

Telephone: _____   _____

Signature: Joint/Second Inventor: _____

Print Name:_____   Date: _____

Residence: _____   Citizen Of:_____

Post Office Address: _____

Telephone: _____   _____

In the United States Patent and Trademark Office

First/Sole Applicant:_____

Joint/Second Applicant:_____

Title:"_____"

Small Entity Declaration—Independent Inventor(s)

As a below-named inventor, I hereby declare that I qualify as an independent inventor as defined in 37 CFR 1.9(c) for purposes of paying reduced fees under Section 41(a) and (b) of Title 35 United States Code, to the Patent and Trademark Office with regard to my above-identified invention described in the specification filed herewith. I have not assigned, granted, conveyed, or licensed—and am under no obligation under any contract or law to assign, grant, convey, or license—any rights in the invention to either (a) any person who could not be classified as an independent inventor under 37 CFR 1.9(c) if that person had made the invention, or (b) any concern which would not qualify as either (i) a small business concern under 37 CFR 1.9(d) or (ii) a nonprofit organization under 37 CFR 1.9(e).

Each person, concern, or organization to which I have assigned, granted, conveyed, or licensed—or am under an obligation under contract or law to assign, grant, convey, or license—any rights in the invention is listed below:

[ ] There is no such person, concern, or organization.

[ ] Any applicable person, concern, or organization is listed below:[*]
Full Name: _____

Address:

_____

I acknowledge a duty to file, in the above application for patent, notification of any change in status resulting in loss of entitlement to small entity status prior to paying, or at the time of paying, the earliest of the issue fee or any maintenance fee due after the date on which status as a small entity is no longer appropriate. (37 CFR 1.28(b)).

I hereby declare that all statements made herein of my own knowledge are true and that all statements made on information and belief are believed to be true; and further that these statements were made with the knowledge that willful false statements and the like so made are punishable by fine or imprisonment or both, under Section 1001 of Title 18 of the United States Code, and that such willful false statements may jeopardize the validity of the application, any patent issuing thereon, or any patent to which this verified statement is directed.

_____   _____
Signature of Sole/First Inventor   Signature of Joint/Second Inventor

_____   _____
Print Name of Sole/First Inventor   Print Name of Second/Joint Inventor

_____   _____
Date of Signature: _____   Date of Signature: _____

_____

[*]Note: A separate Small Entity Statement is required from any listed entity.

In The United States Patent And Trademark Office

First/Sole Applicant: _____

Joint/Second Applicant: _____

Title: " _____ "

Small Entity Declaration—Non-Inventor Individual

I hereby declare that I am making this verified statement to support a claim by

_____

for small entity status for purposes of paying reduced fees under 35 USC 41(a) & (b) with regard to the above-entitled invention of the above applicants and described in the specification filed herewith.

I hereby declare that I would qualify as an independent inventor as defined in 37 CFR 1.9(c) for the purpose of paying reduced fees under 35 USC 41(a) & (b) if I had made the above-identified invention.

I have not assigned, granted, conveyed, or licensed—and am under no obligation under any contract or law to assign, grant, convey, or license—any rights in the invention to either (a) any person who could not be classified as an independent inventor under 37 CFR 1.9(c) if that person had made the invention, or (b) any concern which would not qualify as either (i) a small business concern under 37 CFR 1.9(d) or (ii) a nonprofit organization under 37 CFR 1.9(e).

I have assigned, granted, conveyed, or licensed—and am not under any obligation under contract or law to assign, grant, convey, or license—any rights in the invention to any person, concern, or organization.

I acknowledge a duty to file, in the above application for patent, notification of any change in status resulting in loss of entitlement to small entity status prior to paying, or at the time of paying, the earliest of the issue fee or any maintenance fee due after the date on which status as a small entity is no longer appropriate (37 CFR 1.28(b)).

I hereby declare that all statements made herein of my own knowledge are true and that all statements made on information and belief are believed to be true; and further that these statements were made with the knowledge that willful false statements and the like so made are punishable by fine or imprisonment, or both, under Section 1001 of Title 18 of the United States Code, and that such willful false statements may jeopardize the validity of the application, any patent issuing thereon, or any patent to which this verified statement is directed.

_____   _____

Signature of Non-Inventor   Date of Signature

_____

Print Name and Address of Non-Inventor

_____

In the United States Patent and Trademark Office

First/Sole Applicant:_____

Joint/Second Applicant:_____

Title: " _____ "

Small Entity Declaration—Small Business Concern

I hereby declare that I am

[] the owner of the small business concern identified below:

[] an officer of the small business concern empowered to act on behalf of the concern identified below:

NAME OF CONCERN _____

ADDRESS OF CONCERN _____

_____

I hereby declare that the above identified small business concern qualifies as a small business concern as defined in 13 CFR 121.3-18, and reproduced in 37 CFR 1.9(d), for purposes of paying reduced fees under section 41(a) and (b) of Title 35, United States Code, in that the number of employees of the concern, including those of its affiliates, does not exceed 500 persons. For purposes of this statement, (1) the number of employees of the business concern is the average over the previous fiscal year of the concern of the persons employed on a full-time, part-time or temporary basis during each of the pay periods of the fiscal year, and (2) concerns are affiliates of each other when either, directly or indirectly, one concern controls or has the power to control the other, or a third party or parties controls or has the power to control both.

I hereby declare that rights under contract or law have been conveyed to and remain with the small business concern identified above with regard to the above entitled invention of the above applicants and the specification filed herewith.

I acknowledge a duty to file, in the above application for patent, notification of any change in status resulting in loss of entitlement to small entity status prior to paying, or at the time of paying, the earliest of the issue fee or any maintenance fee due after the date on which status as a small entity is no longer appropriate. (37 CFR 1.28(b)).

I hereby declare that all statements made herein of my own knowledge are true and that all statements made on information and belief are believed to be true; and further that these statements were made with the knowledge that willful false statements and the like so made are punishable by fine or imprisonment or both, under Section 1001 of Title 18 of the United States Code, and that such willful false statements may jeopardize the validity of the application, any patent issuing thereon, or any patent to which this verified statement is directed.

_____   _____

Signature of Officer of Small Business Concern   Date

_____

Name and Title of Officer

_____

Address of Officer

_____

In the United States Patent and Trademark Office

First/Sole Applicant: _____

Joint/Second Applicant: _____

Title:" _____ "

Small Entity Declaration—Nonprofit Organization

I hereby declare that I am an official empowered to act on behalf of the nonprofit organization identified below:

NAME OF ORGANIZATION _____

ADDRESS OF ORGANIZATION _____

_____

TYPE OF ORGANIZATION

[ ] UNIVERSITY OR OTHER INSTITUTION OF HIGHER EDUCATION

[ ] TAX EXEMPT UNDER INTERNAL REVENUE SERVICE CODE (26 USC 501(a) and 501(c) (3))

[ ] NONPROFIT SCIENTIFIC OR EDUCATIONAL UNDER STATUTE OF STATE OF THE THE UNITED STATES OF AMERICA
    (NAME OF STATE_____)
    (CITATION OF STATUTE_____)

[ ] WOULD QUALIFY AS TAX EXEMPT UNDER INTERNAL REVENUE SERVICE CODE
    (26 USC 501(a) and 501(c) (3)) IF LOCATED IN THE UNITED STATES OF AMERICA

[ ] WOULD QUALIFY AS NONPROFIT SCIENTIFIC OR EDUCATIONAL UNDER STATUTE OF STATE OF THE UNITED STATES OF AMERICA IF LOCATED IN THE UNITED STATES OF AMERICA
    (NAME OF STATE_____)
    (CITATION OF STATUTE_____)

I hereby declare the the nonprofit organization identified above qualifies as a nonprofit organization as defined in 37 CFR 1.9(e) for purposes of paying reduced fees under section 41(a) and (b) of Title 35 , United States Code with regard to the above-entitled invention of the above applicant(s) and the specification filed herewith.

I hereby declare that rights under contract or law have been conveyed to and remain with the nonprofit organization with regard to the above identified invention.

I acknowledge a duty to file, in the above application for patent, notification of any change in status resulting in loss of entitlement to small entity status prior to paying, or at the time of paying, the earliest of the issue fee or any maintenance fee due after the date on which status as a small entity is no longer appropriate. (37 CFR 1.28(b)).

I hereby declare that all statements made herein of my own knowledge are true and that all statements made on information and belief are believed to be true; and further that these statements were made with the knowledge that willful false statements and the like so made are punishable by fine or imprisonment or both, under Section 1001 of Title 18 of the United States Code, and that such willful false statements may jeopardize the validity of the application, any patent issuing thereon, or any patent to which this verified statement is directed.

_____   _____

Signature or Officer of Non-Profit Organization   Date

_____

Name and Title of Officer

_____

Address of Officer

In the United States Patent and Trademark Office

Serial Number: _____

Appn. Filed: _____

Applicant(s): _____

Appn. Title: _____

Examiner/GAU: _____

Mailed: _____

At: _____

Information Disclosure Statement

Commissioner of Patents and Trademarks
Washington, District of Columbia 20231

Sir:

Attached is a completed Form PTO-1449 and copies of the pertinent parts of the references cited thereon.

Following are comments on these references pursuant to Rule 98:

FORM PTO-1449 (REV. 7-80)	U.S. DEPARTMENT OF COMMERCE PATENT AND TRADEMARK OFFICE	ATTY. DOCKET NO.	SERIAL NO.
LIST OF PRIOR ART CITED BY APPLICANT *(Use several sheets if necessary)*		APPLICANT	
		FILING DATE	GROUP

U.S. PATENT DOCUMENTS

*EXAMINER INITIAL		DOCUMENT NUMBER	DATE	NAME	CLASS	SUBCLASS	FILING DATE IF APPROPRIATE
	AA						
	AB						
	AC						
	AD						
	AE						
	AF						
	AG						
	AH						
	AI						
	AJ						
	AK						

FOREIGN PATENT DOCUMENTS

		DOCUMENT NUMBER	DATE	COUNTRY	CLASS	SUBCLASS	TRANSLATION YES	TRANSLATION NO
	AL							
	AM							

OTHER PRIOR ART *(Including Author, Title, Date, Pertinent Pages, Etc.)*

	AR	
	AS	

EXAMINER	DATE CONSIDERED

*EXAMINER: Initial if reference considered, whether or not citation is in conformance with MPEP 609; Draw line through citation if not in conformance and not considered. Include copy of this form with next communication to applicant.

USCOMM-DC 80-3985

Form 10-6

In the United States Patent and Trademark Office

Serial Number: _____

Appn. Filed:_____

Applicant(s):_____

Appn. Title:_____

Examiner/GAU: _____

Mailed: _____

At:_____

Petition to Make Special

Commissioner of Patents and Trademarks
Washington, District of Columbia 20231

Sir:

    Applicant hereby respectfully petitions that the above application be made special under MPEP Sec. 708.02 for the following reason; attached is a declaration in support thereof:

    I. [ ] Manufacturer Available*;   VI. [ ] Energy Savings Will Result;

    II. [ ] Infringement Exists*;   VII. [ ] Recombinant DNA is involved*;

    III. [ ] Applicant's Health Is Poor;   VIII.[ ] Special Procedure: Search Was Made*;

    IV. [ ] Applicant's Age is 65 Or Greater;   IX [ ] Superconductivity is advanced.

    V. [ ] Environmental Quality will Be Enhanced;

*[ ] Also attached, since reason I., II., VII., or VIII. has been checked, is the $_____ Petition Fee pursuant to Rules 102 and 17(i).

Very respectfully,

Applicant(s):_____

_____

Attachment: Supporting Declaration And Fee if Indicated

c/o:_____

_____

_____

Tel.:_____

Design Patent Application—Preamble, Specification, and Claim

Commissioner of Patents and Trademarks
Washington, District of Columbia 20231

Sir:

Preamble:

    The petitioner(s) whose signature(s) appear on the declaration attached respectfully request that Letters Patent be granted to such petitioner(s) for the new and original design set forth in the following specification. The filing fee of $_____, a patent application declaration, a small entity declaration, and a return receipt postcard are attached.

Specification:

    The undersigned has (have) invented a new, original, and ornamental design entitled
"_____"
of which the following is a specification. Reference is made to the accompanying drawings which form a part hereof, the figures of which are described as follows:

    Fig. 1 is a _____ view.
    Fig. 2 is a _____ view.

Claim: I:
(We) Claim:
The ornamental design for a _____ as shown.

In the United States Patent and Trademark Office

Serial Number: _____

Appn. Filed:_____

Applicant(s):_____

Appn. Title:_____

Examiner/GAU: _____

Mailed: _____

At:_____

Amendment _____

Commissioner of Patents and Trademarks
Washington, District of Columbia 20231

Sir:

In response to the Office Letter mailed _____, 19___, please amend the above application as follows:

In the United States Patent and Trademark Office

Serial Number: _____

Appn. Filed: _____

Applicant(s): _____

Appn. Title: _____

Examiner/GAU: _____

Mailed: _____

At: _____

Request for Approval of Proposed Drawing Amendment

Commissioner of Patents and Trademarks
Washington, District of Columbia 20231

Sir:

    Applicant(s) respectfully request(s) permission to amend the drawing(s) of the above application after allowance. The proposed changes are indicated in red on the photocopy(ies) of Fig.(s) _____ or sheets _____ thereof attached hereto.

Very respectfully,

Applicant(s): _____

_____

c/o: _____

_____

_____

Tel.: _____

Certificate of Mailing

    I certify that this correspondence will be deposited with the United States Postal Service as first class mail with proper postage affixed in an envelope addressed to: "Commissioner of Patents and Trademarks, Washington, DC 20231" on the date below.

Date: 198 _____   _____ , Applicant

In the United States Patent and Trademark Office

Serial Number: _____

Appn. Filed:_____

Applicant(s):_____

Appn. Title:_____

Examiner/GAU: _____

Mailed: _____

At:_____

Submission of Corrected Drawings

Commissioner of Patents and Trademarks
Washington, District of Columbia 20231
        Attn: Chief Draftsperson

Sir:

    New drawing sheet(s)_____for the above application is/are enclosed, corrected as necessary. Please substitute this/these for the corresponding sheet(s) on file.

Very respectfully,

        Applicant(s):_____

_____

c/o:_____

_____

_____

Tel.:_____

Certificate of Mailing

    I certify that this correspondence will be deposited with the United States Postal Service as first class mail with proper postage affixed in an envelope addressed to: "Commissioner of Patents and Trademarks, Washington, DC 20231" on the date below.

Date: 198 _____   _____ , Applicant

In the United States Patent and Trademark Office

Serial Number: _____

Appn. Filed: _____

Applicant(s): _____

Appn. Title: _____

Examiner/GAU: _____

Mailed: _____

At: _____

Supplemental Declaration

(for Use After Close of Prosecution or
With Continuation-In-Part Application)

As an applicant in the above-identified application, I declare as follows:

1. If only one inventor is named below, I am a sole inventor, and if more than one inventor is named below, I am a joint inventor with the inventor(s) named below of the subject matter of the above-identified application.

2. I have reviewed and understand the contents of the specification and claims, as originally filed, and as amended by the amendment(s) dated _____

3. I believe that I, and the other inventor(s) named below if more than one inventor is named below, am the original and first inventor or inventors of the subject matter which is claimed and for which a patent is sought.

4. I acknowledge the duty to disclose information which is material to the examination of the application in accordance with 37 C.F.R. Section 1.56(a), and if this oath accompanies or refers to a continuation-in-part application, I acknowledge the duty to disclose material information as defined in 37 C.F.R. Section 1.56(a) which occurred between the filing date of the prior application and the national or PCT international filing date of the continuation-in-part application.

5. I hereby declare that all statements made herein of my own knowledge are true and that all statements made on information and belief are believed to be true, and further that these statements were made with the knowledge that willful false statements and the like so made are punishable by fine or imprisonment, or both, under Section 1001 of Title 18 of the United States Code, and that such willful false statements may jeopardize the validity of the application, any patent issuing thereon, or any patent to which this verified statement is directed.

_____   _____
Signature of Inventor   Signature of Joint Inventor

_____   _____
Printed Name of Inventor   Printed Name of Joint Inventor

Date_____   Date_____

In the United States Patent and Trademark Office

Serial Number: _____

Appn. Filed: _____

Applicant(s): _____

Appn. Title: _____

Examiner/GAU: _____

Mailed: _____

At: _____

Petition for Extension of Time
(Rules 136 and 17(a)-(d))

Outstanding Office Action Mailed 198 _____

Original Period for Response Expired 198 _____

Request for Extension of _____ Month(s) to 198 _____

Sml. Ent. Petn. Fee Enc.: [ ] $____ (1 mo.); [ ] $___ (2 mo.); [ ] $___ (3 mo.); [ ] $___(4 mo.)

Commissioner of Patents and Trademarks
Washington, District of Columbia 20231

Sir:

In the above application, applicant(s) respectfully petition that the period for response to the outstanding Office Action indicated above be extended for the additional month(s) also indicated above. A response to such Office Action and the above Petition Fee (Small Entity) are enclosed herewith. (This extension will not extend the time over the statutory period of six months from the date of the Office Action.)

Very respectfully,

Applicant(s): _____

_____

c/o: _____

_____

_____

Tel.: _____

Encs.

Certificate of Mailing

I certify that this correspondence will be deposited with the United States Postal Service as first class mail with proper postage affixed in an envelope addressed to: "Commissioner of Patents and Trademarks, Washington, DC 20231" on the date below.

Date: 198 _____   _____ , Applicant

In the United States Patent and Trademark Office

Serial Number: _____

Serial Filed: _____

Applicant(s): _____

Appn. Title: _____

Examiner/GAU: _____

Mailed: _____

At: _____

Request for File-Wrapper-Continuing Application

Box FWC
Commissioner of Patents and Trademarks
Washington, District of Columbia 20231

Sir:

Pursuant to Rule 62, please file a file-wrapper-continuing application of the above pending complete application. This Rule 62 continuing application is being actually filed in the PTO or express mailed during the pedency of the above application. It should be a ( ) Continuation-In-Part ( ) Division ( ) Continuation of the above application. Please use the specification (including the claims and abstract) of the above application and also use the Declaration therefrom if this is a Continuation or Division.

( ) Enclosed is a Preliminary Amendment for this Rule 62 Application

( ) Please enter the Amendment Under Rule 116 in the parent application.

( ) Enclosed is a new Declaration if this is a CIP application.
Please use ( ) the drawings of the above application ( ) the new drawing(s) enclosed(____ sheets).

After entry of this Preliminary Amendment (or any Amendments Under Rule 116 in the above application) there will be not more than 3 independent claims and 20 total claims. A small entity declaration ( ) was enclosed for the above application ( ) is enclosed. Thus the filing fee for the Rule 62 Application will be $_____ a check for this amount is enclosed.

Very respectfully,

Applicant(s): _____

Enc(s): Filing Fee; Also New Declaration and/or Amendment, if indicated

c/o: _____

_____

_____

Tel.: _____

Express Mail Label # _____ ; **Date of Deposit 198** _____

I hereby certify that this paper or fee is being deposited with the United States Postal Service using "Express Mail Post Office To Addressee" service under 37 CFR 1.10 on the date indicated above and is addressed to "Commissioner of Patents and Trademarks, Washington, DC 20231."

Signed: _____

Form 14-1

In the United States Patent and Trademark Office

Patent No.: _____

Issued: _____

Patentee(s): _____

Ser. Nr: _____

Filed: _____

Request for Certificate of Correction

Date: _____

Commissioner of Patents and Trademarks
Washington, District of Columbia 20231

Sir:

1. The above patent contains significant error, as indicated on the attached Certificate of Correction form (submitted in duplicate). These errors arose at the respective places in the application file indicated below.

( ) 2. Since such error arose through the fault of the Patent and Trademark Office, it is requested that the Certificate be issued at no cost to applicant.

( ) 3. Such error arose through the fault of applicant(s). A check for $ _____ for the fee is enclosed. Such error is of a clerical or minor nature and occurred in good faith and therefore issuance of the Certificate of Correction is respectfully requested.

4. Specifically,

Very respectfully,

_____   _____

Patentee   Co-Patentee

Encs.

_____

_____

(_____) ____-_____

UNITED STATES PATENT AND TRADEMARK OFFICE

Patent No.:

Dated:

Inventor(s):

It is certified that error appears in the above-identified patent and that said Letters Patent are hereby corrected as shown below:

Mailing Address of Sender:

Patent No. _____

_____

_____

_____

In the United States Patent and Trademark Office

Patent No.: _____

Issued: _____

Patentee(s): _____

Ser. Nr: _____

Filed: _____

Submission of Maintenance Fee

Commissioner of Patents and Trademarks
Washington, District of Columbia 20231

Sir:

Enclosed is the following maintenance fee for the above patent: this fee is for a ( ) large entity
( ) small entity since a small-entity declaration was filed in connection with the above application.

[]   3.5 yr fee; $_____; due 3.0 to 3.5 yrs after issue; covers yrs 4.0 thru 8.0
[]   7.5 yr fee; $_____; due 7.0 to 7.5 yrs after issue; covers yrs 8.0 thru 12.0
[] 11.5 yr fee; $_____; due 11.0 to 11.5 yrs after issue; covers yrs 12.0 thru 17.0

[] Also enclosed is a surcharge of $_____ (total enclosed $_____) since this fee is being filed in the six-month grace period after the above due period.

Very respectfully,

_____

Either Patentee/ Assignee

_____

_____

(_____) ____-_____

Certificate of Mailing

I hereby certify that this correspondence will be placed in an envelope marked "First Class Mail", addressed to "Commissioner of Patents and Trademarks, Washington DC 20231", and having adequate first class postage affixed, and that such envelope will then be sealed and deposited in an approved United States Postal Service depository on the date below.

Date: 198_____   _____

Joint Owners' Agreement

This agreement is made by and between the following parties who, by separate assignment or as joint applicants, own the following respective shares of the invention and patent application identified below:

_____ of _____ , _____ %,

_____ of _____ , _____ %,

_____ of _____ , _____ %,

Invention Title:_____

Patent Application Ser. Nr.:_____ , Filed: _____

Applicants: _____

_____

The above patent application data is to be filled in as soon as it becomes available if the application has not yet been filed.

The parties desire to stipulate the terms under which they will exploit this invention and patent application and therefore agree as follows:

1. No Action Without Everyone's Consent: None of the parties to this agreement shall license, use, make, or sell the invention or application, or take any other action, other than normal prosecution, without the written consent and cooperation of the other party or parties (hereinafter "parties") to this agreement, except as provided below. Any action so taken shall be committed to a writing signed by all of the parties, or as many parties as consent, with copies to all other parties.

2. Decisions: In case any decision must be made in connection with the invention or the patent application, including foreign filing, appealing from an adverse decision in the Patent and Trademark Office, or any opportunity to license, sell, make, or use the invention or application, the parties shall consult on such opportunity and a majority decision shall control. In the event the parties are equally divided, the matter shall be submitted to an impartial, mutually-acceptable arbiter whose decision shall control. If no arbiter can be agreed upon, then the parties shall each select a representative and the parties' representative shall select the arbiter. After a decision is so made, all parties shall abide by the decision and shall cooperate fully by whatever means are necessary to implement and give full force to such decision. However, if there is time for any parties to obtain a better or different offer, they shall be entitled to do so and the decision shall be postponed for up to one month to allow such other parties to act.

3. Proportionate Sharing: The parties to this agreement shall share, in the percentages indicated above, in all income from, liabilities, and expenditures agreed to be made by any decision under Part 2 above in connection with the invention or patent application. In case a decision is made to make any expenditure, as for foreign patent application filing, exploitation, etc., and a minority or other parties opposes such expenditure or is unable to contribute his or her proportionate share, then the others shall advance the minority or other parties' share of the expenditure. Such others shall be reimbursed by the minority or other parties by double the amount so advanced from the minority or other parties' proportionate share of any income received, provided such income has some reasonable connection with the expenditure. No party shall be entitled to reimbursement or credit for any labor unless agreed to in advance by all of the parties hereto.

4. If Any Parties Desire to Manufacture, Etc.: If any parties who do not constitute all of the parties to this agreement desire to manufacture, distribute, or sell any product or service embodying the above invention, they may do so with the written consent of the other parties under Part 1 above. The cost of the product or service shall include, in addition to normal profit, labor, commission, and/or overhead, etc., provision for a reasonable royalty which shall be paid for the term of the above patent application and any patent which may issue thereon. Such royalty shall be determined before any action is taken under this part and as if a valid patent on the invention had been licensed to an unrelated exclusive licensee (or a nonexclusive licensee if the patent is licensed to others) in an arm's length transaction. Such royalty shall be distributed to all of the parties hereto according to their proportionate shares and on a quarterly basis, accompanied by a written royalty report and sent within one month after the close of each calendar quarter.

5. In Case of Dispute: In case any dispute or disagreement arises out of this agreement or in connection with the invention or patent application, the parties shall confer as much as necessary to settle the disagreement; all parties shall act and compromise to at least the degree a reasonable person would act. If the parties cannot settle their differences on their own, they shall submit the dispute to mediation by an impartial third party or professional mediator agreed to by all of the parties. If the parties cannot agree on a mediator, or cannot come to an agreement after mediation, then they shall submit the matter to binding arbitration with a mutually-acceptable arbitrator or the American Arbitration Association. The arbitrator shall settle the dispute in whatever manner he or she feels will do substantial justice, recognizing the rights of all parties and commercial realities of the marketplace. The parties shall abide by the terms of the arbitrator's decision and shall cooperate fully and do any acts necessary to implement such decision. The costs of the arbitrator shall be advanced by all of the parties or in accordance with Part 3 above and the arbitrator may make any allocation of arbitration costs he or she feels is reasonable.

_____   _____

Date: _____   Date:_____

_____

Date: _____

Assignment of Invention and Patent Application

For value received, _____,

_____

of_____

(hereinafter **Assignor**), hereby sells, assigns, transfers, and sets over unto

_____

of_____

and her or his successors or assigns (hereinafter **Assignee**) ____ % of the following: (A) **Assignor's** right, title and interest in and to the invention entitled "_____

_____ "

invented by **Assignor**; (B) the application for Untied States patent therefor, signed by **Assignor** on

_____,U.S. Patent and Trademark Office Serial Number _____;

Filed _____; (C) any patent or reissues of any patent that may be granted thereon; and (D) any applications which are continuations, continuations-in-part, substitutes, or divisions of said application. **Assignor** authorizes **Assignee** to enter the date of signature and/or Serial Number and Filing Date in the spaces above. **Assignor** also authorizes and requests the Commissioner of Patents andTrademarks to issue any resulting patent(s) as follows: _____ % to **Assignor** and _____ % to **Assignee**. (The singular shall include the plural and vice-versa herein.)

    Assignor hereby further sells, assigns, transfers, and sets over unto **Assignee**, the above percentage of **assignor's** entire right, title and interest in and to said invention in each and every country foreign to the United States; and **Assignor** further conveys to **Assignee** the above percentage of all priority rights resulting from the above-identified application for United States patent. **Assignor** agrees to execute all papers, give any required testimony and perform other lawful acts, at **Assignees** expense, as **Assignee** may require to enable **Assignee** to perfect **Assignee's** interest in any resulting patent of the United States and countries foreign thereto, and to acquire, hold, enforce, convey, and uphold the validity of said patent and reissues and extensions thereof, and **Assignee's** interest therein.

    In testimony whereof **Assignor** has hereunto set its hand and seal on the date below.

_____

_____

State: _____:

                        :ss

County:_____:

Subscribed and sworn to before me_____

                              (date)

_____.

             Notary Public

SEAL

Universal License Agreement

1. Parties, Terms, and Parameters:

This agreement is between:

Licensor: _____ , of

_____ .

Licensee: _____ , of

_____ .

Patent Royalty Rate (%): _____ % x Est. 1st Yr's Sales (units): _____ x

Estimated Unit Price $ _____ = Resultant Licensing Fee $ _____

Type of License: [ ] Exclusive   [ ] Nonexclusive

Invention Title: _____ .

Patent Application Ser. Nr.: _____ , Filing Date: _____

Minimum Nr. of Units to be Sold to Compute Min. Annual Royalty: _____

Minimum Annual Royalties Start Year Commencing 19 _____ .

   [ ] Option Granted: Premium $ _____ For Term Of (months): _____

   [ ] Know-How Licensed: Know-How Royalty Rate (%): _____

Running Royalty (Patent Royalty and Know-How Royalty, if applicable) (%): _____

2. Effective Date:
This agreement shall be effective as of the latter of the signature dates below written and shall be referred to as the Agreement of such date.

3. Recitals:

A. LICENSOR has developed an invention having the above title and warrants that LICENSOR has filed a patent application on such invention in the U.S. Patent and Trademark Office, which patent application is identified by the above title, Serial Number, and Filing Date. LICENSOR warrants that licensor has full and exclusive right to grant this license on this invention and LICENSOR'S patent application. If the "Know-How" block above is checked, LICENSOR has also developed know-how in connection with said invention and warrants that LICENSOR owns and has the right to license said know-how.

B. LICENSEE desires, if the "Option Granted" block above is checked, to exclusively investigate LICENSOR'S above invention for the term indicated. If said "Option Granted" box is not checked, or if said box is checked and LICENSEE investigates LICENSOR'S invention for the term indicated and such investigation is favorable, LICENSEE desires to make, use and sell the products embodying such invention and covered by the claims of LICENSOR'S patent application and any patent(s) issuing thereon (hereinafter "Licensed Product").

4. If Option Granted:
If the "Option Granted" box above is checked, then (A) the patent license grant of Part 5 below shall not take effect except as defined in this part, and (B) LICENSOR hereby grants LICENSEE, for the option premium stated above, an exclusive option to investigate LICENSOR'S invention for the term indicated above, such term to commence from the date of this Agreement. LICENSOR will furnish LICENSEE with all information and know-how (if any) concerning LICENSOR'S invention in LICENSOR'S possession. LICENSEE will investigate LICENSOR'S invention for operability, costing, marketing, etc. LICENSEE shall report the results of its investigation to LICENSOR at any time before the end of the option term. If LICENSEE'S determination is favorable, it may thereupon exercise this option and the patent license grant of Part 5 below shall become effective. If LICENSEE'S determination is unfavorable, then said option shall not be exercised and no patent license grant shall take effect and all rights hereunder shall revert to LICENSOR and LICENSEE shall deliver to LICENSOR all results of its investigations for LICENSOR'S benefit.

5. Patent License if Option Exercised or if Option Not Granted:
If the "Option Granted" box above is checked and LICENSEE has investigated LICENSOR'S invention and such investigation is favorable and LICENSEE has exercised its option, or if said box is not checked, then LICENSOR hereby grants to LICENSEE, subject to the terms and conditions herein, a patent license of the type (Exclusive or Nonexclusive) checked above. Such patent license shall include the right to grant sublicenses, to

make, have made, use, and sell the Licensed Product throughout the United States, its territories, and possessions. Such patent license shall be under LICENSOR'S patent application, any continuations, divisions, continuations-in-part, substitutes, reissues of any patent from any of such applications (hereinafter and hereinbefore LICENSEE'S patent application), any patent(s) issuing thereon, and if the "Know-How Licensed" box is checked above, any know-how transferred to LICENSEE.

6. If Know-How Licensed: If the "Know-How" box above is checked, LICENSOR shall communicate to LICENSEE all of LICENSOR'S know-how in respect of LICENSOR'S invention within one month after the date of this Agreement and shall be available to consult with LICENSEE, for up to 80 hours, with respect to the licensed invention and know-how. All travel and other expenses of LICENSOR for such consultation shall be reimbursed by LICENSEE within one month after LICENSOR submits its voucher therefor. LICENSOR makes no warranty regarding the value, suitability, or workability of such know-how. The royalty applicable for such know-how shall be paid, at the rate indicated above, for a minimum of three years from the date of this Agreement if no option is granted, or for three years from the date of exercise if an option is granted and exercised by LICENSOR, and thereafter for so long as LICENSEE makes, uses, or sells Licensed Products and has a share in the United States for of at least 15% of the competitive market for Licensed Products.

7. Royalties:

A. Licensing Fee: Unless the "Option Granted" box above is checked, LICENSEE shall pay to LICENSOR, upon execution of this Agreement, a non-refundable Licensing Fee. This Licensing Fee shall also serve as an advance against future royalties. Such Licensing Fee shall be computed as follows: (A) Take the Running Royalty Rate in percent, as stated above. (B) Multiply by LICENSEE'S Estimate Of Its First Year's Sales, in units Of Licensed Product, as stated above. (C) Multiply by LICENSEE'S Estimated Unit Price Of Licensed Product, in dollars, as stated above. (D) The combined product shall be the Resultant Licensing Fee, in dollars, as stated above. When licensee begins actual sales of the Licensed Product, it shall certify its Actual Net Factory Sales Price of Licensed Product to LICENSOR in writing and shall either (1) simultaneously pay LICENSOR any difference due if the Actual Net Factory Sales Price of Licensed Product is more than the Estimated Unit Price, stated above, or (2) advise LICENSOR of any credit to which LICENSEE is entitled if the Actual Net Factory Sales Price of Licensed Product is less than the above Estimated Unit Price. In the latter case, LICENSEE may deduct such credit from its first royalty remittance to LICENSOR, under subpart B below. If an option is granted and exercised under Part 4 above, then LICENSEE shall pay this Resultant Licensing Fee to LICENSOR if and when LICENSEE exercises its option.

B. Running Royalty: If the "Option Granted" box above is not checked, or if said box is checked and LICENSEE has exercised its option under Part 4, LICENSEE shall also pay to LICENSOR a Running Royalty, at the rate stated above. Such royalty shall be at the Patent Royalty Rate stated in Part 1C above, plus, if the "Know-How Licensed" box above is checked, a Know-How Royalty at the Know-How Royalty Rate stated above. Said Running Royalty shall be computed on LICENSEE'S Net Factory Sales Price of Licensed Product. Such Running Royalty shall accrue when the Licensed Products are first sold or disposed of by LICENSEE, or by any sublicensee of LICENSEE. LICENSEE shall pay the Running Royalty due to LICENSOR within one month after the end of each calendar quarter, together with a written report to LICENSOR of the number of units, respective sales prices, and total sales made in such quarter, together with a full itemization of any adjustments made pursuant to subpart F below. LICENSEE'S first report and payment shall be made within one month after the end of the first calendar quarter following the execution of this Agreement. No royalties shall be paid by LICENSEE to LICENSOR until after the Licensing Fee under subpart A above has been earned, but LICENSEE shall make a quarterly report hereunder for every calendar quarter after the execution hereof, whether or not any royalty payment is due for such quarter, except that if an option is granted, LICENSEE shall not make any royalty reports until and if LICENSEE exercises its option.

C. Minimum Annual Royalties: If the "Exclusive" box above is checked, so that this is an exclusive license, then this subpart C and subpart D shall be applicable. But if the "Nonexclusive" box is checked above, then these subparts C and D shall be inapplicable. There shall be no minimum annual royalties due under this Agreement until the "Year Commencing," as identified in Part 1 above. For the exclusivity privilege of the patent license grant under Part 5 above, a minimum annual royalty shall be due beginning with such royalty year and for each royalty year ending on the anniversary of such royalty year thereafter. Such minimum annual royalty shall be equal to the royalty which would have been due if the "Minimum Number Of Units [of Licensed Product] To Be Sold To Compute Minimum Annual Royalty" identified Part 1 above were sold during such royalty year. If less than such number of units of Licensed Product are not sold in any royalty year, then the royalty payable for the fourth quarter of such year shall be increased so as to cause the royalty payments for such year to equal said minimum annual royalty. If an option is granted under Parts 1 and 4, then no minimum annual royalties shall be due in any case until and if LICENSEE exercises its option.

D. If Minimum Not Paid: If this part is applicable and if sales of Licensed Product in any royalty year do not equal or exceed the minimum number of units identified in Part 1 above, LICENSEE may choose not to pay the minimum annual royalty under subpart C above. In this case, LICENSEE shall so notify LICENSOR by the date on which the last royalty for such year is due, i.e., within one month after any anniversary of the date identified in Part 1 above. Thereupon the license grant under Part 4 above shall be converted to a non-exclusive grant, and LICENSOR may immediately license others under the above patent.

E. Most Favored Licensee: If this license is nonexclusive, or if it becomes nonexclusive under subpart D above, then (a) LICENSOR shall not grant any other license under the above patent to any other party under any terms which are more favorable than those which LICENSEE pays or enjoys under this Agreement, and (b) LICENSOR shall promptly advise LICENSEE of any such other grant and the terms thereof.

F. When No Royalties Due: No Patent Royalties shall be due under this Agreement after the above patent expires or if it is declared invalid by a court of competent jurisdiction from which no appeal can be taken. Also, if LICENSOR'S patent application becomes finally abandoned without any patent issuing, then the Patent Royalty under this Agreement shall be terminated as of the date of abandonment. Any Know-How Royalties under Part 6 above shall continue after any Patent Royalties terminate, provided such Know-How Royalties are otherwise due under such Part 6.

G. Late Payments: If any payment due under this Agreement is not timely paid, then the unpaid balance shall bear interest until paid at an annual rate of 10% until the delinquent balance is paid. Such interest shall be compounded monthly.

H. Net Factory Sales Price: "Net Factory Sales Price" is defined as the gross factory selling price of Licensed Product, or the U.S. importer's gross selling price if Licensed Product is made abroad, less usual trade discounts actually allowed, but not including advertising allowances or fees or commissions paid to employees or agents of LICENSEE. The Net Factory Sales Price shall not include (1) pack-ing costs, if itemized separately, (2) import and export taxes, excise and other sales taxes, and customs duties, and (3) costs of insurance and transportation, if separately billed, from the place of manufacture if in the U.S., or from the place of importation if manufactured abroad, to the customer's premises or next point of distribution or sale. Bona fide returns may be deducted from units shipped in computing the royalty payable after such returns are made.

8. Records: LICENSEE and any of its sublicensees shall keep full, clear, and accurate records with respect to sales subject to royalty under this Agreement. The records shall be made in a manner such that the royalty reports made pursuant to Part 7B can be verified. LICENSOR, or its authorized agent, shall have the right to examine and audit such records upon reasonable notice during normal business hours, but not more than twice per year. In case of any dispute as to the sufficiency or accuracy of such records, LICENSOR may have any independent auditor examine and certify such records. LICENSEE shall make prompt adjustment to compensate for any errors or omissions disclosed by any such examination and certification of LICENSEE'S records. If LICENSOR does not examine LICENSOR'S records or question any royalty report within two years from the date thereof, then such report shall be considered final and LICENSOR shall have no further right to contest such report.

9. Sublicensees: If LICENSEE grants any sublicenses hereunder, it shall notify LICENSOR within one month from any such grant and shall provide LICENSOR with a true copy of any sublicense agreement. Any sublicensee of LICENSEE under this Agreement shall be bound by all of the terms applying to LICENSEE uhereunder and LICENSEE shall be responsible for the obligations and duties of any of its sublicensees.

10. Patent Prosecution:

A. Domestic: LICENSOR shall, at LICENSOR'S own expense, prosecute its above U.S. patent application, and any continuations, divisions, continuations-in-part, substitutes, and reissues of such patent application or any patent thereon, at its own expense, until all applicable patents issue or any patent application becomes finally abandoned. LICENSOR shall also pay any maintenance fees which are due on any patent(s) which issue on said patent application. If for any reason LICENSOR intends to abandon any patent application hereunder, it shall notify LICENSEE at least two months in advance of any such abandonment so as to give LICENSEE the opportunity to take over prosecution of any such application and maintenance of any patent. If LICENSEE takes over prosecution, LICENSOR shall cooperate with LICENSEE in any manner LICENSEE requires, at LICENSEE'S expense.

B. Foreign: LICENSOR shall have the opportunity, but not the obligation, to file corresponding foreign patent applications to any patent application under subpart A above. If LICENSOR files any such foreign patent applications, LICENSOR may license, sell, or otherwise exploit the invention, Licensed Product, or any such foreign application in any countries foreign to the United States as it chooses, provided that LICENSOR must give LICENSEE a right of first refusal and at least one month to exercise this right before undertaking any such foreign exploitation. If LICENSOR chooses not to file any corresponding foreign applications under this part, it shall notify LICENSEE at least one month prior to the first anniversary of the above patent application so as to give LICENSEE the opportunity to file corresponding foreign patent applications if it so chooses.

C. If Licensee Acts: If LICENSEE takes over prosecution of any U.S. patent application under subpart A above, and LICENSEE is successful so that a patent issues, then LICENSEE shall pay LICENSOR royalties thereafter at a rate of 75% of the royalty rate and any applicable minimum under Part 7C above and LICENSOR shall be entitled to deduct prosecution and maintenance expenses from its royalty payments. If LICENSEE elects to prosecute any foreign patent applications under subpart B above, then LICENSEE shall pay LICENSOR royalties of 50% of the royalty rate under Part 7 above for any applicable foreign sales, less all foreign prosecution and maintenance expenses incurred by LICENSEE.

11. Marking: LICENSEE shall mark all units of Licensed Product, or its container if direct marking is not feasible, with the legend "Patent Pending" until any patent(s) issue from the above patent application. When any patent(s) issue, LICENSOR shall promptly notify LICENSEE and thereafter LICENSEE shall mark all units of Licensed Product which it sells with proper notice of patent marking under 35 U.S.C. Section 287.

12. If Infringement Occurs: If either party discovers that the above patent is infringed, it shall communicate the details to the other party. LICENSOR shall thereupon have the right, but not the obligation, to take whatever action it deems necessary, including the filing of lawsuits, to protect the rights of the parties to this Agreement and to terminate such infringement. LICENSEE shall cooperate with LICENSOR if LICENSOR takes any such action, but all expenses of LICENSOR shall be borne by LICENSOR. If LICENSOR recovers any damages or compensation for any action it takes hereunder, LICENSOR shall retain 100% of such damages. If LICENSOR does not wish to take any action hereunder, LICENSEE shall also have the right, but not the obligation, to take any such action, in which case LICENSOR shall cooperate with LICENSEE, but all of LICENSEE'S expenses shall be borne by LICENSEE. LICENSEE shall receive 75% of any damages or compensation it recovers for any such infringement and shall pay 25% of such damages or compen-sation to LICENSOR, after deducting its costs, including attorney fees.

13. Disclaimer of Warranty: Nothing herein shall be construed as a warranty or representation by LICENSOR as to the scope or validity of the above patent application or any patent issuing thereon.

14. Term: The term of this Agreement shall end with the expiration of the last of any patent(s) which issues on LICENSOR'S patent application, unless terminated sooner for any reason provided herein, or unless know-how is licensed, in which case the terms of Part 6 shall cover the term of this Agreement.

15. Termination: This Agreement may be terminated under and according to any of the following contingencies:

A. Default: If LICENSEE fails to make any payment on the date such payment is due under this Agreement, or if LICENSEE makes any other default under or breach of this Agreement, LICENSOR shall have the right to terminate this Agreement upon giving three months' written Notice Of Intent To Terminate, specifying such failure, breach, or default to LICENSEE. If LICENSEE fails to make any payment in arrears, or otherwise fails to cure the breach or default within such three-month period, then LICENSOR may then send a written Notice Of Termination to LICENSEE, whereupon this Agreement shall terminate in one month from the date of such Notice of Termination. If this Agreement is terminated hereunder, LICENSEE shall not be relieved of any of its obligations to the date of termination and LICENSOR may act to enforce LICENSEE'S obligations after any such termination.

B. Bankruptcy, Etc.: If LICENSEE shall go into receivership, bankruptcy, or insolvency, or make an assignment for the benefit of creditors, or go out of business, the Agreement shall be immediately terminable by LICENSOR by written notice, but without prejudice to any rights of LICENSOR hereunder.

C. Antishelving: If LICENSEE discontinues its sales or manufacture of Licensed Product without intent to resume, it shall so notify LICENSOR within one month of such discontinuance, whereupon LICENSOR shall have the right to terminate this Agreement upon one month's written notice, even if this Agreement has been converted to a non-exclusive grant under Part 5D above. If LICENSEE does not begin manufacture or sales of Licensed Product within one and one-half years from the date of this Agreement or the date of its option exercise if an option is granted, or, after commencing manufacture and sales of Licensed Product, discontinues its manufacture and sales of Licensed Product for one and one-half years, LICENSOR shall have the right to terminate this Agreement upon one months' written notice, unless LICENSEE can show that it in good faith intends and is actually working to resume or begin manufacture or sales, and has a reasonable basis to justify its delay. In such case LICENSEE shall advise LICENSOR in writing, before the end of such one and one-half year period, of the circumstances involved and LICENSEE shall thereupon have up to an additional year to resume or begin manufacture or sales. It is the intent of the parties hereto that LICENSOR shall not be deprived of the opportunity, for an unreasonable length of time, to exclusively license its patent if LICENSEE has discontinued or has not commenced manufacture or sales of Licensed Product. In no case shall LICENSOR have the right to terminate this Agreement if and so long as LICENSEE is paying LICENSOR minimum annual royalties under Part 7C above.

16. Notices: All notices, payments, or statements under this Agreement shall be in writing and shall be sent by first-class certified mail, return receipt requested, postage prepaid, to the party concerned at the above address, or to any substituted address given by notice hereunder. Any such notice, payment, or statement shall be considered sent or made on the day deposited in the mails.

17. Mediation and Arbitration: If any dispute arises under this Agreement, the parties shall negotiate in good faith to settle such dispute. If the parties cannot resolve such dispute themselves, then either party may submit the dispute to mediation by a mediator approved by both parties. The parties shall both cooperate with the mediator. If the parties cannot agree to any mediator, or if either party does not wish to abide by any decision of the mediator, then they shall submit the dispute to arbitration by any mutually-acceptable arbitrator. If no arbitrator is mutually acceptable, then they shall submit the matter to arbitration under the rules of the American Arbitration Association (AAA). Under any arbitration, both parties shall cooperate with and agree to abide finally by any decision of the arbitration proceeding. If the AAA is selected, the arbitration shall take place under the auspices of the nearest branch of the AAA such to the other party. The costs of the arbitration proceeding shall be born according to the decision of the arbitrator, who may apportion costs equally, or in accordance with any finding of fault or lack of good faith of either party. The arbitrator's award shall be non-appealable and enforceable in any court of competent jurisdiction.

18. Assignment: The rights of LICENSOR under this Agreement shall be assignable or otherwise transferrable, in whole or in part, by LICENSOR and shall vest LICENSOR'S assigns or transferrees with the same rights and obligations as were held by LICENSOR. This Agreement shall be assignable by LICENSEE to any entity that succeeds to the business of LICENSEE to which Licensed Products relate or to any other entity if LICENSOR'S permission is first obtained in writing.

19. Jurisdiction: This Agreement shall be interpreted under the laws of LICENSOR'S state, as given in Part 1 above.

20. Non-Frustration: Neither party to this Agreement shall commit any act or take any action which frustrates or hampers the rights of the other party under this Agreement. Each party shall act in good faith and engage in fair dealing when taking any action under or related to this Agreement.

21. Signatures: The parties, having carefully read this Agreement and having consulted or have been given an opportunity to consult counsel, have indicated their agreement to all of the above terms by signing this Agreement on the respective dates below indicated. LICENSEE and LICENSOR have each received a copy of this Agreement with both LICENSEE'S and LICENSOR'S original ink signatures thereon.

Date: 19 _____   _____

Print Licensor's Name _____

Date: 19 _____   _____

Print Licensee's Name _____

INDEX

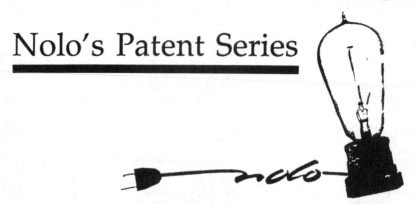

ACCESS TO LAW CATALOG

FAMILY MATTERS

A Legal Guide for Lesbian and Gay Couples

ATTORNEYS HAYDEN CURRY & DENIS CLIFFORD
NATIONAL 5TH ED.

Laws designed to regulate and protect married couples don't apply to lesbian and gay couples. This book shows you, step-by-step, how to write a living-together contract, plan for medical emergencies (using durable powers of attorney), and plan your estates (wills and probate avoidance techniques). It also discusses legal aspects of having and raising children and relating to ex-spouses and children of former marriages. Complete with forms, sample agreements and lists of both national lesbian and gay legal organizations, and AIDS organizations.

$17.95 / LG

The Guardianship Book

BY LISA GOLDOFTAS &
ATTORNEY DAVID BROWN
CALIFORNIA 1ST ED.

Thousands of children in California are left without a guardian because their parents have died, abandoned them or are unable to care for them. *The Guardianship Book* provides step-by-step instructions and the forms needed to obtain a legal guardianship without a lawyer.

$19.95 / GB

How to Do Your Own Divorce

ATTORNEY CHARLES E. SHERMAN
(TEXAS ED. BY SHERMAN & SIMONS)
CALIFORNIA 15TH ED. & TEXAS 2ND ED.

This is the book that launched Nolo Press and advanced the self-help law movement. During the past 18 years, over 500,000 copies have been sold, saving consumers at least $50 million in legal fees (assuming 100,000 have each saved $500—certainly a conservative estimate). Contains all the forms and instructions you need to do your divorce without a lawyer.

CALIFORNIA $18.95 / CDIV
TEXAS $14.95 / TDIV

Practical Divorce Solutions

ATTORNEY CHARLES E. SHERMAN
CALIFORNIA 1ST ED.

This book is a valuable guide to the emotional aspects of divorce as well as an overview of the legal and financial decisions that must be made.

$12.95 / PDS

The Living Together Kit

ATTORNEYS TONI IHARA & RALPH WARNER
NATIONAL 6TH ED.

"Written in plain language, free of legal mumbo jumbo, and spiced with witty personal observations."

—**Associated Press**

The Living Together Kit is a detailed guide designed to help the increasing number of unmarried couples living together understand the laws that affect them. *The Living Together Kit* contains comprehensive information on estate planning, paternity agreements, living together agreements, buying real estate, and much more. Sample agreements and instructions are included.

$17.95 / LTK

How to Adopt You Stepchild in California

FRANK ZAGONE & ATTORNEY MARY RANDOLPH
CALIFORNIA 3RD ED.

For many families that include stepchildren, adoption is a sure-fire way to avoid confusion over inheritance or guardianship. This book provides sample forms and complete step-by-step instructions for completing a simple uncontested stepparent adoption in California.

$19.95 / ADOP

How to Modify and Collect Child Support In California

ATTORNEYS JOSEPH MATTHEWS, WARREN SIEGEL & MARY WILLIS
CALIFORNIA 3RD ED.

Using this book, parents can determine the level of child support they are entitled to receive, or obliged to pay, and can go to court to modify existing support to the appropriate level.

$17.95 / SUPP

California Marriage & Divorce Law

ATTORNEYS RALPH WARNER, TONI IHARA &
STEPHEN ELIAS
CALIFORNIA 10TH ED.

This practical handbook is for the Californian who wants to understand marriage and divorce laws. It explains community property, pre-nuptial contracts, foreign marriages, buying a house, the steps for getting a divorce, dividing property, and much more.

$17.95 / MARR

PATENT, COPYRIGHT & TRADEMARK

Patent It Yourself

ATTORNEY DAVID PRESSMAN
NATIONAL 2ND ED.

Every step of the patent process is presented in order in this gem of a book, complete with official forms…"

—**San Francisco Chronicle**

This state-of-the-art guide is a must for all inventors interested in obtaining patents—from the patent search to the actual application. Patent attorney and former patent examiner David Pressman covers use and licensing, successful marketing, and how to deal with infringement.

$29.95 / PAT

The Inventor's Notebook

GRISSOM & ATTORNEY PRESSMAN
NATIONAL 1ST ED.

The best protection for your patent is adequate records. *The Inventor's Notebook* helps you document the process of successful independent inventing by providing forms, instructions, references to relevant areas of patent law, a bibliography of legal and non-legal aids, and more.

$19.95 / INOT

How to Copyright Software

ATTORNEY M.J. SALONE
NATIONAL 3RD ED.

"Written by practicing lawyers in the straightforward and informative Nolo style, the book covers just about everything that might be of interest to a software developer or publisher. Even those who are employed by a company on a full-time or contractual basis will find much to ponder here."

—**PC Week**

Now that you've spent hours of time and sleepless nights perfecting your software creation, learn how to protect it from plagiarism. Copyright laws give you rights against those who use your work without your permission. This book tells you how to enforce those rights, how to register your copyright for maximum protection, and discusses who owns a copyright on software developed by more than one person.

$34.95 / COPY

Legal Care for Your Software

ATTORNEYS DANIEL REMER & STEPHEN ELIAS
NATIONAL

Legal Care for Your Software is out of print. Nolo authors are in the process of writing a new 4th edition of this book.

The Independent Paralegal's Handbook

ATTORNEY RALPH WARNER
NATIONAL 1ST ED.

A large percentage of routine legal work in this country is performed by typists, secretaries, researchers and other law office helpers generally labeled paralegals. For those who want to take these services out of the law office and offer them for a reasonable fee in an independent business, attorney Ralph Warner provides legal and business guidelines.

$12.95 / PARA

Getting Started as an Independent Paralegal

(TWO AUDIO TAPES)
ATTORNEY RALPH WARNER
NATIONAL 1ST ED.

Approximately three hours in all, these tapes are a carefully edited version of a seminar given by Nolo Press founder Ralph Warner. They are designed to be used with *The Independent Paralegal's Handbook.*

$24.95 / GSIP

Marketing Without Advertising

MICHAEL PHILLIPS & SALLI RASBERRY
NATIONAL 1ST ED.

"There are good ideas on every page. You'll find here the nitty gritty steps you need to—and can—take to generate sales for your business, no matter what business it is."

—Milton Moskowitz, syndicated columnist and author of *The 100 Best Companies to Work For in America*

The best marketing plan encourages customer loyalty and personal recommendation. Phillips and Rasberry outline practical steps for building and expanding a small business without spending a lot of money on advertising.

$14.00 / MWA

The Partnership Book

ATTORNEYS CLIFFORD & WARNER
NATIONAL 3RD ED.

Lots of people dream of going into business with a friend. The best way to keep that dream from turning into a nightmare is to have a solid partnership agreement. This book shows you, step-by-step, how to write an agreement that meets your need. It covers initial contributions to the business, wages, profit-sharing, buy-outs, death or retirement of a partner, and disputes.

$18.95 / PART

How to Write a Business Plan

MIKE MCKEEVER
NATIONAL 3RD ED.

"...outlines the kinds of credit available... shows how to prepare cashflow forecasts, capital spending plans, and other vital ideas. An attractive guide for would-be entrepreneurs."

—ALA Booklist

If you're thinking of starting a business or raising money to expand an existing one, this book will show you how to write the business plan and loan package necessary to finance your business and make it work.

$17.95 / SBS

How to Form Your Own Corporation

ATTORNEY ANTHONY MANCUSO
CALIFORNIA 7TH ED.
TEXAS 4TH ED.
NEW YORK 2ND ED.
FLORIDA 2ND ED.

Incorporating your small business lets you take advantage of tax benefits, limited liability and benefits of employee status, and financial flexibility. These books contain the forms, instructions and tax information you need to incorporate a small business yourself and save hundreds of dollars in lawyers' fees. Each contains up-to-date corporate and tax information.

CALIFORNIA $29.95 / CCOR
TEXAS $24.95 / TCOR
NEW YORK $24.95 / NYCO
FLORIDA $24.95 / FLCO

The California Professional Corporation Handbook

ATTORNEY ANTHONY MANCUSO
CALIFORNIA 4TH ED.

Health care professionals, lawyers, accountants and members of certain other professions must fulfill special requirements when forming a corporation in California. Professional corporations offer liability protection, the financial benefits of a corporate retirement plan, and lower tax rates on the first $75,000 of taxable income. This edition contains up-to-date tax information plus all the forms and instructions necessary to form a California professional corporation. An appendix explains the special rules that apply to each profession.

$34.95 / PROF

The California Nonprofit Corporation Handbook

ATTORNEY ANTHONY MANCUSO
CALIFORNIA 5TH ED.

This book shows you step-by-step how to form and operate a nonprofit corporation in California. It includes the latest corporate and tax law changes, including expanded protection from personal liability for corporate directors. Includes forms for the Articles, Bylaws and Minutes you need. Contains complete instructions for obtaining federal 501(c)(3) tax exemptions and benefits, which may be used in any state.

$29.95 / NON

Elder Care: Choosing and Financing Long-Term Care

JOSEPH L. MATTHEWS
NATIONAL 1ST ED.

Until recently, the only choice for the elderly in deteriorating health was to enter a nursing home. Now older people who need care and their families are faced with many more choices, ranging from care at home to residential facilities to complete care homes. This book will guide you in choosing and paying for long-term care, alerting you to practical concerns and explaining laws that may affect your decisions.

$16.95 / ELD

Social Security, Medicare & Pensions

ATTORNEY JOSEPH MATTHEWS & DOROTHY MATTHEWS BERMAN
NATIONAL 5TH ED.

When the Catastrophic Coverage Act was repealed recently, it drastically changed the kinds and amounts of government benefits for older Americans. While assistance with income and healthcare is still available, it requires more perseverance and understanding to get your due.

This new edition includes invaluable guidance through the current maze of rights and benefits for those 55 and over, including Medicare, Medicaid and Social Security retirement and disability benefits and age discrimination protections.

$15.95 / SOA

Plan Your Estate

ATTORNEY DENIS CLIFFORD
NATIONAL 1ST ED.

"One of the best personal finance books of 1989." —**Money Magazine**

This book covers every significant aspect of estate planning, and gives detailed, specific instructions for preparing a living trust. *Plan Your Estate* shows how to prepare an estate plan without the expensive services of a lawyer and ncludes all the tear-out forms and step-by-step instructions to let people with estates under $600,000 do the job themselves.

$19.95 / NEST

Nolo's Simple Will Book

ATTORNEY DENIS CLIFFORD
NATIONAL 2ND ED.

It's easy to write a legally valid will using this book. The instructions and forms enable people to draft a will for all needs, including naming a personal guardian for minor children; leaving property to minor children or young adults; and updating a will when necessary. This edition also contains a discussion of estate planning basics with information on living trusts, death taxes, and durable powers of attorney. Good in all states except Louisiana.

$17.95 / SWIL

How To Probate an Estate

JULIA NISSLEY
CALIFORNIA 5TH ED.

If you find yourself responsible for winding up the legal and financial affairs of a deceased family member or friend, you can often save costly attorneys' fees by handling the probate process yourself. *How to Probate an Estate* shows you, step-by-step, how to actually settle an estate. It also covers the simple procedures you can use to transfer assets that don't require probate, including property held in joint tenancy or living trusts or as community property.

$29.95 / PAE

The Power of Attorney Book

ATTORNEY DENIS CLIFFORD
NATIONAL 3RD ED.

Who will take care of your affairs, and make your financial and medical decisions if you can't? With this book you can appoint someone you trust to carry out your wishes and stipulate exactly what kind of care you want or don't want. Includes Durable Power of Attorney and Living Will forms.

$19.95 / POA

Everybody's Guide to Small Claims Court

ATTORNEY RALPH WARNER
NATIONAL 4TH ED.
CALIFORNIA 8TH ED.

So, the dry cleaner ruined your good flannel suit. Your roof leaks every time it rains, and the contractor who supposedly fixed it won't call you back. The landlord won't return your security deposit. This book will help you decide if you should sue in small claims court, show you how to file and serve papers, tell you what to bring to court, and how to collect a judgment.

NATIONAL $14.95 / NSCC
CALIFORNIA $14.95 / CSCC

Collect Your Court Judgment

GINI GRAHAM SCOTT, ATTORNEY STEPHEN ELIAS & LISA GOLDOFTAS
CALIFORNIA 1ST ED.

After you win a judgment in small claims, municipal or superior court, you still have to collect your money. If the debtor doesn't pay up voluntarily, you need to know how to collect your judgment from the debtor's bank accounts, wages, business receipts, real estate or other assets. This book contains step-by-step instructions and all the forms you need.

$24.95 / JUDG

Fight Your Ticket

ATTORNEY DAVID BROWN
CALIFORNIA 3RD ED.

Here's a book that shows you how to fight an unfair traffic ticket—when you're stopped, at arraignment, at trial and on appeal. No wonder a traffic court judge (who must remain nameless) told us that he keeps this book by his bench for easy reference!

$17.95 / FYT

The Criminal Records Book

ATTORNEY WARREN SIEGEL
CALIFORNIA 3RD ED.

We've all done something illegal. If you were one of those who got caught, your juvenile or criminal court record can complicate your life years later. *The Criminal Records Book* shows you, step-by-step, how to seal criminal records, dismiss convictions, destroy marijuana records, and reduce felony convictions.

$19.95 / CRIM

Dog Law

ATTORNEY MARY RANDOLPH
NATIONAL 1ST ED.

Do you own a dog? Do you live down the street from one? If you do, you need *Dog Law*, a practical guide to the laws that affect dog owners and their neighbors. *Dog Law* answers common questions on such topics as biting, barking, veterinarians, leash laws, travel, landlords, wills, guide dogs, pit bulls, cruelty and much more.

$12.95 / DOG

How to Change Your Name

ATTORNEYS DAVID LOEB & DAVID BROWN
CALIFORNIA 4TH ED.

Wish you had gone back to your former name after the divorce? Tired of spelling V-e-n-k-a-t-a-r-a-m-a-n S-u-b-r-a-m-a-n-i-a-m over the phone? This book explains how to change your name legally and provides all the necessary court forms with detailed instructions on how to fill them out.

$19.95 / NAME

How to File For Bankruptcy

ATTORNEYS STEPHEN ELIAS, ALBIN RENAUER & ROBIN LEONARD
NATIONAL 1ST ED.

It's no fun having to think about declaring bankruptcy. But easy credit, high interest rates, unexpected illness, job lay-offs and inflation often conspire to leave people with a satchel full of debts. Here we show you how to decide whether or not filing for bankruptcy makes sense and if it does, we give you forms and step-by-step instructions on how to do it.

$24.95 / HFB

Simple Contracts for Personal Use

ATTORNEY STEPHEN ELIAS
NATIONAL 1ST ED.

If you've ever sold a car, lent money to a relative or friend, or put money down on a prospective purchase, you should have used a contract.

Here are clearly written legal form contracts to: buy and sell property, borrow and lend money, store and lend personal property, make deposits on goods for later purchase, release others from personal liability, or pay a contractor to do home repairs.

$12.95 / CONT

How to Buy A House in California Strategies for Beating the Affordability Gap

ATTORNEY RALPH WARNER, IRA SERKES & GEORGE DEVINE
CALIFORNIA 1ST EDITION

Most potential California home buyers are finding themselves victims of skyrocketing prices, as the demand for housing at all price levels continues to increase relentlessly. This book shows how to find a house, work with a real estate agent, make an offer, and negotiate intelligently. Information on all types of mortgages as well as private financing options place this book in a category by itself.

$18.95 / BHC

Homestead Your House

ATTORNEYS RALPH WARNER, CHARLES SHERMAN & TONI IHARA
CALIFORNIA 7TH ED.

Under California homestead laws, up to $75,000 of the equity in your home may be safe from creditors. But to get the maximum legal protection, you should file a Declaration of Homestead before a creditor gets a court judgment against you and puts a lien (legal claim) on your house. This book shows you how and includes complete instructions and tear-out forms.

$9.95 / HOME

The Deeds Book

ATTORNEY MARY RANDOLPH
CALIFORNIA 1ST ED.

If you own real estate, you'll need to sign a new deed when you transfer the property or put it in trust as part of your estate planning. This book shows you how to choose the right kind of deed, complete the tear-out forms, and record them in the county recorder's public records. Includes real property disclosure requirements and California community property rules.

$15.95 / DEED

For Sale by Owner

GEORGE DEVINE
CALIFORNIA 1ST ED.

If you sell your house at California's median price—$200,000—the standard broker's commission (6%) amounts to $12,000. That's money you could save if you sold your own house. This book provides essential information about pricing your house, marketing it, writing a contract and going through escrow.

$24.95 / FSBO

The Landlord's Law Book: Vol. 1, Rights & Responsibilities

ATTORNEYS DAVID BROWN & RALPH WARNER
CALIFORNIA 2ND ED.

The era when a landlord could substitute common sense for a detailed knowledge of the law is gone forever. Everything from the amount you can charge for a security deposit to terminating a tenancy, to your legal responsibility for the illegal acts of your manager is closely regulated by the law. This volume covers: deposits, leases and rental agreements, inspections (tenants' privacy rights), habitability (rent withholding), ending a tenancy, liability, and rent control.

$24.95 / LBRT

The Landlord's Law Book: Vol. 2, Evictions

ATTORNEY DAVID BROWN
CALIFORNIA 2ND ED.

What do you do if you've got a tenant who won't pay the rent—and won't leave? There's only one choice: go to court and get an eviction. This book takes you through the process step-by-step. It's even got a special section on local rent control laws. All the tear-out forms and instructions you need are included.

$24.95 / LBEV

Tenants' Rights

ATTORNEYS MYRON MOSKOVITZ & RALPH WARNER
CALIFORNIA 10TH ED.

Your "security building" doesn't have a working lock on the front door. Is your landlord liable? How can you get him to fix it? Under what circumstances can you withhold rent? When is an apartment not "habitable?" Moskovitz and Warner explain the best way to handle your relationship with your landlord and your legal rights when you find yourself in disagreement. A special section on rent control cities is included.

$15.95 / CTEN

Legal Research

ATTORNEY STEPHEN ELIAS
NATIONAL 2ND ED.

A valuable tool on its own, or as a companion to just about every other Nolo book. *Legal Research* gives easy-to-use, step-by-step instructions on how to find legal information. The legal self-helper can find and research a case, read statutes and administrative regulations, and make Freedom of Information Act requests. A great resource for paralegals, law students, legal secretaries and social workers.

$14.95 / LRES

Family Law Dictionary

ATTORNEYS ROBIN LEONARD & STEPHEN ELIAS
NATIONAL 1ST ED.

Finally, a legal dictionary that's written in plain English, not "legalese"! The *Family Law Dictionary* is designed to help the nonlawyer who has a question or problem involving family law—marriage, divorce, adoption, or living together. The book contains many examples as well as definitions, and extensive cross-references to help you find the information you need.

$13.95 / FLD

Patent, Copyright & Trademark: The Intellectual Property Law Dictionary

ATTORNEY STEPHEN ELIAS
NATIONAL 1ST ED.

...uses simple language free of legal jargon to define and explain the intricacies of items associated with trade secrets, copyrights, trademarks and unfair competition, patents and patent procedures, and contracts and warranties.

—IEEE Spectrum

If you're dealing with any multi-media product, a new business product or trade secret, you need this book.

$19.95 / IPLD

Legal Research Made Easy: A Road Map Through the Law Library Maze

2 1/2 hr. Videotape and 40 page manual
NOLO PRESS-LEGAL STAR COMMUNICATIONS

If you're a law student, paralegal or librarian—or just want to look up the law for yourself—this video is for you. From statutes to court cases and agency regulations, University of California law professor, Bob Berring explains how to use all the basic legal research tools in your local law library. Berring comes armed with an easy-to-follow six-step research plan and a sense of humor.

$89.95 / LRME

WillMaker

NOLO PRESS & LEGISOFT, INC.
NATIONAL 4TH ED.

"A well crafted document. That's even what my lawyer said ...noting that a few peculiar twists in my state's law were handled nicely by the computerized lawyer."

—Peter H. Lewis, The New York Times

"An excellent addition to anyone's home-productivity library."

—Home Office Computing

"A fertile hybrid that I expect to see more of: can-do software that lives inside a how-to-book. In this case, the book itself is one of the better ones on preparing your own will."

—Whole Earth Review

Recent statistics say chances are better than 2 to 1 that you haven't written a will, even though you know you should. *WillMaker* makes the job easy, leading you step-by-step in a question and answer format. Once you've gone through the program, you print out the will and sign it in front of witnesses. *WillMaker* comes with a 200-page manual which tracks the program and provides the legal background necessary to make sound choices about how and to whom you should leave property. Good in all states except Louisiana.

IBM PC 3 1/2 & 5 1/4	$69.95	MI4

AVAILABLE FALL 1990:

MACINTOSH	$69.95	WM4

For the Record

CAROL PLADSEN & ATTORNEY RALPH WARNER
NATIONAL 1ST ED.

This easy-to-use software program provides a single place to keep a complete inventory of all your important legal, financial, personal, and family records. Having accurate and complete records facilitates tax preparation and helps loved ones manage your affairs if you become incapacitated or die. The detailed manual offers an overview of how to reduce estate taxes and avoid probate, and tells you what records you need to keep.

MACINTOSH	$49.95	FRM
IBM PC 3 1/2	$49.95	FR3I
IBM PC 5 1/4	$49.95	FRI

California Incorporator

ATTORNEY ANTHONY MANCUSO & LEGISOFT, INC.
CALIFORNIA 1ST ED.

"...easy to use...the manual consists of far more than instructions for using the software...[it is] a primer that provides a great deal of background, including detailed explanations of the legal implications of each decision you make."

—Los Angeles Times

About half of the small California corporations formed today are done without the services of a lawyer. Now, this easy-to-use software program makes the job even easier.

Just answer the questions on the screen, and *California Incorporator* will print out the 35-40 pages of documents you need to make your California corporation legal.

Comes with a 200-page manual that explains the incorporation process.

IBM PC 3 1/2 & 5 1/4	$129.00	INCI

The California Nonprofit Corporation Handbook: computer edition with disk

ATTORNEY ANTHONY MANCUSO
CALIFORNIA 1ST ED.

This is the standard work on how to form a nonprofit corporation in California. Included on the disk are the forms for the Articles, Bylaws and Minutes you will need, as well as regular and special director and member minute forms. Also included are several chapters with line-by-line instructions explaining how to apply for and obtain federal tax-exempt status. This is a critical step in the incorporation of any nonprofit organizaton and applies to incorporating in any state.

IBM PC 5 1/4	$69.95	NPI
IBM PC 3 1/2	$69.95	NP3I
MACINTOSH	$69.95	NPM

How to Form Your Own New York Corporation: computer edition with disk

How to Form Your Own Texas Corporation: computer edition with disk

ATTORNEY ANTHONY MANCUSO

More and more businesses are incorporating to qualify for tax benefits, limited liability status, the benefit of employee status and financial flexibility. This software package contains all the instructions, tax information and forms you need to incorporate a small business, including the Certificate of Incorporation, Bylaws, Minutes and Stock Certificates. The 250-page manual includes instructions on how to incorporate a new or existing business; tax and securities law information; information on S corporations; Federal Tax Reform Act rates and rules; and the latest procedures to protect your directors under state law. All organizational forms are on disk.

NEW YORK 1ST ED.

IBM PC 5 1/4	$69.95	NYCI
IBM PC 3 1/2	$69.95	NYC3I
MACINTOSH	$69.95	NYCM

TEXAS 1ST ED.

IBM PC 5 1/4	$69.95	TCI
IBM PC 3 1/2	$69.95	TC3I
MACINTOSH	$69.95	TCM

GET A TWO-YEAR SUBSCRIPTION TO THE NOLO NEWS FREE!

(normally $12)

Nolo Press wants you to have top quality, up-to-date legal information. The *Nolo News,* our "Access to Law" quarterly newspaper, contains an update section which will keep you abreast of any changes in the law relevant to *How to Buy a House in California.* You'll find interesting articles on a number of legal topics, book reviews and our ever-popular lawyer joke column. Send in the registration card below and receive FREE a two-year subscription to the *Nolo News* (normally $12.00). Your subscription will begin with the first quarterly issue published after we receive your card. (This offer is good in the U.S. only.)

REGISTRATION CARD

How to Buy a House in California

Fill out and return this postage paid card for a FREE two-year subscription to the *Nolo News.* We'll also notify you when we publish a new edition of *How to Buy a House in California.* If you're already a subscriber, we will extend your subscription for two more years.

BONUS! When you send in this card we'll send you a discount coupon good on your next Nolo Press book or software purchase.

Please print clearly, or type.

YOUR NAME

STREET ADDRESS

CITY   STATE   ZIP

COMMENTS

Because we respect your privacy and hate all the junk mail we get in our mailboxes too, we pledge never to sell, rent or give your name and address to any other organization. (This offer available to U.S. residents only.)

[Nolo books are]..."written in plain language, free of legal mumbo jumbo, and spiced with witty personal observations."

—ASSOCIATED PRESS

"Well-produced and slickly written, the [Nolo] books are designed to take the mystery out of seemingly involved procedures, carefully avoiding legalese and leading the reader step-by-step through such everyday legal problems as filling out forms, making up contracts, and even how to behave in court."

—SAN FRANCISCO EXAMINER

"...Nolo publications...guide people simply through the how, when, where and why of law."

—WASHINGTON POST

"Increasingly, people who are not lawyers are performing tasks usually regarded as legal work... And consumers, using books like Nolo's, do routine legal work themselves."

—NEW YORK TIMES

"...All of [Nolo's] books are easy-to-understand, are updated regularly, provide pull-out forms...and are often quite moving in their sense of compassion for the struggles of the lay reader."

—SAN FRANCISCO CHRONICLE

NO POSTAGE
NECESSARY
IF MAILED
IN THE
UNITED STATES

BUSINESS REPLY MAIL
FIRST-CLASS MAIL PERMIT NO 3283 BERKELEY CA

POSTAGE WILL BE PAID BY ADDRESSEE

NOLO PRESS
950 Parker Street
Berkeley CA 94710-9867